Entropy Application for Forecasting

Entropy Application for Forecasting

Special Issue Editors

Ana Jesús López Menéndez
Rigoberto Pérez-Suárez

MDPI • Basel • Beijing • Wuhan • Barcelona • Belgrade • Manchester • Tokyo • Cluj • Tianjin

Special Issue Editors

Ana Jesús López Menéndez
University of Oviedo
Spain

Rigoberto Pérez-Suárez
University of Oviedo
Spain

Editorial Office
MDPI
St. Alban-Anlage 66
4052 Basel, Switzerland

This is a reprint of articles from the Special Issue published online in the open access journal *Entropy* (ISSN 1099-4300) (available at: https://www.mdpi.com/journal/entropy/special_issues/ entropy_forecasting).

For citation purposes, cite each article independently as indicated on the article page online and as indicated below:

LastName, A.A.; LastName, B.B.; LastName, C.C. Article Title. *Journal Name* **Year**, *Article Number*, Page Range.

ISBN 978-3-03936-487-9 (Hbk)
ISBN 978-3-03936-488-6 (PDF)

Contents

About the Special Issue Editors

Ana Jesús López Menéndez works as a full professor in the Department of Applied Economics at the University of Oviedo. Her research activities are related to the regional modeling and forecasting, the measurement of economic inequality, and the socioeconomic impact of ICT. She has supervised several Ph.D. candidates and research projects. She has published in high-impact international journals, including *Economics Letters*, *Regional Studies*, *Applied Economics Letters*, *TEST*, *Journal of Forecasting*, *Social Indicators Research*, *Empirical Economics*, *Information & Management*, and *Entropy*. She has been a visiting fellow at universities in the UK, Hungary, Cuba, and Portugal, and she has worked as an expert evaluator for Spanish National Agencies such as OAPEE, SEPIE, ANECA, and CNEAI.

Rigoberto Pérez Suárez works as a full professor in the Department of Applied Economics at the University of Oviedo. He has published several books and articles in high-impact journals, such as *Kybernetes*, *Metrika*, *Empirical Economics*, *Regional Studies*, *Technological Forecasting and Social Change*, *IEEE Transactions on Fuzzy Systems*, and *Entropy*. He has been involved in a wide variety of research projects related to information measures, econometric modeling and forecasting, and ICT impacts, mainly e-learning. He has been Head of the Department of Applied Economics and Director of the University of Oviedo Innovation Center and the G9 Shared Virtual Campus, including nine Spanish universities.

Editorial

Entropy Application for Forecasting

Ana Jesús López-Menéndez * and Rigoberto Pérez-Suárez

Department of Applied Economics, University of Oviedo, Campus del Cristo s/n, 33006 Oviedo, Asturias, Spain; rigo@uniovi.es
* Correspondence: anaj@uniovi.es

Received: 19 May 2020; Accepted: 27 May 2020; Published: 29 May 2020

Keywords: information theory; uncertainty; forecasting methods; forecasting evaluation; accuracy; M-competition; combined forecasts; scenarios

The information theory developed by Shannon [1] defines the entropy for any probabilistic system as a measure of the related uncertainty. This measure, inspired by the entropy defined in thermodynamics by Boltzmann [2], provides a link between uncertainty and probability and opens a wide variety of applications in different fields.

The basic idea of information theory is that the informational content of a message depends on the degree to which it is surprising: if an event is very likely to occur, there is no surprise when this event happens as expected; on the contrary, it is much more informative to know that an unlikely event has taken place.

In this context, entropy can be understood as a measure of unpredictability and therefore it is not surprising that entropy and information theory can be of great help in a broad range of problems related to forecasting, as shown by Theil [3,4].

The contributions to this Special Issue "Entropy Application for Forecasting" show the enormous potential of entropy and information theory in forecasting, including both theoretical developments and empirical applications.

The contents cover a great diversity of topics, such as the aggregation and combination of individual forecasts [5,6], the comparison of forecasting performances [7,8], the analysis of forecasting uncertainty [9], robustness [10] and inconsistency [11], and the proposal of new forecasting approaches [12–14].

A great diversity is also observed in the methods, since the contributions encompass a wide variety of time series techniques (ARIMA, VAR, State Space Models, etc.) as well as econometric methods and machine learning algorithms.

Furthermore, the empiric contents are also diverse including both simulated experiments and real-world applications. More specifically, the contributions provide empirical evidence that refer to the economic growth and gross domestic product (GDP) [5,9], the M4 competition dataset [8], the confidence and industrial trend surveys [9], and some stock exchange composite indices (Taiwan, Shanghai, Hong-Kong) [11], as well as other real data from a Portuguese retailer [7] and a Chinese grid company [12].

In summary, this Special Issue provides an engaging insight into entropy applications for forecasting, offering an interesting overview of the current situation and suggesting possibilities for further research in this field.

Acknowledgments: We want to express our thanks to the authors of the contributions of this Special Issue, and to the journal referees for their valuable comments and suggestions. We also acknowledge the confidence of the journal *Entropy* and its support in the development of this Special Issue.

Conflicts of Interest: The authors declare no conflict of interest.

References

1. Shannon, C.E. A Mathematical Theory of Communication. *Bell Syst. Tech. J.* **1948**, *27*, 379–423. [CrossRef]
2. Boltzmann, L. Über die Mechanische Bedeutung des Zweiten Hauptsatzes der Wärmetheorie. *Wien. Ber.* **1866**, *53*, 195–220.
3. Theil, H. *Applied Economic Forecasting*; North Holland Publishing: Amsterdam, The Netherlands, 1966.
4. Theil, H. *Economics and Information Theory*; North Holland Publishing: Amsterdam, The Netherlands, 1967.
5. Bretó, C.; Espinosa, P.; Hernández, P.; Pavía, J.M. An Entropy-Based Machine Learning Algorithm for Combining Macroeconomic Forecasts. *Entropy* **2019**, *21*, 1015. [CrossRef]
6. Fernández-Vázquez, E.; Moreno, B.; Hewings, G. A Data-Weighted Prior Estimator for Forecast Combination. *Entropy* **2019**, *21*, 429. [CrossRef]
7. Oliveira, J.M.; Ramos, P. Assessing the Performance of Hierarchical Forecasting Methods on the Retail Sector. *Entropy* **2019**, *21*, 436. [CrossRef]
8. Ponce-Flores, M.; Frausto-Solís, J.; Santamaria-Bonfil, G.; Pérez-Ortega, J.; González-Barbosa, J.J. Time Series Complexities and Their Relationship to Forecasting Performance. *Entropy* **2020**, *22*, 89. [CrossRef]
9. López-Menéndez, A.J.; Pérez-Suárez, R. Acknowledging Uncertainty in Economic Forecasting. Some Insight from Confidence and Industrial Trend Surveys. *Entropy* **2019**, *21*, 413. [CrossRef]
10. Mei, W.; Liu, Z.; Su, L.; Du, L.; Huang, J. Evolved-Cooperative. *Entropy* **2019**, *21*, 912. [CrossRef]
11. Guan, H.; Dai, Z.; Guan, S.; Zhao, A. A Neutrosophic Forecasting Model for Time Series Based on First-Order State and Information Entropy of High-Order Fluctuation. *Entropy* **2019**, *21*, 455. [CrossRef]
12. Lei, M.; Ming, S.; Yu, S. Demand Forecasting Approaches Based on Associated Relationships for Multiple Products. *Entropy* **2019**, *21*, 874. [CrossRef]
13. Vanhoucke, M.; Batselier, J. A Statistical Method for Estimating Activity Uncertainty Parameters to Improve Project Forecasting. *Entropy* **2019**, *21*, 952. [CrossRef]
14. Popkov, Y.S. Soft Randomized Machine Learning Procedure for Modeling Dynamic Interaction of Regional Systems. *Entropy* **2019**, *21*, 424. [CrossRef]

Article

Acknowledging Uncertainty in Economic Forecasting. Some Insight from Confidence and Industrial Trend Surveys

Ana Jesús López-Menéndez * and Rigoberto Pérez-Suárez

Department of Applied Economics, University of Oviedo, Campus del Cristo s/n, 33006 Oviedo, Asturias, Spain; rigo@uniovi.es
*   Correspondence: anaj@uniovi.es; Tel.: +34-985103759

Received: 18 March 2019; Accepted: 12 April 2019; Published: 18 April 2019

Abstract: The role of uncertainty has become increasingly important in economic forecasting, due to both theoretical and empirical reasons. Although the traditional practice consisted of reporting point predictions without specifying the attached probabilities, uncertainty about the prospects deserves increasing attention, and recent literature has tried to quantify the level of uncertainty perceived by different economic agents, also examining its effects and determinants. In this context, the present paper aims to analyze the uncertainty in economic forecasting, paying attention to qualitative perceptions from confidence and industrial trend surveys and making use of the related ex-ante probabilities. With this objective, two entropy-based measures (Shannon's and quadratic entropy) are computed, providing significant evidence about the perceived level of uncertainty. Our empirical findings show that survey's respondents are able to distinguish between current and prospective uncertainty and between general and personal uncertainty. Furthermore, we find that uncertainty negatively affects economic growth.

Keywords: uncertainty; qualitative surveys; Shannon's entropy; quadratic entropy; VAR; impulse-response analysis

---

1. Introduction

In the context of a complex world characterized by high levels of uncertainty, several works have emphasized the need of acknowledging uncertainty in economic modeling and forecasting [1–3], also suggesting the convenience of complementing the predictions with the surrounding levels of uncertainty [4,5].

The controversial debate about the effects of uncertainty in consumers, managers, investors, … is not easy to solve due both to the lack of data and to methodological difficulties. Although the traditional practice consisted of reporting point predictions without specifying the attached probabilities, uncertainty about the prospects deserves increasing attention, and recent literature has tried to quantify the level of uncertainty perceived by different economic agents also examining its effects and determinants.

Within this context, the present paper aims to analyze the uncertainty around economic forecasts, paying attention to qualitative perceptions. With this purpose, the next section briefly describes the role of uncertainty in economic forecasting and the main difficulties that should be addressed in order to approach the level of uncertainty from surveys.

The materials and methods are presented in Section 3 where we set three different hypotheses referred to the measurement of forecasting uncertainty and its impact on economic growth. Since the estimation of uncertainty is closely related to the available information, this section also describes the statistical sources (barometers of the Spanish Center for Sociological Research and regional Industrial Trend Surveys) and the proposed measures (Shannon's and quadratic entropy).

The empirical results are described in Section 4, where we summarize the main findings on the proposed hypotheses based on Confidence and Industrial Trend Surveys. Finally, section five contains the discussion of the obtained results and some concluding remarks.

2. Uncertainty in Economic Forecasting

In spite of the wide consensus on the main role of uncertainty in economic forecasting, it appears not to receive the academic attention it deserves, as emphasis is often made in best estimates and predictions without paying attention to the surrounding uncertainty. However, uncertainty has become increasingly important in economic forecasting due to both theoretical and empirical reasons and recent literature has tried to quantify the level of uncertainty perceived by different economic agents also examining its effects and determinants.

Different approaches can be used in the measurement of uncertainty, including statistical models and human judgement. While ex-post uncertainty has been usually studied by looking at forecasting errors, ex-ante uncertainty—which is particularly interesting from the economic point of view—could be estimated from survey data, as we intend in this work. With regard to the ex-post approach, empirical evidence including the M-competitions [6,7] shows that neither forecasting errors nor uncertainty are reduced with more sophisticated forecasting techniques or higher level of respondents' expertise. From the ex-ante perspective, as explained by [8] the methodology is evolving with the types of surveys and datasets. Different proxies have been proposed to approach forecast uncertainty being one of the most popular disagreement, usually measured through the variance of the point forecasts. However, several authors [8–10] have emphasized the limitations of this approach, since disagreement between forecasters only considers the between component, and its reliability as a proxy for uncertainty will depend on several factors, as the stability and length of the forecasting horizon. In this context, the use of entropy-based measures seems to be a good option to take advantage of the information provided by forecasts surveys, including both the expected economic outcomes and the surrounding uncertainty levels. Unfortunately, as pointed out in [9] most of the professional surveys lack quantitative measures of uncertainty as they only aggregate the information of individuals' assessment on the economic variables.

Measuring the level of uncertainty greatly depends on the information available to estimate probabilities that appear in uncertainty measures. A wide variety of existing surveys are summarized in Table 1, taking into account their size, level of expertise and information content.

Table 1. Main typologies of forecasting surveys.

Survey	Size	Level of Expertise	Information
Surveys of professional forecasters	Medium	High	Detailed (Density forecasts)
Panels of professional forecasters	Medium	High	Reduced (consensus forecasts)
Expert elicitations	Small	Very high	Detailed (subjective probabilities)
Confidence surveys	High	Low/Medium	Medium (frequency probabilities)

The first category considered corresponds to surveys of professional forecasters (SPF), provided quarterly by the Federal Reserve Bank of Philadelphia, the European Central Bank and some other institutions, such as the Bank of England. Although the antecedents of SPF date from 1968 when the American Statistical Association and the National Bureau of Economic Research jointly started a quarterly survey of macroeconomic forecasters, the Federal Reserve Bank of Philadelphia assumed the responsibility for the survey and named it SPF in 1990. Similar investigations have been developed by the European Central Bank since 1999 (Survey of Professional Forecasters) and by the Bank of England since 1996 (Survey of External Forecasters). These highly specialized panels have an intermediate size (around 36 forecasters in the US-SPF and 75 forecasters in the EU-SPF) and collect forecasters' expectations on key economic variables, such as inflation and GDP growth and unemployment rate, also including a particularly interesting feature: forecasters are asked to provide their subjective probabilities that a variable will fall into each of the predefined forecasting intervals, thus allowing the estimation of uncertainty from density forecasts as shown in [9,10]. With this aim, different approaches have been proposed to handle density functions, assuming some specific probability models such as the uniform [11], normal [12,13] or generalized beta [14].

Despite their success, surveys of professional forecasters also have some important limitations such as the difficulty of response and the lack of homogeneity, due to methodological changes and the replacement of forecasters.

The second category refers to panels of institutional or professional forecasters that are available for different countries, providing short-term predictions referred to the main economic aggregates (GDP and its components, employment, prices, etc.). These panels usually comprise a moderate number of recognized institutions including universities, research services of banks and economic analysis institutes. In the Spanish context, the private non-profit organization FUNCAS (a think tank dedicated to social and economic research, https://www.funcas.es). publishes the Spanish economy forecast panel, a survey carried out every two months among a panel of 19 institutions that has been studied in [15,16]. Although this kind of panel usually includes a consensus forecast (computed as the average) and some measures of dispersion (rank, variance, etc.) they do not allow the estimation of probabilities and uncertainty measures.

Expert elicitations are another interesting source of specialized information referring to future prospects and associated uncertainties, usually collected through subjective probabilities. This third category has been increasingly used in order to obtain experts judgments from scientists, engineers, and other analysts who are knowledgeable about particular issues and variables of interest, as described in [17] among others. Obviously, the size of these panels is quite small due to the required level of expertise and the difficulty of assigning the required probabilities.

Finally, the fourth category corresponds to confidence surveys, comprising a wide variety of initiatives performed for different countries and sectors, where a high number of economic agents (consumers, managers, etc.) show their positive or negative attitudes with regard to the current, previous or future economic activity. In the European framework, regular harmonized surveys are conducted for the member countries under the Joint Harmonized EU Program of Business and Consumer Surveys. The information provided by business and consumer confidence surveys has been proven to be extremely useful for short-term forecasting, detection of turning points and economic analysis [18,19]. Confidence survey data are generally presented as balances between the percentage of positive and negative answers to each question and their results are mainly used to compute synthetic indicators built on selected questions (confidence indicators, economic sentiment indicators, business climate indicators, etc.).

Furthermore, the vast amount of information provided by the participants in these surveys allows the estimation of frequentist probabilities and uncertainty measures, as we will show in the next sections of this paper.

3. Materials and Methods

Although the previously described surveys provide a huge amount of information, many empirical studies make exclusive use of consensus forecasts rather than analyzing individual forecasts and examining the surrounding level of uncertainty. Moreover, the estimation of uncertainty has mainly been based on subjective probabilities provided by the surveys of professional forecasters or the experts' elicitations, while this approach has scarcely been used in the case of confidence surveys. In this paper we aimed to fill this gap, approaching the economic uncertainty with probabilities estimated from confidence and industrial trend surveys. More specifically, we focused on the barometers developed by the Spanish Center for Sociological Research (CIS) and the regional Industrial trend Surveys (ECI), referred to as Asturias, providing significant evidence about both the economic situation and the encompassing uncertainty.

3.1. Hypotheses

Three hypotheses have been proposed referred to the informational content of the considered surveys and the relationship between uncertainty and economic growth:

1. Confidence surveys allow an adequate estimate of the economic situation and the surrounding uncertainty.
2. A survey's respondents can properly distinguish between current and prospective uncertainty and between general and personal uncertainty.
3. Uncertainty negatively affects economic growth.

With the aim of testing the proposed hypotheses we firstly describe the available information, respectively provided by the barometer of the Spanish Center of Sociological Research and the regional Industrial Trend Survey. Besides supplying synthetic indicators, both sources allow the estimation of probabilities and uncertainty levels through entropy-based measures. More specifically in this paper we used Shannon's and quadratic Indexes, thus allowing a comparison of the uncertainty levels estimated by both expressions.

Furthermore, the estimation of econometric models allows a more detailed analysis about the causal relationship and the impact of uncertainty on economic growth. Thus, vector autoregresive (VAR) models were estimated, and their results are described in Section 4.

3.2. Data Description: Confidence Barometers and Industrial Trend Surveys

CIS is an independent entity assigned to the Ministry of the Presidency, and gathers the necessary data for research in very different fields, carrying out a wide variety of surveys, whose data is in the public domain. The CIS databank includes confidence barometers, polls carried out since 1994 on a monthly basis (except in August), with the aim of measuring Spanish public opinion. As described in the CIS website [20] these polls involve interviews with around 2500 randomly-chosen people from all over the country, including a block of variable questions which focuses on the assessment of both the economic situation in Spain and the personal economic situation, as described in Table 2.

Table 2. Spanish Center for Sociological Research (CIS) confidence barometer.

Items	Options
Assessment of the current economic situation in Spain	Very Good, Good, Intermediate, Bad, Very Bad
Retrospective assessment of the economic situation in Spain (one year before)	Better. Equal, Worse
Prospective assessment of the economic situation in Spain (one year)	Better. Equal, Worse
Assessment of the current personal economic situation	Very Good, Good, Intermediate, Bad, Very Bad
Prospective assessment of the personal economic situation (one year)	Better, Equal, Worse

Microdata provided by the monthly polls can be downloaded from the CIS website www.cis.es and allow the calculation of probabilities based on relative frequencies assigned to the alternative options.

Regarding the Spanish industrial trend surveys, the Ministry of Industry, Trade and Tourism, and also some regional statistical offices develop qualitative surveys with the aim of catching the opinion of industrial managers about the current situation and future prospects. More specifically, the questionnaire is directed to the management industrial personnel and compiles qualitative information referred to the present levels of the portfolio orders and the production, sale prices and employment expected for the next months.

Three alternative answers (high, normal or low) are provided for those questions reflecting the present level, while the options to increase, to stay or to diminish can be selected if the questions refer to the expected tendency. The individual answers given to the different questions are aggregated in order to obtain series by classes and categories and the balance between the extreme options provides an indicator with values oscillating between $+100$ and -100 (totally' optimistic and pessimistic situations). The results for each variable can also be summarized through the industrial climate indicator (ICI) computed as an arithmetic mean of the balances of the portfolio orders, the production expectations and, with the opposite sign, the level of finished product stocks. This composite indicator is widely used to provide a global vision of the industrial confidence in relation to the conjunctural evolution. In fact, as the leading indicator signals summarized in the ICI are assumed to happen before the economy turning points, this index can be used as a leading indicator of economic activity allowing the obtention of economic turning point forecasts as shown in [16].

Since the estimation of uncertainty requires detailed information about individuals perceptions we focus on the regional industrial trend survey referred to Asturias, whose databank is fully available from [21] allowing the estimation of the corresponding probabilities.

3.3. Shannon's and Quadratic Entropy Measures

Although qualitative surveys have been extensively used to obtain synthetic indicators, few attempts have been made in order to quantify the uncertainty level perceived by the respondents. In this paper we aim at filling this gap, and also analyzing to which extent the level of uncertainty perceived by the experts is related with the economic situation.

Entropy measures provide a suitable framework for our goal, as entropy is a function of the probability distribution and not a function of the actual values taken by the random variable. Since microdata of qualitative surveys allow the estimation of the probabilities assigned to each possible outcome, entropy measures can also be estimated. Thus, given the set of n distinct mutually exclusive options for a specific question, the individual responses allow the estimation of frequency probabilities $p_i, \forall i = 1, \ldots n$ such that $p_i \geq 0, \sum_i p_i = 1$. Shannon [22] defines the information content of a single outcome as $h(p_i) = \log\left(\frac{1}{p_i}\right)$. According to this definition, observing a rare event provides much more information than observing another, more probable outcome.

In this context, Shannon's entropy is defined as the expected amount of information and can be computed as $H = -\sum_i p_i \log(p_i)$. This expression plays a central role since it fulfills a number of interesting properties which, as shown in [22] substantiate it as a reasonable measure of information, choice or uncertainty:

1. $H = 0$ if and only if all the p_i but one are zero, this one having the value unity. Thus the result of H is null only when we are certain about the outcome, and otherwise H is positive.
2. For a given n, H is a maximum and equal to $\log(n)$ when all the p_i are equal $p_i = \frac{1}{n}, \forall i = 1, 2, \ldots, n$. This is also intuitively the most uncertain situation.
3. Any change toward equalization of the probabilities $p_1, p_2, \ldots, p_n$ increases H. Thus, if $p_1 < p_2$ and we increase p_1 decreasing p_2 an equal amount so that p_1 and p_2 are more nearly equal, then H increases. More generally, if we perform any averaging operation on the p_i of the form $p_i' = \sum_j a_{ij} p_j$ where $\sum_i a_{ij} = \sum_j a_{ij} = 1$ and $a_{ij} \geq 0, \forall i, j = 1, \ldots, n$ then H increases, except in the case where this transformation amounts to no more than a permutation of the p_i with H remaining the same.

Following a similar approach, Pérez [23] proposes the individual quadratic entropy, which can be computed for a single outcome as $h^2(p_i) = 2(1 - p_i)$. According to this proposal, the quadratic entropy is quantified as twice the distance of the probability of an event from the true outcome, and similarly to Shannon's measure, the information provided by a rare event is higher than the information corresponding to a more likely one.

Given a set of probabilities $p_i, \forall i = 1, \ldots, n$ such that $p_i \geq 0, \sum_i p_i = 1$, the quadratic entropy is defined in [23] as the expected value of the individual quadratic entropies, given by the expression $H^2 = 2 \sum_i p_i(1 - p_i)$. This is a suitable measure of uncertainty since it fulfils the requirements proposed by Shannon. More specifically:

1. $H^2 = 0$ if and only if all the p_i but one are zero, this one having the value unity.
2. For a given n, H^2 is a maximum when all the p_i are equal $p_i = \frac{1}{n}, \forall i = 1, 2, \ldots, n$. This maximum value, corresponding to the most uncertain situation, is given by the expression $2(1 - \frac{1}{n})$ and in the limit it takes a value of two.
3. Any change toward equalization of the probabilities $p_1, p_2, \ldots, p_n$ increases the quadratic entropy H^2. Thus, if we perform any averaging operation on the p_i of the form $p_i' = \sum_j a_{ij} p_j$ where $\sum_i a_{ij} = \sum_j a_{ij} = 1$ and $a_{ij} \geq 0, \forall i, j = 1, \ldots, n$ then H^2 increases, except if this transformation is only a permutation of the p_i (in this case H^2 does not change, since the quadratic entropy fulfils the property of symmetry).

The quadratic measure has been successfully used in different economic applications, including the evaluation of forecasts [24,25]. Taking into account its suitable behavior, in this paper we propose the joint use of Shannon's and quadratic entropy to approach the level of uncertainty.

4. Results

This section summarizes the results obtained from the CIS barometer and the industrial confidence survey, providing empirical evidence referred to the three proposed hypotheses. As previously described, the available information allows us to compute uncertainty levels through Shannon's and quadratic entropy measures, respectively given by the expressions:

$$H = -\sum_{i=1}^{n} p_i \log p_i \tag{1}$$

$$H^2 = 2 \sum_{i=1}^{n} p_i(1 - p_i). \tag{2}$$

As these expressions verify the reasonable properties to be considered as suitable measures of uncertainty they have been used in a complementary way.

4.1. Hypothesis 1

According to the first proposed hypothesis, confidence surveys allow an adequate estimate of the economic situation and the surrounding uncertainty. With the aim of testing this assumption we first consider the CIS Confidence barometers collecting extremely interesting information referred to respondents' perception about both the economic situation in Spain and their personal situation. Since the CIS survey is not available in august, we have used quarterly series. The results of both entropy measures are represented in Figure 1, showing a very similar evolution. As expected, Shannon's and quadratic entropy appear to be highly correlated (the linear correlation coefficient between them reaches the value 0.91) and the level of uncertainty significantly increases between 2005 and 2007 according to both measures. Subsequently, since the end of 2007, a decreasing pattern is observed until the first quarter of 2013 when both indicators reach their minimum value and the uncertainty starts a new rise.

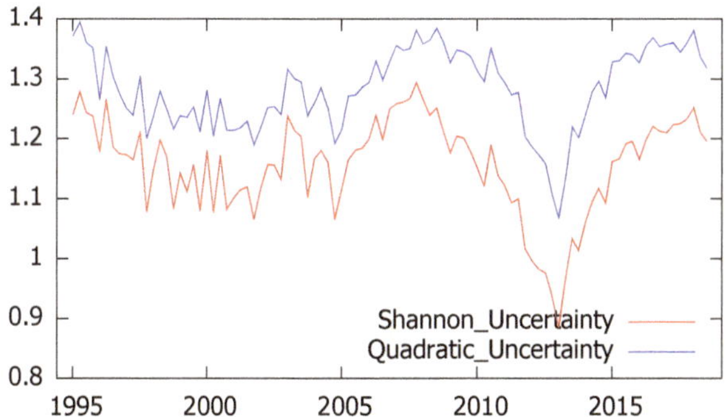

Figure 1. Evolution of Shannon's and quadratic uncertainty associated to current economic situation in Spain.

The analysis of these time series confirms that seasonality does not affect the levels of perceived uncertainty (the Kruskal–Wallis test fails to reject the null hypothesis of non seasonality and the same conclusion is obtained through an OLS regression with periodic dummy variables, that are found to be non-significant). It is also interesting to remark that the "herding effect" which has been largely studied in panels of forecasters does not appear in this case, as the respondents have been randomly selected and there is no influence among them.

A similar analysis has been performed on the industrial trend survey that, as we have previously described, aims at catching industrial managers' opinions about the present and future economic situation. In this case we analyze the information referred to the region of Asturias from January 1990 to December 2018 and, although the questionnaire includes qualitative information related to several variables, we mainly focus on industrial production.

Experts' answers were used to compute the probabilities associated to the three alternative options for the current output level (high, normal and low), leading to the estimation of monthly series for Shannon and quadratic uncertainty whose results are plot in Figure 2.

As expected, both entropy measures provide quite similar results when measuring uncertainty referred to the current industrial production, leading to a linear correlation coefficient of 0.98.

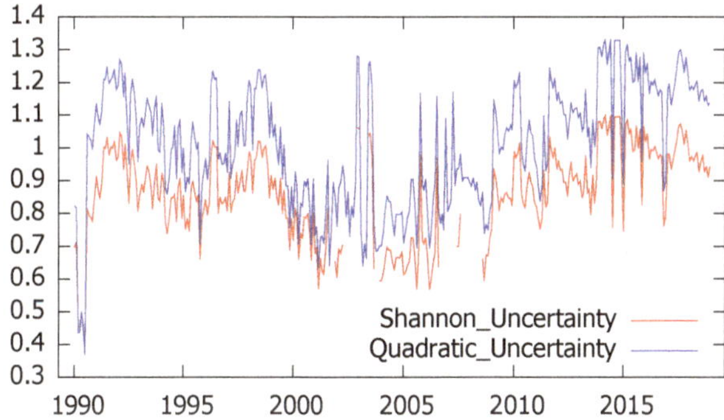

Figure 2. Evolution of Shannon's and quadratic uncertainty associated to current industrial production in Asturias.

4.2. Hypothesis 2

The second hypothesis refers to the ability of survey's respondents to distinguish between current and prospective uncertainty and between general and personal uncertainty. Since the CIS barometers include current, retrospective and prospective assessments of the economic situation in Spain, we have compared the corresponding levels of Shannon's and quadratic uncertainty, represented in Figures 3 and 4. As it can be seen, according to both entropy measures current uncertainty is found to be higher than prospective uncertainty, which generally exceeds past uncertainty. However, some exceptions are found, corresponding to years 2012 and 2013 when the present uncertainty reaches its minimum values and is exceeded by prospective uncertainty.

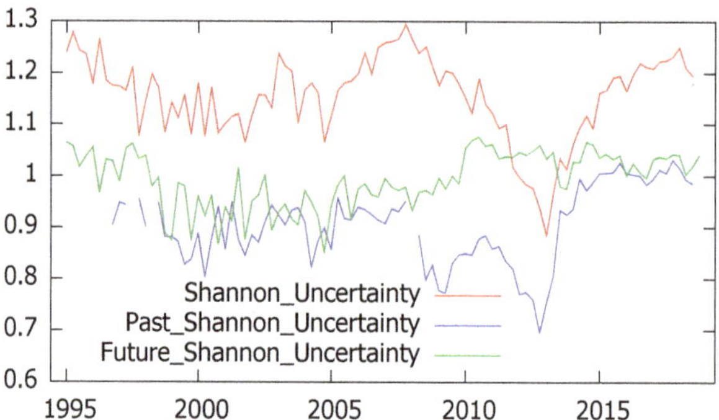

Figure 3. Shannon's uncertainty for current, retrospective and prospective economic situation in Spain.

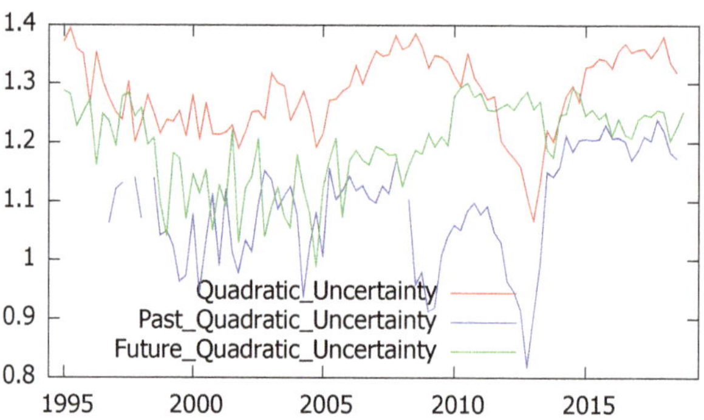

Figure 4. Quadratic uncertainty for current, retrospective and prospective economic situation in Spain.

As we have seen in the previous figures, Shannon's and quadratic entropy mostly agree in the quantification of uncertainty. No matter if we consider the general or the personal situation or if uncertainty refers to present, past or future periods, the correlation coefficients always exceed 90% as summarized in Table 3.

Table 3. Correlation coefficients between Shannon's and quadratic Uncertainty.

	Spanish Economy	Personal Economy
Current	0.91	0.97
Retrospective (one year before)	0.99	—
Prospective (one year)	0.99	0.97

In order to analyze to which extent survey's respondents can properly distinguish between general and personal uncertainty we have also studied the perceptions about their personal economic situation. Although these series, represented in Figure 5 were quite short (they are only available since 2010) and therefore should be considered cautiously, the results show that until 2015 the level of uncertainty was higher when it refers to the personal situation. However, the perception of personal uncertainty seems to be more stable than that referred to the general economic situation and both measures are negatively correlated (-0.73 and -0.6 for Shannon and quadratic uncertainty respectively).

It is also interesting to mention that this situation changes when we focus on uncertainty about the future. In this case, we find no significant correlation between personal and general uncertainties, measured either with Shannon or quadratic entropy.

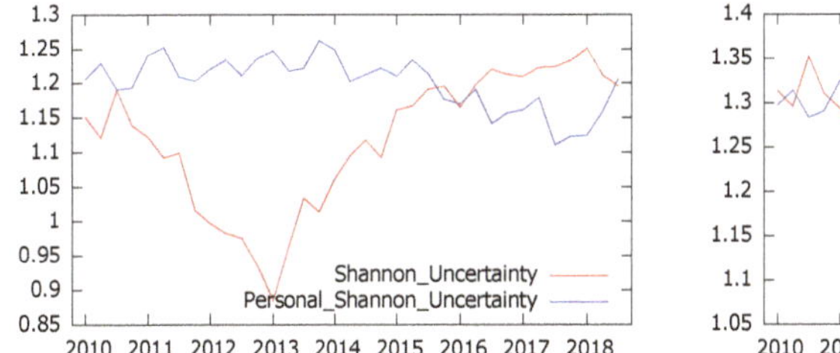

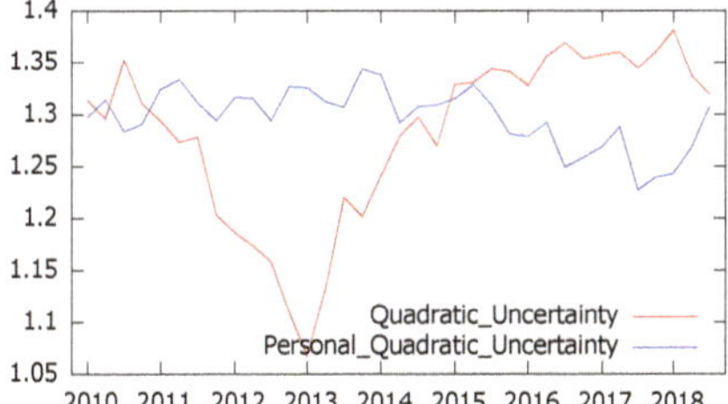

Figure 5. Shannon's (**left**) and quadratic (**right**) uncertainty for personal and Spanish economic situation.

Regarding the relationship between current and prospective uncertainty, the findings differ from personal to country's uncertainty (Table 4). It is interesting to remark that—independently of the measure of entropy used—when we pay attention to the personal situation there is a strong relationship between current and prospective uncertainty while this correlation does not exist when we focus on the assessment of the general economic situation. These findings confirm that the respondents were able to properly distinguish the perceptions related to their own economic situation and prospects from those referred to the country as a whole.

Table 4. Correlation coefficients between current and prospective uncertainty.

	Spanish Economy	Personal Economy
Shannon's Entropy	-0.023	0.816
Quadratic Entropy	0.19	0.769

With regard to the industrial trend surveys, the experts' answers referred to future prospects (whose alternative options are to increase, to stay and to decrease) allow the estimation of future uncertainty, leading to similar results for Shannon's and quadratic entropy (the correlation coefficient amounted to 0.99). As in the previous application, the obtained results show that the respondents clearly distinguished between present and prospective uncertainty. In fact, regardless of the entropy measure considered, uncertainty referred to the present industrial output is found to be higher and more stable than uncertainty referred to the future industrial production.

These findings, represented in Figure 6 for the quadratic entropy, have been corroborated through paired difference tests, leading to the conclusion that the expected current uncertainty significantly exceeds prospective uncertainty.

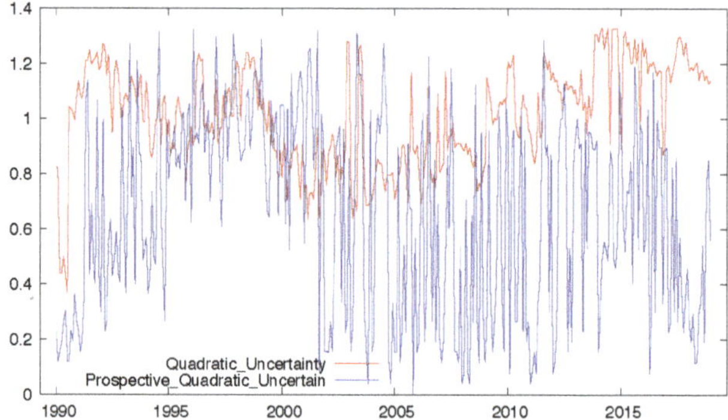

Figure 6. Evolution of quadratic uncertainty associated to current and prospective industrial production in Asturias.

4.3. Hypothesis 3

According to the third hypothesis, which we consider especially interesting, uncertainty negatively affects economic growth. In order to analyze this assumption we first focus on the CIS barometer, considering the estimated Shannon's and quadratic entropy together with two additional quarterly series: the annual GDP growth rate and a synthetic indicator.

Denoting by X_t the quarterly GDP, the related annualized growth rate is given by the expression $g = \frac{X_t}{X_{t-4}} - 1$.

Furthermore, following a widely extended practice in this kind of surveys, a synthetic index can be computed in order to summarize the answers. Focusing on the assessment of the current economic situation in Spain, this indicator can be easily obtained as follows: $SI = 2p_{very\_good} + p_{good} - p_{bad} - 2p_{very\_bad}$, where p_{very_good}, p_{good}, p_{bad}, p_{very_bad} represent the probabilities assigned to each of the considered categories, estimated through the corresponding relative frequencies.

Once this indicator has been computed we can analyze the relationship between the perceived economic situation and the corresponding level of uncertainty. Although these quarterly series appear to be contemporaneously uncorrelated, the scatter diagram represented in Figure 7 provides some interesting hints about the parabolic pattern of uncertainty regarding the synthetic index.

As it can be seen in this graph, low uncertainty with low dispersion is associated with very negative perceptions of the economic situation, whilst as perceptions of economic situation increase, so too do measures of uncertainty with associated increasing dispersion.

With the aim of examining how uncertainty impacts on economic activity, a more detailed analysis has been developed through VAR models. More specifically, we propose VAR models involving the economic growth, the synthetic index and the uncertainty measure, and we run two versions by using either Shannon's entropy or quadratic entropy as the measure of uncertainty. We estimated both VAR models on quarterly data from 1996 to 2018 (T = 89) and, following the commonly used information criteria (Akaike, Schwartz and Hannah–Quinn), we considered two lags (p = 2). Tables A1 and A2 in the Appendix A collect the VAR estimation results.

It is interesting to notice that the Granger causality test (whose null hypothesis is "no Granger causality") leads to the p-values collected in Tables 5 and 6, showing that variations in GDP are explained by both the synthetic index and the level of uncertainty, regardless of the entropy measure used. Moreover, uncertainty was found to Granger cause the synthetic index at the 10% level.

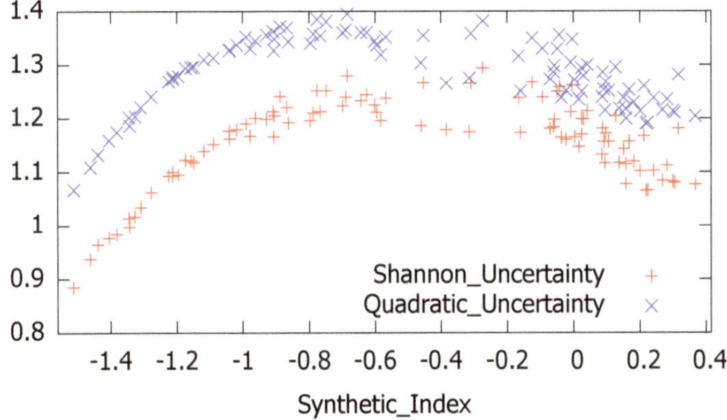

Figure 7. Shannon's and quadratic uncertainty versus synthetic index.

Table 5. p-values for the Granger causality tests (F-test of zero restrictions) in vector autoregresive (VAR) 1.

	GDP Growth	Synthetic Index	Shannon Uncertainty
All lags of GDP growth	0.0000	0.0633	0.1369
All lags of synthetic index	0.0059	0.0000	0.6183
All lags of Shannon's uncertainty	0.0059	0.1200	0.0000

Table 6. p-values for the Granger causality tests (F-test of zero restrictions) in VAR 2.

	GDP Growth	Synthetic Index	Quadratic Uncertainty
All lags of GDP growth	0.0000	0.0510	0.1571
All lags of synthetic index	0.0031	0.0000	0.5184
All lags of quadratic uncertainty	0.0269	0.2874	0.0000

Since uncertainty causes economic growth, we have also analyzed the impulse responses for GDP growth and the synthetic index to a one standard deviation shock in the uncertainty level, measured both by Shannon and quadratic entropy. The results are plot in Figures 8 and 9, showing that the effects of one standard deviation shock to the uncertainty in economic growth are mostly negative with their largest impacts around 12–15 months.

According to the impulse-response analysis, the behavior is quite robust with regard both to the economic indicator (GDP growth and synthetic index) and the uncertainty measure (Shannon's and quadratic entropy).

Regarding the impact of the synthetic index on GDP growth, Figure 10 represents the impulse-response analysis for one standard deviation shock in the synthetic index. As expected, the response in this case is positive and faster, with its largest impact taking place around five months.

Following the same method we examine the relationship between uncertainty and industrial production. As in the previous analysis we estimate two VAR models including, in this case, four monthly series, corresponding to the regional IPI, the ICI, the synthetic index (SI) and the level of uncertainty.

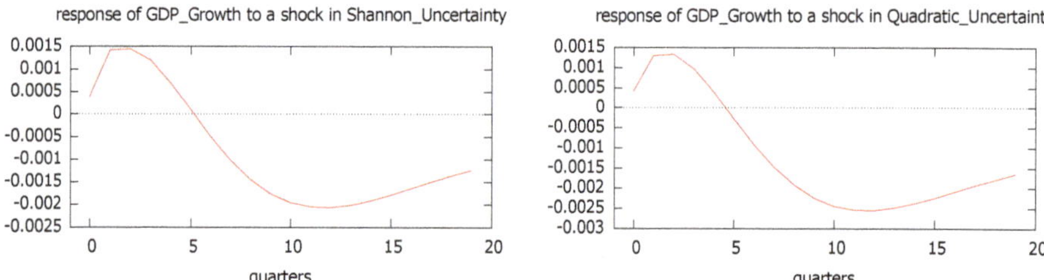

Figure 8. Impulse responses of GDP growth to a shock in Shannon's uncertainty (**left**, VAR 1) and quadratic uncertainty (**right**, VAR 2).

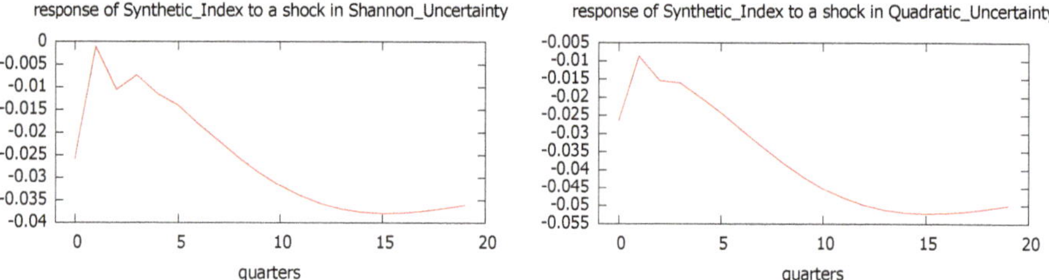

Figure 9. Impulse responses of the synthetic index to a shock in Shannon's uncertainty (**left**, VAR 1) and quadratic uncertainty (**right**, VAR 2).

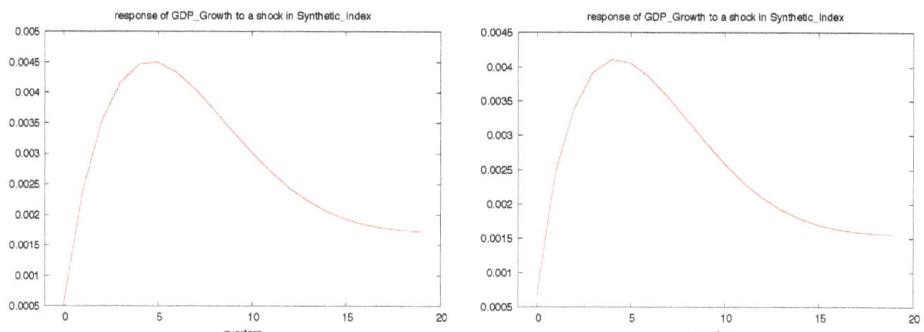

Figure 10. Impulse response of GDP growth to a shock in the synthetic index according to VAR 1 (**left**) and VAR2 (**right**).

These series have been obtained from SADEI [21], the regional statistical office of Asturias which provides monthly information about the industrial production index (currently referred to year 2010) and the ICI, a leading indicator of economic activity [16] computed as an arithmetic mean of the balances of the portfolio orders, the production expectations and—with the opposite sign—the level of stocks. Regarding the Synthetic Index, it has been computed as in the previous subsection from the balance of positive and negative answers referred to industrial output, using the estimated frequency probabilities.

Finally, with regard to the level of uncertainty, two VAR models have been estimated, using Shannon's entropy in the first one and quadratic entropy in the second. Since Shannon's index cannot be computed for some months with null probability in any of the categories we have restricted the sample size in both models (T = 124) in order to provide fully homogeneous results.

It is interesting to remark that, taking into account the series analyzed, VAR specification includes in this case constant, trend and seasonality. Following the information criteria, only one lag was considered.

The estimation results are collected in the Appendix A (Tables A3 and A4) and the conclusions show outstanding similarities for the two uncertainty measures, as it can be seen in Figure 11. As expected, the impulse responses of the regional industrial production index to a one standard deviation shock to the uncertainty level are negative with their largest impacts during the first two periods and a quick recovery in the medium run.

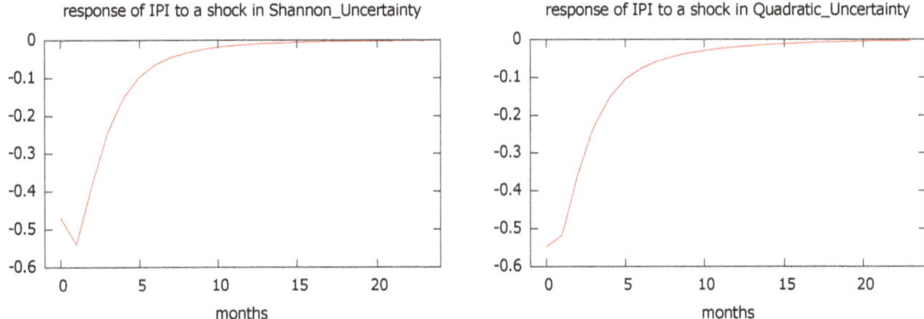

Figure 11. Impulse response of industrial production index (IPI) to a shock in Shannon's (**left**) and quadratic uncertainty (**right**).

5. Discussion and Concluding Remarks

Our empirical results show that qualitative surveys can be successfully used to approach both the economic situation and the surrounding uncertainty, thus agreeing with the first proposed hypothesis. More specifically, the information provided by the respondents to the CIS barometer and the industrial trend survey confirms the usefulness of both sources and the adequacy of entropy-based measures to approach uncertainty. In addition, we find that—as indicated by previous works [2]—the level of expertise does not affect the adequacy of respondents' answers.

According to the two empirical applications, based in confidence barometers and industrial trend surveys, Shannon's and quadratic entropy mostly agree in the quantification of uncertainty, no matter if we consider the Spanish or the personal economic situation or if uncertainty refers to present, past or future periods.

Regarding the second hypothesis, the available information suggests that surveys respondents can properly distinguish between current and prospective uncertainty and between general and personal uncertainty. According to the CIS barometer and the Industrial Trend Survey, current uncertainty is higher than prospective uncertainty, regardless of the measure used. Furthermore, the CIS barometer provides significant evidence about the capability of survey's respondents to distinguish between personal and national uncertainty: first, the perception of personal uncertainty seems to be more stable than that referred to the Spanish economic situation and second, a strong positive correlation is found between current and prospective uncertainty referred to the personal situation, unlike what happens when we focus on the economic situation of the country. Finally, our empirical applications show that uncertainty negatively affects economic growth, providing evidence about the responses of economic growth and industrial production to a shock in the uncertainty measures.

The estimation of VAR models leads to some interesting findings that broadly match with previous works as [26–28]. More specifically, the adverse impacts of uncertainty shocks on economic activity have been documented among others in [26,27] while [28] provides significant evidence about the fall of industrial production as a response to a volatility shock.

Our results based on the CIS barometer are quite robust, since they confirm that uncertainty shocks, regardless of the entropy measure used, have a negative impact on economic activity, whether measured through GDP growth or the synthetic index. According to the impulse-response analysis, the largest impacts take place around 12–15 months, followed by a slow recovery. Similarly, when we focus on the industrial trend survey, we find that one standard deviation shock to the uncertainty level (measured either by Shannon's or quadratic entropy) leads to sharp reductions in the regional industrial production, with a quick recovery in the medium run.

Despite their limitations, these interesting findings confirm the potential of qualitative surveys in the assessment of economic uncertainty also suggesting the need of further research in this field.

Author Contributions: Both authors contributed to the design and implementation of the research, to the analysis of the results and to the writing of the manuscript.

Funding: This research was funded by Consejería de Hacienda y Sector Público (Treasury and Public Sector Counseling, Government of the Principalty of Asturias, Spain), Grant SV-PA-18-03.

Acknowledgments: The authors would like to thank two anonymous referees for very helpful comments. We would also like to acknowledge the financial support of the Government of the Principalty of Asturias.

Conflicts of Interest: The authors declare no conflict of interest.

Appendix A

Table A1. Results of VAR estimation for GDP growth, synthetic index and Shannon's uncertainty.

	Equation (1): GDp Growth	Equation (2): Synthetic Index	Equation (3): Shannon's Uncertainty
const	0.0075 (0.0077)	0.0417 (0.1992)	0.1898 ** (0.0784)
GDP_Growth_1	1.4824 *** (0.0792)	4.6636 ** (2.0422)	1.0144 (0.8033)
GDP_Growth_2	−0.6090 *** (0.0826)	−3.7670 * (2.1275)	−0.3986 (0.8369)
Synthetic_Index_1	0.0159 *** (0.0044)	0.8687 *** (0.1129)	0.0225 (0.0444)
Synthetic_Index_2	−0.0123 *** (0.0043)	0.0896 (0.1098)	−0.0319 (0.0431)
Shannon_Uncertainty_1	0.0305 *** (0.0105)	0.4737 * (0.2713)	0.5175 *** (0.1067)
Shannon_Uncertainty_2	−0.0332 *** (0.0102)	−0.5452 ** (0.2626)	0.3018 *** (0.1033)

Note: Standard deviation in parenthesis; $*p < 0.1, **p < 0.05, ***p < 0.01$.

Table A2. Results of VAR estimation for GDP growth, synthetic index and quadratic uncertainty.

	Equation (1): GDP Growth	Equation (2): Synthetic Index	Equation (3): Quadratic Uncertainty
const	0.0101 (0.0099)	0.1756 (0.2523)	0.1781 ** (0.0797)
GDP_Growth_1	1.4832 *** (0.0810)	4.7214 ** (2.0709)	0.8938 (0.6543)
GDP_Growth_2	−0.6036 *** (0.0848)	−3.5476 (2.1687)	−0.4385 (0.6852)
Synthetic_Index_1	0.0143 *** (0.0044)	0.8280 *** (0.1132)	−0.0048 (0.0358)
Synthetic_Index_2	−0.0111 ** (0.0043)	0.1169 (0.1093)	−0.0086 (0.0345)
Shannon_Uncertainty_1	0.0315 ** (0.0134)	0.3344 (0.3447)	0.6423 *** (0.1089)
Shannon_Uncertainty_2	−0.0362 *** (0.0132)	−0.5131 (0.3375)	0.2058 * (0.1066)

Note: Standard deviation in parenthesis; $*p < 0.1, **p < 0.05, ***p < 0.01$.

Table A3. Results of VAR estimation for IPI, industrial climate indicator (ICI), synthetic index and Shannon's uncertainty.

	Equation (1): IPI	Equation (2): ICI	Equation (3): Synthetic Index	Equation (4): Shannon's Uncertainty
const	78.2967 *** (13.5687)	−7.7117 (21.9591)	−50.1833 (30.3293)	0.4149 (0.2651)
IPI_1	0.4264 *** (0.0892)	−0.2030 (0.1444)	0.3989 ** (0.1995)	−0.0008 (0.0017)
ICI_1	0.0655 (0.0539)	0.5432 *** (0.0872)	0.5030 *** (0.1204)	0.0023 ** (0.0011)
Synthetic_Index_1	0.0253 (0.0221)	0.0512 (0.0358)	0.6958 *** (0.0495)	−0.0010 ** (0.0004)
Shannon_Uncertainty_1	−5.2332 (4.0951)	3.3735 (6.6274)	−14.5058 (9.1536)	0.5159 *** (0.0800)
S1	5.6129 *** (1.7703)	−5.9058 ** (2.8650)	4.8208 (3.9570)	−0.0036 (0.0346)
S2	2.1790 (1.6913)	−8.3030 *** (2.7372)	1.7326 (3.7805)	0.0462 (0.0330)
S3	9.2197 *** (1.7113)	−6.2971 ** (2.7695)	2.6542 (3.8251)	0.0175 (0.0334)
S4	−0.6028 (1.7058)	−5.3639 * (2.7606)	2.9481 (3.8129)	0.0333 (0.0333)
S5	7.2989 *** (1.7316)	−11.0793 *** (2.8024)	3.8561 (3.8706)	0.0151 (0.0338)
S6	4.0416 ** (1.7026)	−7.6083 *** (2.7554)	6.5073 * (3.8057)	0.0112 (0.0332)
S7	3.1868 * (1.7073)	−9.1271 *** (2.7631)	3.4087 (3.8163)	−0.0083 (0.0333)
S8	−7.7571 ** (1.7324)	1.3488 (2.8037)	1.8777 (3.8724)	0.0934 *** (0.0338)
S9	8.9359 *** (2.0608)	−12.5681 *** (3.3351)	6.5238 (4.6064)	−0.0287 (0.0403)
S10	8.6702 *** (1.6487)	−5.2552 * (2.6682)	4.2867 (3.6852)	0.0222 (0.0322)
S11	3.7736 ** (1.6597)	−5.0736 * (2.6860)	1.2455 (3.7099)	−0.0004 (0.0324)
time	−0.0860 *** (0.0230)	0.1221 *** (0.0372)	0.0567 (0.0514)	0.0002 (0.0004)

Note: Standard deviation in parenthesis; $*p < 0.1, **p < 0.05, ***p < 0.01$.

Table A4. Results of VAR estimation for IPI, ICI, synthetic index and quadratic uncertainty.

	Equation (1): IPI	Equation (2): ICI	Equation (3): Synthetic Index	Equation (4): Quadratic Uncertainty
const	11.9661 ** (5.0065)	28.2318 *** (8.1289)	−34.1409 *** (12.0858)	0.6871 *** (0.1196)
IPI_1	0.8428 *** (0.0357)	−0.1463 ** (0.0580)	0.2226 ** (0.0861)	−0.0032 *** (0.0009)
ICI_1	−0.0067 (0.0201)	0.8077 *** (0.0326)	0.1322 *** (0.0485)	0.0011 ** (0.0005)
Synthetic_Index_1	0.0130 (0.0130)	0.0231 (0.0211)	0.8162 ** (0.0314)	−0.0002 ((0.0003))
Quadratic_Uncertainty_1	−2.2750 (1.8431)	2.3620 (2.9924)	−4.8080 (4.4491))	0.6057 *** (0.0440)
S1	11.9399 *** (1.3078)	−11.0163 *** (2.1234)	11.9912 *** (3.1570)	−0.0490 (0.0312))
S2	5.5726 *** (1.2800)	−9.5696 *** (2.0783)	9.7667 *** ((3.0900))	−0.0202 (0.0306)
S3	14.2356 *** (1.2811)	−6.1080 ** (2.0801)	12.7661 *** (3.0926)	−0.0735 ** (0.0306)
S4	2.1922 * (1.2881)	−7.6131 *** (2.0914)	5.9668 * (3.1094)	0.0042 (0.0308)
S5	10.7613 *** (1.2791)	−10.7401 *** ((2.0768)	6.8457 ** (3.0877)	−0.0209 (0.0305)
S6	5.4104 *** (1.2813)	−8.4761 *** (2.0804)	8.3448 *** (3.0931)	0.0002 (0.0306)
S7	3.7122 *** (1.2791)	−7.3248 *** (2.0768)	5.6735 * (3.0878)	−0.0321 (0.0305)
S8	−7.40764 *** (1.2834)	−2.4897 (2.0838)	4.3521 (3.0981)	0.0105 (0.0307)
S9	20.0910 *** (1.3870)	−12.7784 *** (2.2520)	17.4210 *** (3.3482)	−0.1031 *** (0.0331)
S10	11.8102 *** (1.2800)	−7.2666 *** (2.0783)	10.8198 *** (3.0900)	−0.0278 (0.0306)
S11	5.9073 *** (1.284)	−10.3945 *** (2.0840)	8.6740 *** (3.0984)	−0.0396 (0.0307)
time	−0.0041 (0.0038)	−0.0227 *** (0.0062)	0.0294 *** (0.0092)	0.0002 ** (0.0001)

Note: Standard deviation in parenthesis; $*p < 0.1, **p < 0.05, ***p < 0.01$.

References

1. Ericsson, N.R. Forecast Uncertainty in Economic Modeling. *FRB Int. Financ. Discuss. Pap. No.697*. 2001. Available online: http://dx.doi.org/10.2139/ssrn.266494 (accessed on 13 April 2019). [CrossRef]
2. Makridakis, S.; Hogarth, R.M.; Gaba, A. Forecasting and uncertainty in the economic and business world. *Int. J. Forecast.* **2009**, *25*, 794–812. [CrossRef]
3. Bloom, N. Fluctuations in Uncertainty. *J. Econ. Perspect.* **2014**, *28*, 153–176. [CrossRef]
4. Makridakis, S.; Taleb, N. Living in a world of low levels of predictability. *Int. J. Forecast.* **2009**, *25*, 840–844. [CrossRef]
5. Pain, N.; Lewis, C.; Dang, T.T.; Jin, Y.; Richardson, P. OECD Forecasts during and after the Financial Crisis: A Post Mortem. *OECD Econ. Dep. Work. Pap.* **2014**. [CrossRef]
6. Green, K.C.; Armstrong, J.S. Simple versus complex forecasting: The evidence. *J. Bus. Res.* **2015**, *68*, 1678–1685. [CrossRef]
7. Makridakis, S.; Hibon, M. The M3-competition: Results, conclusions and implications. *Int. J. Forecast.* **2000**, *16*, 451–476. [CrossRef]
8. Lahiri, K.; Sheng, X. Measuring Forecast Uncertainty by Disagreement: The Missing Link. *J. Appl. Econ.* **2010**, *25*, 514–538. [CrossRef]
9. Conflitti, C. Measuring Uncertainty and Disagreement in the European Survey of Professional Forecasters. *OECD J. J. Bus. Cycle Meas. Anal.* **2012**, *2011*, 69–103. [CrossRef]
10. Lahiri, K.; Wang, W. Estimating Macroeconomic Uncertainty Using Information Measures from SPF Density Forecasts. Available online: https://www.albany.edu/economics/images/jobmarket/2017-18/Wang/Wuwei_Wang_JMP.pdf (accessed on 13 April 2019).
11. Abel, J.; Rich, R.; Song, J.; Tracy, J. The Measurement and Behavior of Uncertainty: Evidence from the ECB Survey of Professional Forecasters. *J. Appl. Econ.* **2016**, *31*, 533–550. [CrossRef]
12. Giordani, P.; Soderlind, P. Inflation Forecast Uncertainty. *Eur. Econ. Rev.* **2003**, *47*, 1037–1059. [CrossRef]
13. Boero, G.; Smith, J.; Wallis, K.F. Uncertainty and disagreement in economic prediction: The Bank of England Survey of External Forecasters. *Warwick Econ. Res. Pap.* **2013**, *811*, 451–476. [CrossRef]
14. Engelberg, J.; Manski, C.F.; Williams, J. Comparing the Point Predictions and Subjective Probability Distributions of Professional Forecasters. *NBER Work. Pap. No. w11978*; 2006. Available online: https://ssrn.com/abstract=878065 (accessed on 13 April 2019).
15. Gadea, M.D. Las Previsiones Económicas en España. Estudio Comparativo del Panel de FUNCAS. *Fund. Cajas Ahorr.* 2014. Available online: https://www.funcas.es/publicaciones/Sumario.aspx?IdRef=7-05749 (accessed on 13 April 2019). (In Spanish)
16. Moreno, B.; López-Menéndez, A.J. Combining economic forecasts through information measures. *Appl. Econ. Lett.* **2007**, *14*, 899–903. [CrossRef]
17. Anadón, L.D.; Baker, E.D.; Bosetti, V. Integrating Uncertainty into Public Energy Research and Development Decisions. *Nat. Energy* **2017**, *2*, 17071. [CrossRef]
18. Diebold, F.X.; Rudebusch, G.D. Forecasting Output with the Composite Leading Index: A real-Time Analysis. *J. Am. Stat. Assoc.* **1991**, *86*, 603–610. [CrossRef]
19. Moreno, B.; López-Menéndez, A.J. Las opiniones empresariales como predictores de los puntos de giro del ciclo industrial. *Estud. Econ. Appl.* **2007**, *25*, 511–528. (In Spanish)
20. Centro de Investigaciones Sociológicas (CIS). Nota de Investigación Sobre la Metodología General de los BaróMetros Mensuales del Centro de Investigaciones Sociológicas. 2019. Available online: www.cis.es (accessed on 13 April 2019). (In Spanish)
21. Sociedad Asturiana de Estudios Económicos e Industriales (SADEI). Encuesta Industrial de Coyuntura. 2019. Available online: www.sadei.es (accessed on 13 April 2019). (In Spanish)
22. Shannon, C.E. A Mathematical Theory of Communication. *Bell Syst. Tech. J.* **2000**, *27*, 379–423. [CrossRef]
23. Pérez, R. Estimación de la incertidumbre, la incertidumbre útil y la inquietud en poblaciones finitas: Una aplicación a las medidas de desigualdad. *Revista de la Real Academia de Ciencias Exactas, Físicas y Naturales* **1985**, *LXXIX*, 651–654. (In Spanish)
24. Pérez, R.; López-Menéndez, A.J.; Caso, C.; Alvargonzález, M.; Río, M.J. On Economic Applications of Information Theory. In *The Mathematics of the Uncertain*; Springer: Oviedo, Spain, 2018; pp. 515–525.

25. López-Menéndez, A.J.; Pérez, R. Forecasting Performance and Information Measures. Revisiting the M-Competition. *Estud. Econ. Appl.* **2017**, *35*, 299–314.
26. Jurado, K.; Ludvigson, S.C.; Ng, S. Measuring uncertainty. *Am. Econ. Rev.* **2015**, *105*, 1177–1216. [CrossRef]
27. Sheen, J.; Wang, B.Z. Estimating macroeconomic uncertainty from surveys. A mixed frequency approach. In Proceedings of the International Conference on Time Series and Forecasting (ITISE 2018), Granada, Spain, 19–21 September 2018; pp. 197–226.
28. Bloom, N. The impact of uncertainty shocks. *Econometrica* **2009**, *77*, 623–685.

Article

Soft Randomized Machine Learning Procedure for Modeling Dynamic Interaction of Regional Systems

Yuri S. Popkov [1,2,3,4]

[1] Federal Research Center "Computer Science and Control" of Russian Academy of Sciences, 119333 Moscow, Russia; popkov@isa.ru
[2] Institute of Control Sciences of Russian Academy of Sciences, 117997 Moscow, Russia
[3] Department of Software Engineering, ORT Braude College, 216100 Karmiel, Israel
[4] Yugra Research Institute for Information Technologies, 628011 Khanty-Mansiysk, Russia

Received: 19 March 2019; Accepted: 11 April 2019; Published: 20 April 2019

Abstract: The paper suggests a randomized model for dynamic migratory interaction of regional systems. The locally stationary states of migration flows in the basic and immigration systems are described by corresponding entropy operators. A soft randomization procedure that defines the optimal probability density functions of system parameters and measurement noises is developed. The advantages of soft randomization with approximate empirical data balance conditions are demonstrated, which considerably reduces algorithmic complexity and computational resources demand. An example of migratory interaction modeling and testing is given.

Keywords: soft randomization; entropy; entropy operator; migration; immigration; empirical balance; empirical risk

---

1. Introduction

The mutual influence of migratory processes in regional systems is a problem of growing significance in the modern world. The socioeconomic statuses of different regions demonstrate higher heterogeneity in response to rising political and military tension. All these factors cause an abrupt redistribution of migration flows and regional population variations, thereby increasing the cost of regional population maintenance [1–4]. Therefore, it is important to develop different tools (mathematical models, algorithms, and software) for forecasting the distribution of migration flows with adaptation to their dynamics considering available resources.

The authors of [5] suggested a dynamic entropy model for the migratory interaction of regional systems. In comparison with biological reproduction, migration mobility is a rather fast process [1,6]. Thus, the short-term dynamics of regional population size are described by the locally stationary state of a migratory process [7]. The latter can be simulated under the hypothesis that all migrants have a random and independent spatial distribution over interacting regional systems with given prior probabilities. The mathematical model of a locally stationary state is given by a corresponding entropy operator that maps the space of admissible resources into the space of migratory processes [8].

Mathematical modeling and analysis of interregional migration is considered in numerous publications. First, it seems appropriate to mention the monographs [9,10] that are dedicated to a wide range of interregional migration problems, including mathematical modeling of migration flows. Note that the problem of migration touches upon many aspects of socioeconomic, psychological and political status of the space of migratory movements. Thus, of crucial role is the structural analysis of inter- and intraregional migration flows [4] and motivations that generate them [2,11]. The results of structural and motivational analysis of migratory processes are used for computer simulation. There exist three directions of research in this field, each relying on some system of hypotheses. One of the directions involves the stochastic hypothesis about the origin of migratory motivations [12], which is simulated using agent technologies [13,14]. This direction is adjoined by investigations based on the thermodynamic model of migration flows [3,8]. Of course, the short list above does not exhaust the whole variety of migration studies, merely outlining some topics of research.

This paper studies a stochastic version of the model in [5], in which random parameters and measurement noises are characterized by probability density functions (PDFs). These functions are estimated using retrospective information on the real dynamics of regional population size with "soft" randomized machine learning [15]. The learned model was implemented in the form of computer simulations, i.e., generation of an ensemble of random trajectories with the entropy-optimal PDFs of the model parameters and measurement noises. The resulting ensemble was used for testing of the model and also for short-term forecasting.

The method developed below is illustrated by an example of the randomized modeling and forecasting of the migratory interaction among three EU countries (Germany, France, and Italy—the system $\mathcal{GFI}$) and two countries as sources of immigration (Syria and Libya—the system $\mathcal{SL}$).

2. Randomized Model of Migratory Interaction

Consider the dynamic discrete-time model of migratory interaction with shared resource constraints that is presented in [5]. The first sub-model represents migration flows within the system $\mathcal{GFI}$ and is described by the dynamic regression equation

$$\mathbf{K}[(s+1)h] = (A - E)\mathbf{K}[sh] + \mathbf{F}(z[sh]), \qquad (\mathbf{K}, \mathbf{F}) \in R^N, \; s = \overline{0, K-1}, \tag{1}$$

where

$$A = h \begin{pmatrix} 1 & \alpha_2 a_{21} & \cdots & \alpha_N a_{N1} \\ \alpha_1 a_{12} & 1 & \cdots & \alpha_N a_{N2} \\ \cdots & \cdots & \cdots & \cdots \\ \alpha_1 a_{1N} & \alpha_2 a_{2N} & \cdots & 1 \end{pmatrix}, \tag{2}$$

$$E = h \, \mathrm{diag}[\alpha_n, \; n = \overline{1, N}]. \tag{3}$$

In these equations, $\mathbf{K}[sh]$ denotes the population distribution in the regional system $\mathcal{GFI}$ at a time sh.

At a time sh, the distribution of immigration flows from the regional system $\mathcal{SL}$ to the regional system $\mathcal{GFI}$ in terms of an entropy operator is modeled by the second sub-model, which can be described by a vector function $\mathbf{F}(z[sh])$ with the components

$$f_n[sh] = h \sum_{j=1}^{M} b_{jn}(z[sh])^{c_{jn}}, \qquad n = \overline{1, N}, \; s = \overline{0, K-1}, \tag{4}$$

The variable z, which is the exponential Lagrange multiplier in the entropy-optimal distribution problem of immigration flows, satisfies the equation

$$\sum_{k=1}^{M} \sum_{n=1}^{N} c_{kn} b_{kn}(z[sh])^{c_{kn}} = T[sh], \tag{5}$$

where $T[sh]$ is the amount of a shared resource used by all regions from the system $\mathcal{GFI}$ to maintain immigrants.

In this model, the input data are the amounts $T[0], T[h], \ldots, T[(K-1)h]$; and the output data are the regional population distributions $\mathbf{K}[0], \mathbf{K}[h], \ldots, \mathbf{K}[(K-1)h]$.

The dynamic model in Equations (1)–(5) contains the following parameters:

- $\alpha_n \in [0,1]$, $n = \overline{1, N}$, as the shares of mobile population in system regions;
- $a_{in} \in [0,1]$, $(i,n) = \overline{1, N}$, as the prior probabilities of individual migration in the system $\mathcal{GFI}$;

- b_{kn}, $k = \overline{1, M}$, $n = \overline{1, N}$, as the prior probabilities of individual immigration from region k of the system $\mathcal{SL}$ to region n of the system $\mathcal{GFI}$; and
- c_{kn}, $k = \overline{1, M}$, $n = \overline{1, N}$, as the normalized 1 specific generalized cost of immigration maintenance.

Normalization means that $0 < c_{kn} < 1$, $k = \overline{1, M}$, $n = \overline{1, N}$.

The parameters form three groups: mobility, migratory movements within the system $\mathcal{GFI}$, and immigratory movements from the system $\mathcal{SL}$ to the system $\mathcal{GFI}$. All these characteristics are specified by the regions of both systems. The dimensionality of the parametric space is reduced using the same approach as in [5]. The whole essence is to assign a relative regional differentiation of all parameters except for the weights b_1 (mobility) and b_2 (internal migration) of these groups, which are considered as model variables.

This approach leads to the parametric transformation

$$\alpha_n = b_1 m_n, \quad a_{in} = b_2 h_{in}, \tag{6}$$
$$(i, n) = \overline{1, N}; \; k = \overline{1, M},$$

where m_n and h_{in} are given parameters which characterize the relation of variables.

Then, the dynamic model of migratory interaction in Equations (1)–(5) takes the form

$$\mathbf{K}[(s+1)h] = (b_1 b_2 \tilde{A} - b_1 \tilde{E})\mathbf{K}[sh] + \tilde{\mathbf{F}}(z[sh]), \tag{7}$$

with the matrix

$$\tilde{A} = h \begin{pmatrix} 1 & m_2 h_{21} & \cdots & m_N h_{N1} \\ m_1 h_{12} & 1 & \cdots & m_N h_{N2} \\ \cdots & \cdots & \cdots & \cdots \\ m_1 h_{1N} & m_2 h_{2N} & \cdots & 1 \end{pmatrix} \tag{8}$$

and the diagonal matrix

$$\tilde{E} = h \operatorname{diag}[m_n, \, n = \overline{1, N}]. \tag{9}$$

The vector $\tilde{\mathbf{F}}(\mu, z)[sh]$ consists of the components

$$\tilde{f}_n(z[sh]) = h \sum_{k=1}^{M} q_{kn}(z[sh])^{c_{kn}}, \quad n = \overline{1, N}, \; s = \overline{0, K-1}. \tag{10}$$

For each time sh, the variable z satisfies the equation

$$\sum_{k=1}^{M} \sum_{n=1}^{N} c_{kn} q_{kn}(z[sh])^{c_{kn}} = T[sh], \qquad s = \overline{0, K-1}, \tag{11}$$

i.e., there exist K values $z = z^*[sh]$, $s = \overline{0, K-1}$.

The randomized version of this model is described by Equations (7)–(11) but some parameters (variables) have random character. These are two randomized parameters, b_1 and b_2, as well as the variable $z = b_3$, all of the interval type. More specifically, the parameters b_1 and b_2 belong to the intervals

$$\mathcal{B}_1 = [b_1^-, b_1^+], \; \mathcal{B}_2 = [b_2^-, b_2^+]. \tag{12}$$

The interval $\mathcal{B}_3$ of the variable b_3 is given by Equation (11).

Theorem 1. *Let the parameters b_{kn} and c_{kn} in Equation (11) be positive and $c_{kn} \in [0, 1]$. Then, the solution b_3^* of this equation belongs to the interval*

$$\mathcal{B}_3 = [b_3^-, b_3^+], \tag{13}$$

where

$$b_3^- = \left(\frac{T[sh]}{MNc_{max}b_{max}}\right)^{1/c_{max}}; \quad b_3^+ = \left(\frac{T[sh]}{MNc_{min}b_{min}}\right)^{1/c_{min}}; \tag{14}$$

$$c_{min} = \min_{kn} c_{kn}, \ c_{max} = \max_{kn} c_{kn}; \ b_{min} = \min_{kn} b_{kn}, \ b_{max} = \max_{kn} b_{kn}.$$

The proof is postponed to the Appendix A.

Therefore, the randomized dynamic model in Equations (7)–(11) includes three random parameters $\mathbf{b} = \{b_1, b_2, b_3\}$ of the interval type that are defined over the three-dimensional cube with faces (Equations (12) and (13)), i.e.,

$$\mathcal{B} = \bigotimes_{j=1}^{3} \mathcal{B}_j. \tag{15}$$

The probabilistic properties of the randomized parameters are described by a continuously differentiable PDF $W(\mathbf{b})$.

By assumption, real distributions of regional population sizes contain errors that are simulated by a random vector $\tilde{\xi}[sh] \in R^N$ with the interval components

$$\tilde{\xi}[sh] \in \Xi_s = \left[\tilde{\xi}^-[sh], \tilde{\xi}^+[sh]\right]. \tag{16}$$

The probabilistic properties of this vector are described by a continuously differentiable PDF $Q(\tilde{\xi})$. The measured output of the randomized model has an additive noise,

$$\mathbf{v}[sh] = \mathbf{K}[sh] + \tilde{\xi}[sh]. \tag{17}$$

3. Characterization of Empirical Risk and Measurement Noises

Construct a synthetic functional $J[W(\mathbf{b}), Q(\tilde{\xi})]$ that depends on the PDFs of the model parameters and measurement noises for assessing in quantitative terms the empirical risk (the difference between the regional population distribution generated by the model in Equations (7)–(11) and the real counterpart) and the guaranteed power of these noises. The functional must have components characterizing an intrinsic uncertainty of randomized machine learning (RML) procedures, the approximation quality of empirical balances (the empirical risk) and the worst properties of the corresponding random interval-type noises.

1. Uncertainty. In accordance with the general concept of RML, the first component among the listed ones is *an entropy functional* that describes the level of uncertainty:

$$\mathcal{H}[\mathbf{b}), Q(\tilde{\xi})] = - \int_B W(\mathbf{b}) \ln W(\mathbf{b}) d\mathbf{b} - \int_\Xi Q(\tilde{\xi}) \ln Q(\tilde{\xi}) d\tilde{\xi}. \tag{18}$$

The two other functional components are constructed using Hölder's vector and matrix norms (The vector norm has the form $\|\mathbf{a}\|_\infty = \max_n |a_n|$; the matrix norm, the form $\|A\|_\infty = \max_{ij} |a_{ij}|$.) [16].

2. Approximate empirical balances. First, consider a characterization of *the empirical risk*. For the model in Equations (7)–(11), the deviation between the output and real data vectors is given by

$$\bar{\varepsilon}[sh] = \left(b_1 b_2 \tilde{A} - b_1 \tilde{E}\right) \mathbf{Y}[sh] + \mathbf{F}(b_3[sh]) - \mathbf{Y}[sh], \quad s = \overline{0, K-1}. \tag{19}$$

Using well-known inequalities for the matrix and vector norms, it is possible to write

$$\|\bar{\varepsilon}[sh]\|_\infty \leq \|\left(b_1 b_2 \tilde{A} - b_1 \tilde{E}\right)\|_\infty \|\mathbf{Y}[sh]\|_\infty + \|\mathbf{F}(b_3[sh])\|_\infty + \|\mathbf{Y}[(s+1)h]\|_\infty =$$
$$= \varphi(b_1, b_2, b_3, s), \quad s = \overline{0, K-1}. \tag{20}$$

Introducing the average matrix and vector norms over the observation interval,

$$\varphi(b_1, b_2, b_3) \le h \left(\frac{1}{K} \sum_{s=0}^{K-1} \max_n y_n[sh] \right) \left(b_1 \max_n m_n + b_1 b_2 \max_{i,j} h_{ij} \right) +$$

$$+ \quad \frac{1}{K} \sum_{s=0}^{K-1} \max_n y_n[(s+1)h] + MN c_{max} b_{max} (b_3)^{c_{max}}. \tag{21}$$

The parameters b_1 and b_2 take values within the intervals $\mathcal{B}_1$ and $\mathcal{B}_2$ (Equation (12)) while the parameter b_3 within the interval

$$\mathcal{B}_3 = \left[\left(\frac{T_{max}}{MN c_{max} q_{max}} \right)^{1/c_{max}} , \left(\frac{T_{max}}{MN c_{min} q_{min}} \right)^{1/c_{min}} \right], \tag{22}$$

where

$$T_{max} = \max_s T[sh]. \tag{23}$$

Denote

$$U_1 \quad = \quad h \left(\frac{1}{K} \sum_{s=0}^{K-1} \max_n y_n[sh] \right) \max_n m_n; \quad U_2 = h \left(\frac{1}{K} \sum_{s=0}^{K-1} \max_n y_n[sh] \right) \max_{i,j} h_{ij};$$

$$U_3 \quad = \quad MNh c_{max} b_{max}; \quad U_4 = \frac{1}{K} \sum_{s=0}^{K-1} \max_n y_n[(s+1)h]. \tag{24}$$

Then, the function $\varphi(b_1, b_2, b_3)$ takes the form

$$\varphi(b_1, b_2, b_3) = b_1 U_1 + b_1 b_2 U_2 + (b_3)^{c_{max}} U_3 + U_4. \tag{25}$$

Note that the coefficients $U_1, \ldots, U_4$ are determined by real data on regional population distributions and also by the characteristics of internal migration within the system $\mathcal{GFI}$ and immigration flows from the system $\mathcal{SL}$.

The equality in Equation (25) defines a function $\varphi(b_1, b_2, b_3)$ of random variables. Let its expectation be the characteristic of the empirical risk, i.e.,

$$r[W(\mathbf{b})] = \int_{\mathcal{B}} W(\mathbf{b}) \varphi(\mathbf{b}) d\mathbf{b}, \tag{26}$$

where $\mathcal{B} = \mathcal{B}_1 \otimes \mathcal{B}_2 \otimes \mathcal{B}_3$ and the intervals $\mathcal{B}_1$ and $\mathcal{B}_2$ have given limits. At the same time, the limits of the interval $\mathcal{B}_3$ are specified by the equalities in Equation (22).

Power of noises. The measurement noises are simulated by random vectors $\bar{\xi}[sh] \in R^N$, $s = \overline{0, K-1}$. The components of these vectors may have different domains (ranges of values) at different times $s = \overline{0, K-1}$. For each time, introduce the Euclidean norm $\|\bar{\xi}[sh]\|_N^2$ and its expectation

$$n_s[Q(\bar{\xi}[sh])] = \int_{\Xi} Q(\bar{\xi}[sh]) \|\bar{\xi}[sh]\|_N^2 d\bar{\xi}[sh]. \tag{27}$$

The average expectation of this norm over the time interval has the form

$$\bar{n}_s[Q(\bar{\xi}[sh])] = \frac{1}{K} \sum_{s=0}^{K-1} n_s[Q(\bar{\xi}[sh])]. \tag{28}$$

If the measurement noises are the same on the observation interval, then the noise power functional can be written as

$$\bar{n}_s[Q(\bar{\xi}[sh])] = n[Q(\bar{\xi})] = \int_{\Xi} Q(\bar{\xi})\|\bar{\xi}\|_N^2 d\bar{\xi}. \tag{29}$$

This formula involves the Euclidean norm for a quantitative characterization of the noise power. However, it is possible to choose other norms depending on problem specifics.

4. Soft Randomized Estimation of Model Parameters

The model characteristics and measurement noises are estimated using a learning data collection: the real cost of immigrants maintenance $T[0], \ldots, T[(K-1)h]$ (input data) and the real distributions of regional population sizes $\mathbf{Y}[0], \ldots, \mathbf{Y}[(K-1)h]$ (output data).

In accordance with the general procedure of soft randomized machine learning [15], the optimal probability density functions $W(\mathbf{b})$ (model parameters) and $Q(\bar{\xi})$ (measurement noises) are calculated by the constrained minimization of the synthetic functional $J[W(\mathbf{b}), Q(\bar{\xi})]$ that contains the following functionals:

- the entropy

$$\mathcal{H}[W(\mathbf{b})] = -\int_{\mathcal{B}} W(\mathbf{b}) \ln W(\mathbf{b}) d\mathbf{b} - \int_{\Xi} Q(\bar{\xi}) \ln Q(\bar{\xi}) d\bar{\xi}; \tag{30}$$

- the average empirical risk over the observation interval

$$r[W(\mathbf{b})] = \int_{\mathcal{B}} W(\mathbf{b})(b_1 U_1 + b_1 b_2 U_2 + (b_3)^{c_{max}} U_3 + U_4) d\mathbf{b}; \tag{31}$$

and
- the average error norm

$$n[Q(\bar{\xi})] = \int_{\Xi} Q(\bar{\xi}) \sum_{i=1}^{N} \xi_i^2 d\bar{\xi}. \tag{32}$$

The soft randomized learning algorithm has the form

$$J[W(\mathbf{b}), Q(\bar{\xi})] = \mathcal{H}[W(\mathbf{b})] - r[W(\mathbf{b})] - n[Q(\bar{\xi})] \Rightarrow \max, \tag{33}$$

$$\int_{\mathcal{B}} W(\mathbf{b}) d\mathbf{b} = 1, \qquad \int_{\Xi} Q(\bar{\xi}) d\bar{\xi} = 1.$$

The solution of this problem is the optimal PDFs under maximal uncertainty, for the model parameters of the form

$$W^*(\mathbf{b}) = \frac{\exp\left(b_1 U_1 - b_1 b_2 U_2 - (b_3)^{c_{max}} U_3 - U_4\right)}{\mathcal{P}}, \tag{34}$$

where

$$\mathcal{P} = \int_{\mathcal{B}} \exp\left(b_1 U_1 - b_1 b_2 U_2 - (b_3)^{c_{max}} U_3 - U_4\right) d\mathbf{b}, \tag{35}$$

and for the measurement noises of the form

$$Q^*(\bar{\xi}) = \frac{\exp\left(-\sum_{i=1}^{N} \xi_i^2\right)}{\mathcal{Q}}, \tag{36}$$

where

$$\mathcal{Q} = \int_{\Xi} \exp\left(-\sum_{i=1}^{N} \xi_i^2\right) d\bar{\xi}. \tag{37}$$

In the case of soft randomization, there is no need for solving the empirical balance equations, which have high complexity and computational intensiveness due to the presence of integral components. Here, computational resources are required for the normalization procedure of the resulting PDFs. On the other hand, the morphology of the optimal PDFs depends on a specific choice of the approximate data balancing criterion and a numerical characterization of the measurement noises.

5. Randomized Forecasting of Dynamic Migratory Interaction

Consider randomized forecasting of dynamic migratory interaction using the principle of soft randomization. Let $\mathcal{T}_{pr} = [s_0 h, s_{pr} h]$ be the forecasting interval and assume the initial state (the regional population distribution at the initial time $s_0 h$) coincides with the real distribution, i.e., $\mathbf{K}[s_0 h] = \mathbf{Y}[s_0 h]$. The shared cost of the system $\mathcal{GFI}$ to maintain immigrants is distributed in accordance with a given scenario. For each scenario, the value T_{max} and also the interval $\mathcal{B}_3$ in Equations (12), (22), and (23) are determined.

The forecasted trajectories are constructed using the randomized model in Equations (7), (10), and (11)

$$
\begin{aligned}
\mathbf{K}[(s+1)h] &= \left(b_1 b_2 \tilde{A} - b_1 \tilde{E}\right) \mathbf{K}[sh] + \mathbf{F}[sh \mid b_3], \\
\mathbf{F}[sh \mid b_3] &= \{\sum_{k=1}^{M} b_{kn} (b_3)^{c_{kn}}, \quad n = \overline{1, N}\}, \\
s &= \overline{s_0, s_{pr}}, \quad \mathbf{K}[s_0 h] = Y[s_0 h].
\end{aligned}
\tag{38}
$$

The randomized parameters b_1, b_2, and b_3 take values within the corresponding intervals with the probability density function $W^*(\mathbf{b})$ (Equation (34)).

An ensemble of the forecasted trajectories for the model's output is obtained taking into account a random vector $\bar{\xi} \in \Xi$ with the PDF $Q^*(\bar{\xi})$ (Equation (36)):

$$
\mathbf{v}[sh] = \mathbf{K}[sh] + \bar{\xi}, \quad s = \overline{s_0, s_{pr}}.
\tag{39}
$$

For each scenario $T[s_0 h], \ldots, T[s_{p}, h]$, an ensemble $\mathcal{K}$ of random forecasting trajectories is generated via sampling (the transformation of a PDF into a corresponding sequence of random vectors of length I) of the optimal PDFs of the model parameters and measurement noises for each time sh. The resulting ensemble allows deriving empirical estimates of different numerical characteristics as follows:

- the average trajectory

$$
\bar{\mathbf{K}}[sh] = \frac{1}{I} \sum_{i=1}^{I} \mathbf{K}^{(i)}[sh], \qquad s = \overline{s_0, s_{pr}};
\tag{40}
$$

- the variance trajectory

$$
\bar{\sigma}^2[sh] = \frac{1}{I-1} \sum_{i=1}^{I} \|\mathbf{K}^{(i)}[sh] - \bar{\mathbf{K}}[sh]\|^2, \qquad s = \overline{s_0, s_{pr}};
\tag{41}
$$

- the variance pipe, i.e., the set of random trajectories that almost surely (since an ensemble consists of a finite number of trajectories, the matter concerns not probability but its empirical estimate) belong to the domain

$$
\mathcal{D} = \{\mathbf{K}[sh] : \bar{\mathbf{K}}[sh] - \bar{\sigma}^2[sh] \le \mathbf{K}[sh] \le \bar{\mathbf{K}}[sh] + \bar{\sigma}^2[sh], \quad s = \overline{s_0, s_{pr}}\};
\tag{42}
$$

- the empirical probability distribution and its dynamics on the forecasting interval

$$
\mathbb{P}\left(\mathbf{K}[sh] \le \Delta, \ s = \overline{s_0, s_{pr}}\right) = \frac{I_\Delta}{I},
\tag{43}
$$

where I_Δ denotes the number of vectors $\mathbf{K}[sh]$ whose components are smaller than Δ; and

- the median trajectory $\hat{\mathbf{K}}[sh]$, $s = \overline{s_0, s_{pr}}$, which satisfies the equation

$$\mathbb{P}(\mathbf{K}[sh]) = 0,5; \quad s = \overline{s_0, s_{pr}}. \tag{44}$$

The ensemble $\mathcal{K}$ can be used to calculate other characteristics, e.g., α-quantiles, confidence probabilities, etc.

6. Example

The appearance of territories with low economic status always causes the growth of immigration. The early 2000s were remarkable for the formation of several such territories in Northern and Central Africa, the Near East, Afghanistan, etc. As a result, tens of millions of migrants moved to the EU as the level of life in these territories dropped below the subsistence minimum. The EU countries have to allocate considerable financial resources for their filtering and accommodation, which are often unacceptable. An example below illustrates the use of soft randomization for estimating and forecasting of immigration flows from Syria (1) and Libya (2) (the system $\mathcal{SL}$) to Germany (1), France (2), and Italy (3) (the system $\mathcal{GFI}$).

1. Randomized model, parameters, measurement errors, time intervals, and real data collections. Choose the randomized mathematical model (Equation (25)) with the normalized variables

$$p_n[sh] = \frac{K_n[sh]}{K_{max}}, \quad n = \overline{1,3}. \tag{45}$$

This gives

$$
\begin{aligned}
p_n[(s+1)h] &= (1 - b_1 m_n) p_n[sh] + h b_1 b_2 \sum_{i=1, i \neq n}^{3} m_i h_{in} p_i[sh] + h f_n[sh], \\
f_n[sh] &= \sum_{i=1}^{M} b_{in} b_3^{c_{in}}, \quad n = \overline{1,3}, \\
T[sh] &= \sum_{n=1}^{3} \sum_{i=1}^{2} c_{in} b_{in} b_3^{c_{in}}.
\end{aligned}
\tag{46}
$$

The state variables of the system $\mathcal{GFI}$ and also the immigration flows from the system $\mathcal{SL}$ are normalized, i.e.,

$$0 \leq p_n[sh] \leq 1, \quad 0 \leq f_n[sh] \leq 1, \quad n = \overline{1,N}. \tag{47}$$

The variable z^* characterizes the entropy operator of the immigration process and satisfies the last equation in Equation (46). The values of the parameters m_i, h_{in}, b_{in}, and c_{in} are combined in Table 1, where columns are different values of corresponding parameter. Recall that the two lowest rows of Table 1 indicate the values of the parameters c_{in}. By assumption, the regions of both systems have the same specific cost.

Table 1. Values of relative parameters.

m_n	0.43	0.50	0.40
h_{1n}	0	0.3	0.3
h_{2n}	0.3	0	0.3
h_{3n}	0.5	0.4	0
b_{1n}	0.4	0.3	0.3
b_{2n}	0.3	0.1	0.4
c_{1n}	0.4	0.4	0.3
c_{2n}	0.4	0.4	0.3

In accordance with this table, $m_{max} = 0.5$, $h_{max} = 0.5$, $b_{min} = 0.3$, $b_{max} = 0.4$, and $c_{max} = c_{min} = c = 0.5$. The measurement errors of population sizes $\bar{\xi}[sh] \in R^3$ (in normalized units) belong to the intervals

$$\bar{\xi}[sh] \in \Xi = [\bar{\xi}_-, \bar{\xi}_+], \qquad \xi_n^{\pm} = 0.01, \tag{48}$$

and by assumption they have the same limits for times sh.

The normalized observation (model output) has the form

$$\mathbf{v}[sh] = \mathbf{p}[sh] + \bar{\xi}[sh]. \tag{49}$$

The random parameter model in Equation (46) was employed for estimating parameter characteristics and testing on corresponding time intervals with step $h = 1year$:

- $T_{est} = 2009$–2013 as the estimation interval; and
- $T_{tst} = 2014$–2018 as the testing interval.

2. Entropy estimation of PDFs of model parameters and measurement noises (interval T_{est}). This problem was solved using available data on regional population distribution for Germany ($n = 1$), France ($n = 2$), and Italy ($n = 3$) and also on the shared cost of immigrants maintenance on the estimation interval (see Table 2 and UNdata service at https://data.un.org/).

Table 2. Input and output data collections.

Year	2009	2010	2011	2012	2013
s	0	1	2	3	4
$Y_1[s]$	81.90	81.77	80.27	80.42	80.64
$y_1[s]$	1.00	0.998	0.980	0.982	0.985
$Y_2[s]$	62.47	62.80	63.11	63.41	63.70
$y_2[s]$	0.762	0.767	0.771	0.774	0.778
$Y_3[s]$	59.39	59.53	59.63	59.71	59.75
$y_3[s]$	0.725	0.727	0.728	0.729	0.726
$T[s]$ *(billion)*	0.093	0.094	0.095	0.096	0.097

In this model, the random parameters $b_1, b_2,$ and b_3 take values within the intervals

$$b_1 \in \mathcal{B}_1 = [1.0, 2.5]; \quad b_2 \in \mathcal{B}_2 = [0.5, 1.8], \quad b_3 \in \mathcal{B}_3 = [0.3, 1.5]. \tag{50}$$

In accordance with Equation (24),

$$U_1 = 0.5; \; U_2 = 0.5; \; U_3 = 1.2; \; U_4 = 0.986. \tag{51}$$

Then, the soft RML procedure yields the following optimal PDFs of the model parameters and measurement noises:

$$
\begin{aligned}
W^*(\mathbf{b}) &= \frac{\exp\left(-0.5b_1 - 0.5b_1 b_2 - 1.2b_3^{0.5} - 0.986\right)}{\mathcal{W}}, \\
Q^*(\bar{\xi}) &= \frac{\exp\left(-\sum_{n=1}^{3} \xi_n^2\right)}{\mathcal{Q}},
\end{aligned} \tag{52}
$$

where

$$
\begin{aligned}
\mathcal{W} &= \int_{\mathcal{B}_1} \int_{\mathcal{B}_2} \int_{\mathcal{B}_3} \exp\left(-0.5b_1 - 0.5b_1 b_2 - 1.2b_3^{0.5} - 0.986\right) db_1 db_2 db_3, \\
\mathcal{Q} &= \prod_{n=1}^{3} \int_{-0.01}^{0.01} \exp(-\xi^2) d\xi.
\end{aligned} \tag{53}
$$

The two-dimensional sections of the three-dimensional PDFs of the model parameters are shown in Figure 1a–c, while the graphs of the PDFs of the measurementnoises in Figure 2.

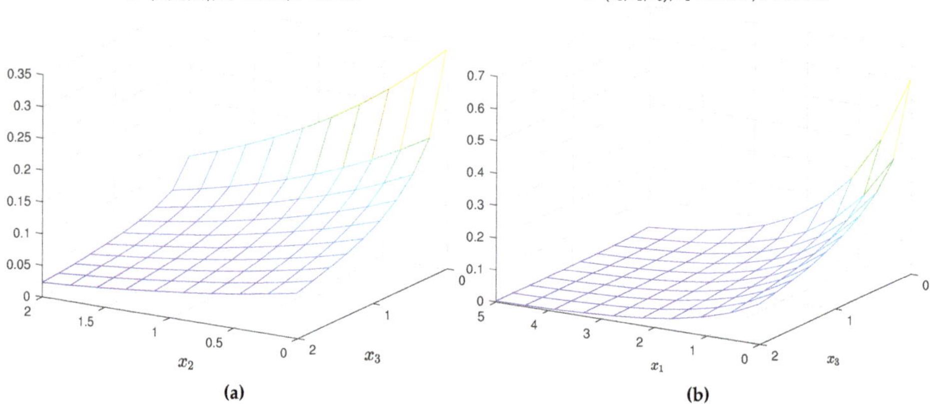

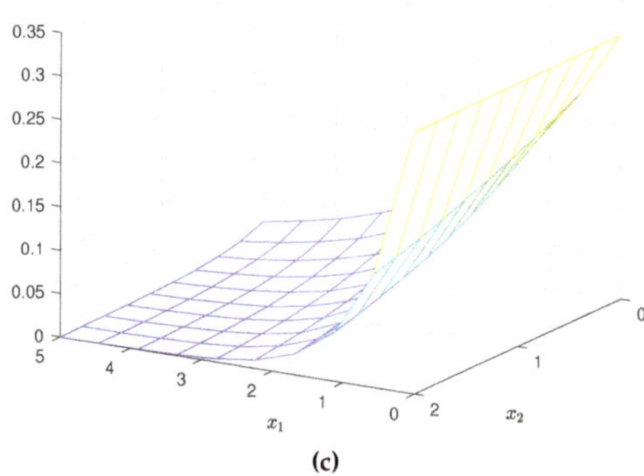

Figure 1. 2-dimensional section of W.

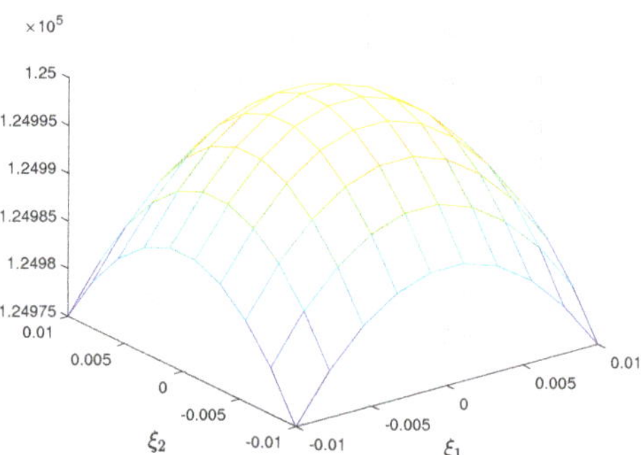

Figure 2. 2-dimensional section of Q.

3. Model testing. The randomized model in Equation (49) with the optimal PDFs in Equations (52) and (53) was tested using the above data on regional population sizes from the UNdata service (https://data.un.org/) (see Table 3). This table also presents the testing results in terms of the ensemble-average trajectories $\bar{p}_1[sh]$, $\bar{p}_2[sh]$, and $\bar{p}_3[sh]$.

Table 3. Input and output data collections.

Year	2014	2015	2016	2017	2018
s	0	1	2	3	4
$Y_1[s]$	81.489	81.707	82.063	82.386	82.674
$y_1[s]$	0.985	0.988	0.993	0.996	1.000
$\bar{p}_1[sh]$	0.986	0.615	0.743	0.639	0.999
$Y_2[s]$	64.190	64.457	64.791	65.134	65.484
$y_2[s]$	0.721	0.472	0.564	0.529	0.708
$\bar{p}_2[sh]$	0.722	0.695	0.707	0.691	0.715
$Y_3[s]$	59.585	59.504	59.504	59.509	59.516
$y_3[s]$	0.775	0.609	0.562	0.699	0.650
$\bar{p}_3[sh]$	0.776	0.617	0.607	0.705	0.628
$T[s]$ (*billion*)	0.097	0.097	0.097	0.098	0.098

Testing was performed via sampling of the randomized interval parameters with the PDFs in Equations (52) and (53) and construction of the corresponding trajectories by Equation (49). Figure 3a–c shows ensembles of such trajectories $v_1[sh]$, $v_2[sh]$, $v[sh]$ as well as the ensemble-average trajectories $\bar{v}_1[sh]$, $\bar{v}_2[sh]$, $\bar{v}_3[sh]$ (Graph 1); the real trajectories $y_1[sh]$, $y_2[sh]$, $y_3[sh]$ of regional population sizes (Graph 2); and the limits of the variance pipes $\bar{p}_1^*[sh] \pm \sigma_1$, $\bar{p}_2^*[sh] \pm \sigma_2$, $\bar{p}_3^*[sh] \pm \sigma_3$ (Graph 3).

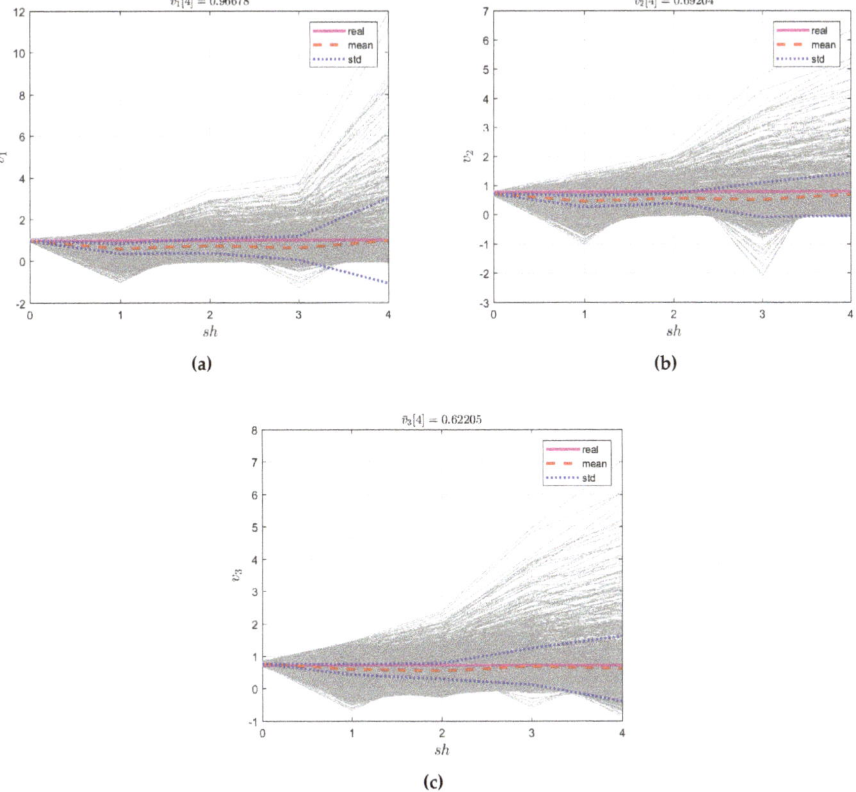

Figure 3. (a) $\bar{v}_1[4]$, (b) $\bar{v}_2[4]$, (c) $\bar{v}_3[4]$.

The testing accuracy was estimated in terms of the relative root-mean-square error

$$\delta_n = \frac{\sqrt{\sum_{s=0}^{4} \left(\bar{p}_n[sh] - y_n[sh]\right)^2}}{\sqrt{\sum_{s=0}^{4}(\bar{p}_n[sh])^2} + \sqrt{\sum_{s=0}^{4}(y_n[sh])^2}}. \tag{54}$$

In the example under study, it constituted 4.6% (Region 1), 3.5% (Region 2), and 2.6% (Region 3).

7. Conclusions

This paper has developed a mathematical model for dynamic migratory interaction of regional systems with locally stationary states described by corresponding entropy operators. The model incorporates random parameters, and their probabilistic characteristics—the probability density functions of system parameters and measurement noises—have been calculated using soft randomized machine learning. An example of migratory interaction modeling and testing has been given.

Funding: This work was supported by Russian Science Foundation (project No. 17-11-01220).

Conflicts of Interest: The author declares no conflict of interest.

Appendix A

Proof of Theorem 1. Consider the function

$$\varphi(z) = \mu \sum_{k=1}^{M} \sum_{n=1}^{N} c_{kn} q_{kn} (z[sh])^{c_{kn}}, \tag{A1}$$

which appears on the left-hand side of Equation (19). Taking advantage of the obvious inequalities,

$$\varphi_-(z) = MN\mu c_{min} q_{min}(z)^{c_{min}} < \varphi(z) < MN\mu c_{max} q_{max}(z)^{c_{max}} = \varphi_+(z). \tag{A2}$$

The variables are $0 < c_{min} < 1$, $0 < c_{max} < 1$, $c_{min} < c_{max}$, and $c_{min} < c_{kn} < c_{max}$. Consider the equations

$$\varphi_-(z) = T[sh], \quad \varphi(z) = T[sh], \quad \varphi_+(z) = T[sh]. \tag{A3}$$

The functions $\varphi_-(z), \varphi(z)$, and $\varphi_+(z)$ are strictly convex. Therefore, the solutions of these equations has the relationship

$$z_- < z^* < z_+, \tag{A4}$$

which concludes the proof of Theorem 1. $\square$

References

1. Bilecen, B.; Van Mol, C. Introduction: International academic mobility and inequalities. *J. Ethic Migr. Stud.* **2017**, *43*, 1241–1255. [CrossRef]
2. Black, R.; Xiang, B.; Caller, M.; Engberson, G.; Heering, L.; Markova, E. Migration and development: Causes and consequences. In *The Dynamics of International Migration and Settlement in Europe: A State of the Art*; Penninx, R., Berger, M., Kraal, K., Eds.; Amsterdam University Press: Amsterdam, The Netherlands, 2006; pp. 41–63.
3. Imel'baev, S.S.; Shmul'yan, B.L. Modeling of stochastic communication systems. In *Entropy Methods for Complex Systems Modeling*; Wilson, A.G., Ed.; Nauka: Moscow, Russian, 1975; pp. 170–234.
4. Van der Knaap, G.A.; Steegers, W.F. Structural analysis of interregional and intraregional migration patterns. In *Demographic Research and Spatial Policy*; Heide, H., Willekens, F., Eds.; Academic Press: Voorburg, The Netherlands, 1984.
5. Popkov, Y.S. Dynamic entropy model for migratory interaction of regional systems. *Tr. Inst. Sist. Analiz. Ross. Akad. Nauk* **2018**, *2*, 3–11.
6. Zelinsky, W. The hypothesis of the mobility transition. *Geogr. Rev.* **1971**, *46*, 219–249. [CrossRef]

7. Popkov, Y.S. *Mathematical Demoeconomy: Integrating Demographic and Economic Approaches*; De Gruyter: Berlin, Germany, 2014.

8. Wilson, A.G. *Entropy in Urban and Regional Modelling*; Routledge: London, UK, 1970.

9. Rogers, A.; Willekens, F.; Raymer, J. Modelling interregional migration flows: continuity and change. *J. Math. Popul. Stud.* **2001**, *9*, 231–263. [CrossRef]

10. Rogers, A.; Little, J.; Raymer, J. *The Indirect Estimation of Migration: Methods for Dealing with Irregular, Inadequate, and Missing Data*; Springer Science & Business Media: Berlin, Germany, 2010.

11. Volpert, V.; Petrovskii, S.; Zincenko, A. Interaction of human migration and wealth distribution. *Nonlinear Anal.* **2017**, *150*, 408–423. [CrossRef]

12. Pan, J.; Nagurney, A. Using Markov chains to model human migration in a network equilibrium framework. *Math. Comput. Model.* **1994**, *19*, 31–39. [CrossRef]

13. Klabunde, A.; Willekens, F. Decision-making in agent-based models of migration: State of the art and challenges. *Eur. J. Popul.* **2016**, *32*, 73–97. [CrossRef] [PubMed]

14. Klabunde, A.; Zinn, S.; Willekens, F.; Leuchter, M. Multistable modelling extended by behavioural rules. An application to migration. *Popul. Stud.* **2017**, *71*, 61–67. [CrossRef] [PubMed]

15. Popkov, Y.S. Soft randomized machine learning. *Doklady Math.* **2018**, *98*, 646–647. [CrossRef]

16. Voevodin, V.V.; Kuznetsov, Y.A. *Matrices and Calculations*; Nauka: Moscow, Russian, 1984. (In Russian)

Article

A Data-Weighted Prior Estimator for Forecast Combination

Esteban Fernández-Vázquez [1],*, Blanca Moreno [1] and Geoffrey J.D. Hewings [2]

[1]   REGIOlab and Department of Applied Economics, University of Oviedo, Faculty of Economics and Business, Avda. del Cristo, s/n, 33006 Oviedo, Spain; morenob@uniovi.es
[2]   Regional Economics Applications Laboratory, University of Illinois at Urbana-Champaign 607 S. Matthew, Urbana, IL 61801-367, USA; hewings@illinois.edu
*   Correspondence: evazquez@uniovi.es

Received: 7 February 2019; Accepted: 18 April 2019; Published: 23 April 2019

Abstract: Forecast combination methods reduce the information in a vector of forecasts to a single combined forecast by using a set of combination weights. Although there are several methods, a typical strategy is the use of the simple arithmetic mean to obtain the combined forecast. A priori, the use of this mean could be justified when all the forecasters have had the same performance in the past or when they do not have enough information. In this paper, we explore the possibility of using entropy econometrics as a procedure for combining forecasts that allows to discriminate between bad and good forecasters, even in the situation of little information. With this purpose, the data-weighted prior (DWP) estimator proposed by Golan (2001) is used for forecaster selection and simultaneous parameter estimation in linear statistical models. In particular, we examine the ability of the DWP estimator to effectively select relevant forecasts among all forecasts. We test the accuracy of the proposed model with a simulation exercise and compare its *ex ante* forecasting performance with other methods used to combine forecasts. The obtained results suggest that the proposed method dominates other combining methods, such as equal-weight averages or ordinal least squares methods, among others.

Keywords: data-weighted prior; generalized maximum entropy method; combined forecast

---

1. Introduction

Forecasting agents can use an ample variety of forecasting techniques and different information sets, thus leading to a wide variety of obtained forecasts. Hence, as each individual forecast captures a different aspect of the available information, a combination of them would be expected to perform better than the individual forecasts. In fact, a growing volume of literature has demonstrated that a combined forecast increases forecast accuracy in several fields (e.g., [1–7]).

The first study about the forecast combination was carried out by [8]. Since their study, several researchers have shown a variety of modeling procedures to estimate the weights of each individual forecast in the combined forecast (a review of the literature can be found in [5,9,10]).

There are several methods for forecast combination that can be classified as variance–covariance methods, probabilistic methods, Bayesian methods, or regression-based methods, among others. The first kind of method allows the calculation of weights of the combined forecast by minimizing the error variance of the combination ([8,11]); Probabilistic methods ([12,13]) weights are linked to the probability that an individual forecast will perform best on the next occasion; Bayesian methods, which were originally put forward by [14], assume that the variable being predicted (y) and the individual forecasts have a random character and the combined forecast is the expected value of the a posteriori distribution of y that is modified from its a priori distribution with the sample information of the individual forecasts ([14–18], among others).

The regression-based methods were introduced by [19]. These methods link the weights of the combined forecasts to the coefficient vector of a linear regression, where individual forecasts are explanatory variables of the variable being predicted. The estimation of the coefficient vector is based on the past available information of individual forecasts and realizations of the variable being predicted. However, when the number of agents providing forecasts increases, the combined regression method involves the estimation of a large number of parameters and a dimensionality problem could arise.

In such a situation, in order to take out relevant information from a large number of forecasts, some procedures can be used, such as the subset selection, factor-based methods ([20,21]), ridge regression [22], shrinkage methods [23], latent root regression [24] or least absolute shrinkage, and the selection operator method ([25,26]), among others. Nevertheless, the simple arithmetic mean of the individual forecasts is the most used strategy to obtain the combined forecast. This strategy could be justified, as some researchers have empirically shown that simple averaging procedures dominate other, more complicated schemes ([2,27–29], among others). Such a phenomenon is usually referred to as the "forecasting combination puzzle" which has been documented by [10], who shows that the simple arithmetic mean constitutes a benchmark. From a theoretical point of view, the simple equal-weight average could be justified when all the forecasters have shown the same forecast performance in the past, or there is not available information about individual forecast's past performance to calibrate them differently.

In such a situation of limited information, the following question arises: Could it be possible to combine individual forecasts differently from the simple average procedure? This drawback of the combination forecast is one of the potential problems which we address in this paper. In fact, under a regression-based combination method framework we propose a procedure that allows for simultaneous parameter estimation and forecast selection in linear statistical models. This procedure is based on the data-weighted prior (DWP) estimator proposed by [30]. This estimator has been previously applied to standard regression analysis, but not specifically to the field of forecast combination. More specifically, we analyze how DWP is able to reduce the number of potential forecasters and estimate a vector of weights different from the simple average in the combined forecast. We use a simulation exercise to compare the ex-ante forecasting performance of the proposed method with other combining methods, such as equal-weight averages or ordinal least square methods, among others. The obtained results indicate that the method based on DWP outperform other examined forecast combination methods.

The paper is organized in five additional sections. Section 2 introduces the framework of the regression-based combination methods. Section 3 presents the data-weighted prior (DWP) estimator. Section 4 shows the simulation experiment and presents the results. Finally, Section 5 summarizes the conclusions of the research.

2. Forecast Combination Methods Based on Regression Methods

There is a large number of individual forecasts to forecast any given variable (y) with forecast horizon h at time t, y_{t+h}. We indicate by x_{it} the forecast referred to $t + h$, given in period t by a forecasting agent or model i ($i = 1, \ldots, K$). The theory of combining forecasts indicates that it could be possible to obtain an aggregated prediction $\hat{y}_t$ that combines the individual forecasts $x = (x_{1t}, \ldots, x_{Kt})$ through a vector of weights $\beta = (\beta_1, \ldots, \beta_K)'$.

The first study about forecast combination focused on the combination of two forecasts whose vector of weights was obtained from the error variances of the individual forecast [8]. Afterward, [11] showed a combined forecast obtained by $\hat{y}_t = x\beta$, with the sum of weights is $l'\beta = 1$, l being a vector ($K \times 1$) of ones and $0 \leq \beta_i \leq 1$. The combined forecast reduces its error variance since:

$$\hat{\beta} = \frac{\left(\Sigma^{-1}l\right)}{\left(l'\Sigma^{-1}l\right)}; \text{ where } \sum = E(e_t e_t') \text{ and } e_t = yl' - x, \tag{1}$$

where e_t is the vector ($K \times 1$) containing the forecast error specific to each forecasting agent or model i.

However, the method does not take into account the possible correlation in the errors of the forecasts being combined. [19] showed that weights of the combined forecasts obtained through conventional methods can be interpreted as the coefficient vector of the linear projection of the variable being predicted from the K individual forecasts as:

$$y_{t+h} = x\beta + e_{t+h}, \tag{2}$$

where y_{t+h} is the variable being predicted (unobservable). The estimation of β is based on the past observations of the variable $y = (y_1, y_2, \ldots, y_T)$ and experts' past performances $X = (x_1, \ldots, x_K)$:

$$y = X\beta + \epsilon, \tag{3}$$

where y is a $(T \times 1)$ vector of observations for y, X is a $(T \times K)$ matrix of experts' past performances, being each x_i a $T \times 1$ vector of individual past forecasts, β is the $(K \times 1)$ vector of unknown parameters $\beta = (\beta_1, \ldots, \beta_K)$ to be estimated, and ϵ is a $(T \times 1)$ vector with the random term of the linear model.

The combining regression-based methods introduced by [19] were extended in several ways. Thus, [31] introduced time varying combining weights and [32] introduced nonlinear specifications in combined regression context. The dynamic combined regressions were introduced by [33] to take into account the serially correlated errors. Moreover, [34,35] considered the problem of non-stationarity.

However, the number of institutions carrying out forecasts has increased considerably in the last few years, thus the projection methodology suggested by Equation (3) would involve the estimation of a large number of weights. Thus a "curse of dimensionality problem" could arise when losing degrees of freedom for the regression estimation. In such cases, it is usual to use the simple mean average of the individual forecasts as a combined forecast.

In this situation of limited information about the past performance of individual forecasts, a question that arises is how to combine individual forecasts differently from the simple mean average. Some authors have shown evidence in support of an alternative that allows the calibration of individual forecasts when the small amount of information available does not allow the use of regression procedures. In a context where entry and exit of individual forecasters makes the regression estimation unfeasible, [36] shows how an affine transformation of the uniform weighted forecast performs reasonably well in small samples. [6] proposes a combination method based on the generalized maximum entropy approach [37]. Through the application of the maximum entropy principle, their method leads the adjustment of a priori weights (which are associated with the simple mean average) into posterior weights by considering a large number of forecasters, for which there is limited available information about their past performances.

3. A Data-Weighted Prior (DWP) Estimator

Generalized cross entropy (GCE) technique has interesting properties when dealing with ill-conditioned datasets (those affected by significant collinearity or small samples) An extensive description of the entropy estimation approach can be found in [37,38]. Thus, in this section we propose the application of an extension of the GCE technique in the context of combining individual predictors.

Let us suppose we are interested in forecasts of a variable y that depends on K explanatory variables x_i:

$$y = X\beta + \epsilon, \tag{4}$$

where y is a $(T \times 1)$ vector of observations for the variable being predicted y, X is a $(T \times K)$ matrix of observations for the x_i variables, β is the $(K \times 1)$ vector of unknown parameters to be estimated $\beta = (\beta_1, \ldots, \beta_K)'$, and ϵ is a $(T \times 1)$ vector containing the random errors. Each unknown parameter β_i is assumed to be a discrete random variable with $M \geq 2$ possible realizations. We suppose that there is some information about those possible realizations based on the researcher's a priori beliefs about the likely values of β_i. That information is included in a support vector $b' = (b_1, \ldots, b_M)$ with corresponding probabilities $p'_i = (p_{i1}, \ldots, p_{iM})$. Although each parameter could have different M values, it is assumed that the M values are the same for every parameter. Thus, vector β can be rewritten as:

$$\beta = \begin{bmatrix} \beta_1 \\ \vdots \\ \beta_K \end{bmatrix} = BP = \begin{bmatrix} b' & 0 & \cdots & 0 \\ 0 & b' & \cdots & 0 \\ \vdots & \vdots & \ddots & \vdots \\ 0 & 0 & \cdots & b' \end{bmatrix} \begin{bmatrix} p_1 \\ p_2 \\ \vdots \\ p_K \end{bmatrix}, \tag{5}$$

where B and P are matrixes with dimensions $(K \times KM)$ and $(KM \times 1)$ respectively. The following expression gives each parameter β_i as:

$$\beta_i = b' p_i = \sum_{m=1}^{M} b_m p_{im}; \ i = 1, \ldots, K \tag{6}$$

A similar approach is followed for ϵ. It is highlighted that, although GCE does not require rigid assumptions about the probability distribution function of the random error, as with other traditional estimation methods, some assumptions are still necessary to be made. It is assumed that ϵ has a mean $E[\epsilon] = 0$ and a finite covariance matrix. Moreover, each element ϵ_t is considered to be a discrete random variable with $J \geq 2$ possible values contained in the vector $v' = \{v_1, \ldots, v_J\}$. Although each ϵ_t could have different J values, it is assumed as common for all of them ϵ_t ($t = 1, \ldots, T$). We also assume that the random errors are symmetric around zero ($-v_1 = v_J$). The upper and lower limits (v_1 and v_J, respectively) are fixed by applying the three-sigma rule (see [37–39]). Thus, vector ϵ can be defined as:

$$\epsilon = \begin{bmatrix} \epsilon_1 \\ \vdots \\ \epsilon_T \end{bmatrix} = VW = \begin{bmatrix} v' & 0 & \cdots & 0 \\ 0 & v' & \cdots & 0 \\ \vdots & \vdots & \ddots & \vdots \\ 0 & 0 & \cdots & v' \end{bmatrix} \tag{7}$$

and each element ϵ_t has the value equals:

$$\epsilon_t = v' w_t = \sum_{j=1}^{J} v_j w_{tj}; \ t = 1, \ldots, T \tag{8}$$

Therefore, model (7) can be transformed into:

$$y = XBP + VW \tag{9}$$

In this context, we need to estimate the elements of matrix P, but also the elements of matrix W (denoted by $\widetilde{w}_{tj}$). The problem of the estimation of the vector of unknown parameters $\beta = (\beta_1, \ldots, \beta_K)'$ for the general linear model is transformed into the estimation of $K + T$ probability distributions. Based on this idea, [30] proposed an estimator that simultaneously allows for the estimation of parameters and the selection of variables in linear regression models. In order to have a basis for extraneous variable identification and coefficient reduction, the estimator uses sample but also non-sample information, as it is related to the Bayesian method of moments (BMOM) (see [40,41]). In other words, this technique allows for classifying some the explanatory variables in the linear model as irrelevant by shrinking the coefficients. Recent empirical applications of this method can also be found in [42–44].

Focusing on the context of combination of predictions, the objective of the DWP estimator is to identify which individual forecaster should receive a weight significantly different from the equal weighting scheme (simple arithmetic mean) and simultaneously to forecast the target variable based on a combination of individual predictors. We begin by specifying a discrete support space b for each β_i symmetric around the value $1/K$ and with large lower and upper limits, so that each β_i is contained in the chosen interval with high probability. The upper and lower bounds for v (v_1 and v_J, respectively) are fixed by applying the three-sigma rule. For the estimation of the β_i parameters, the specification of some a priori distribution q for the values in the supporting vectors is required. Besides fixing a uniform probability distribution that will be used as q in the GCE estimation (i.e., $q_m = \frac{1}{M}$), we also specify a "spike" prior for each β_i, where a very high probability $q_m \cong 1$ is associated with the value $1/K$ for b_m (i.e., $q_m \cong 0$ for the remaining values). Thus, data-based prior is specified so flexibly that for each β_i coordinate either a spike prior at the $b_m = 1/K$, a uniform prior over support space b, or any convex combination of the two, can result. The weight (a weighted formulation in an entropy optimization problem has been also proposed by [45] who proposed a weighted generalized maximum entropy (W-GME) estimator where different weights are assigned to the two entropies (for coefficient distributions and disturbance distributions) in the objective problem. Moreover, under a linear regression model estimation, [46] proposed a streaming generalized cross entropy (Stre-GCE) method to update the estimation of the parameters β_i by combining prior information and new data) given to the spike prior q^s for each parameter β_i is given by γ_i. For each γ_i, a discrete support space b_i^γ is specified with n possible values ($n = 1, \ldots, N$) and corresponding probability distribution p_i^γ. Thus, γ_i is defined as $\gamma_i = \sum_{n=1}^{N} b_{in}^\gamma p_{in}^\gamma$, where $b_{i1}^\gamma = 0$ and $b_{iN}^\gamma = 1$ are, respectively, the lower and upper bounds defined as the support of these parameters.

If q^u and q^s denote the uniform and spike a priori distributions, respectively, we can achieve the objective proposed by minimizing the following constrained problem:

$$
\begin{aligned}
\operatorname*{Min}_{P,P^\gamma,W} D\left(P, P^\gamma, W \| Q, Q^\gamma, W^0\right) = \;& \sum_{i=1}^{K}(1-\gamma_i) \sum_{m=1}^{M} p_{im} ln\left(\frac{p_{im}}{q_{im}^u}\right) \\
& + \sum_{i=1}^{K} \gamma_i \sum_{m=1}^{M} p_{im} ln\left(\frac{p_{im}}{q_{im}^s}\right) \\
& + \sum_{i=1}^{K} \sum_{n=1}^{N} p_{in}^\gamma ln\left(\frac{p_{in}^\gamma}{q_{in}^\gamma}\right) \\
& + \sum_{t=1}^{T} \sum_{j=1}^{J} w_{tj} ln\left(\frac{w_{tj}}{w_{tj}^0}\right)
\end{aligned}
\tag{10}
$$

subject to:

$$
y_t = \sum_{i=1}^{K} \sum_{m=1}^{M} b_m p_{im} x_{it} + \sum_{j=1}^{J} v_j w_{tj}; \quad t = 1, \ldots, T
\tag{11}
$$

$$
\sum_{m=1}^{M} p_{im} = 1; \; i = 1, \ldots, K
\tag{12}
$$

$$
\sum_{j=1}^{J} w_{tj} = 1; \; t = 1, \ldots, T
\tag{13}
$$

$$
\sum_{n=1}^{N} p_{in}^\gamma = 1; \; i = 1, \ldots, K
\tag{14}
$$

$$
\gamma_i = \sum_{n=1}^{N} b_{in}^\gamma p_{in}^\gamma
\tag{15}
$$

The γ_i parameters and the β_i coefficients of the model in (10) are estimated simultaneously. Please note the symmetry between the terms γ and $1 - \gamma$. Permuting the part of the objective function (10) to which they are connected would not change the final result in terms of the weighting scheme estimated.

To understand the logic of the DWP estimator, an explanation regarding the objective function (10) is useful, which is divided into four terms. The first one measures the divergence between the posterior probabilities and the uniform priors for each β_i parameter, this being part of the divergence weighted by $(1 - \gamma_i)$. The second element of (10) measures the divergence between the uniform priors for each β_i with the spike prior and it is weighted by γ_i. The third element in (10) relates to the Kullback divergence of the weighting parameters γ_i. It is highlighted that the a priori probability distribution fixed for each one of those parameters is always uniform ($q_i^\gamma = \frac{1}{N}$ $\forall n = 1, \dots, N$). The last term measures the Kullback divergence between the prior and the posterior probabilities for the random error of the model. The prior distribution of the errors is uniform (again $w_{tj}^0 = \frac{1}{J}$ $\forall t = 1, \dots, T$).

From the recovered $\widetilde{p}_{im}$ probabilities, the estimated value of each parameter β_i is obtained as:

$$\widetilde{\beta}_i = \sum_{m=1}^{M} b_m \widetilde{p}_{im}; \ i = 1, \dots, K \tag{16}$$

Under some mild assumptions (see [30], page 177), there is a guarantee that DWP estimates are consistent and asymptotically normal. Moreover, it is also ensured that the approximate variance of the DWP estimator is lower than the approximate variance of the GCE estimator, where the variance is lower than the approximate variance of an Maximum Likelihood- Least Squares estimator (see [30], page 179).

As it was highlighted, the DWP estimator allows simultaneously the estimation of parameters and the selection of predictors in linear regression models. The strategy to reach this objective has two steps. First, the estimates of the weighting parameters γ_i are obtained as:

$$\widetilde{\gamma}_i = \sum_{n=1}^{N} b_{in}^\gamma \widetilde{p}_{in}^\gamma; \ i = 1, \dots, K \tag{17}$$

which can be used as a tool for this purpose: As $\widetilde{\gamma}_i \to 0$, the prior gets closer to the uniform and the estimated parameters approach those of the GME estimator. This indicates that the parameter associated with this predictor can take values far from the center of the support vector (i.e., $1/K$). On the other hand, for large values of $\widetilde{\gamma}_i$, the part of the objective function with the spike prior on $1/K$ takes over. Consequently, the predictors considered in the combination that should receive a weight equal to those in a simple mean average will be characterized by large values of $\widetilde{\gamma}_i$ ([30] considers sufficiently large values when $\widetilde{\gamma}_{ih} > 0.49$), together with estimates of β_i close to $1/K$.

Moreover, it is possible to test if the estimate for β_i is significantly different from $1/K$ by constructing an χ^2 statistic. In other words, the statistic allows us to test if the estimated $\widetilde{p}_{im}$ is significantly different from the respective spike prior q_{im}^s. The Kullback–Leibler divergence measure between the estimated and the a priori probabilities related to the spike prior is:

$$D_i\left(\widetilde{p}_i \| q_i^s\right) = \sum_{m=1}^{M} \widetilde{p}_{im} ln\left(\frac{\widetilde{p}_{im}}{q_{im}^s}\right) \tag{18}$$

The χ^2 divergence between both probabilities distributions is:

$$\chi_{M-1}^2 = M \sum_{m=1}^{M} \frac{\left(\widetilde{p}_{im} - q_{im}^s\right)^2}{q_{im}^s} \tag{19}$$

A second-order approximation of $D_h\left(\widetilde{p}_h \| q_h^s\right)$ is the entropy-ratio statistic for evaluating $\widetilde{p}_h$ versus q_h^s:

$$D_i\left(\widetilde{p}_i \| q_i^s\right) \cong \frac{1}{2} \sum_{m=1}^{M} \frac{\left(\widetilde{p}_{im} - q_{im}^s\right)^2}{q_{im}^s} \tag{20}$$

Consequently:

$$2MD_i\left(\widetilde{p}_i \| q_i^s\right) \to \chi_{M-1}^2 \tag{21}$$

Thus, the measure $2MD_i\left(\widetilde{p}_i\|q_i^s\right)$ allows us to test the null hypothesis $H_0: \beta_i = 1/K$. If H_0 is not rejected, we conclude that a predictor x_i should be weighted as a simple arithmetic. (We would like to point out that, when computing, $\log(0)$ presents problems in the computation. In order to overcome this, in the empirical application on the next section, the spike priors q_i^u have been specified with a point mass at zero equal to 0.999 and 0.0005 respectively for the other points of the support vectors.) In such a case, the vector of weights of the combined forecast estimated by using the DWP estimator is not different from the simple average. It means that the sample does not contain information providing strong empirical evidence to weigh differently than equal.

4. A Numerical Simulation Study

In this section of the paper, we compare the performance of the proposed DWP estimator with other methods used to combine individual forecasts by carrying out a numerical simulation study. Forecast combinations have been successfully applied in several areas of forecasting, such as economy (gross valued added, inflation, or stock returns), meteorology (wind speed, rainfall, see e.g., [47] in *Entropy* journal), or energy fields (wind power), among others. We focus our empirical exercise in the economic area; in fact, we take variable y as the gross value added being forecasted. (It is supposed that y is measured without error. In a situation in which y was measured with error, [48] proposed a method to extend the simple linear measurement error model through the inclusion of a composite indicator by using the GME estimator.)

The starting point of the numerical simulation is the unknown series y_t $(t = 1, \ldots, T)$ that contains the target variable and a $(T \times K)$ matrix X with K potential unbiased forecasters of this series along the T time periods. The basic idea is that X should contain some imperfect information on the target series. Specifically, in the experiment, the elements of X will be generated in the following way:

$$x_{it} = y_t + u_{it}; \; t = 1, \ldots, T; \; i = 1, \ldots, K \tag{22}$$

where $u_i \sim N(0, \sigma_i)$ is a noise term that reflects the accuracy of x_i as a forecaster of y and σ_i is a scalar that adjusts the variability of this noise. Note that σ_i indicates the degree of information for the target series that is contained in predictor x_i, i.e., the higher the value of σ_i, the less informative x_i is about y.

Given that in our numerical experiment we would like to replicate situations normally observed in the context of forecasting economic series, instead of numerically generating the values of our target variable y, we opted for taking actual values of an economic indicator. More specifically, we have taken the annual Gross Value Added rate of change in the region of Catalonia (Spain) from 1980 to 2013. We have extracted this information (at constant prices of 2008) from the BDmores database. (This database is generated by the Spanish Ministry of Economy, Industry and Competitiveness. More details can be found in: http://www.sepg.pap.minhap.gob.es/sitios/sepg/en-GB/Presupuestos/Documentacion/paginas/base0sdatosestudiosregionales.aspx).

Concerning the configuration of matrix X, we consider different numbers of potential predictors (dimension K) to be combined. Given that, in the context of forecasting regional indicators, the number of forecasters is normally smaller than when national or supra-national variables are predicted, we have set three different values for K, with K set to 6, 12, and 24. Moreover, we have considered that the behavior of these predictors can be heterogeneous when aiming at forecasting variable y. In particular, we have divided our set of K forecasters into two different subsets that can be classified as "good" or "bad" predictors. The logic of this idea is that the information that the predictors provide for forecasting variable y can vary among them, with a "good" predictor preferable to a "bad" one, but with the caveat that the comparatively "bad" forecaster may still contain some potentially useful information to be considered in the combination. In order to reflect this idea, the elements of matrix X will be generated differently in the following two subsets:

$$x_{it} = y_t + u_{it}^g; \; t = 1, \ldots, T; \; i = 1, \ldots, G \tag{23}$$

$$x_{it} = y_t + u_{it}^b; \ t = 1,\ldots,T; \ i = G+1,\ldots,K \tag{24}$$

where u_{it}^g is the noise term for the subset of G "good" predictors and u_{it}^b is the corresponding element for the comparatively "bad" ones. The difference between u_{it}^g and u_{it}^b is on its variability, since:

$$u_i^g \sim N\left(0, \frac{s}{2}\right) \tag{25}$$

$$u_i^b \sim N(0,s) \tag{26}$$

where s is the standard deviation in the sample 1980–2013 of the target variable y. Equation (25) and Equation (26) indicate that the variance of the forecasters classified as "good" presents a variance four times lower than for those classified as "bad".

In the simulation, we have set different proportions between these two subsets of predictors. First, a more realistic situation where 5/6 of the total of K forecasters belong to the group of "good" predictors and only 1/6 are classified as "bad." Additionally, and for comparative purposes, a situation where they are distributed in equal parts (50%) to each group is considered as well.

In the experiment, all the simulated predictors are combined through the regression-based method of combining forecasts:

$$y_t = \sum_{i=1}^{K} \beta_i x_{it} + e_{it}; \ t = 1,\ldots,T \tag{27}$$

with the target of the different methods for combining these forecasters to determine the best possible values for the $\beta's$ parameters.

The benchmark for comparing the competing methods will be the arithmetic mean of the forecasters, where $\beta_i = 1/K$, $\forall i$, which is normally the strategy taken as a valid reference in the literature on combination of forecasters. In fact, it is sometimes considered as the best way of combining information of individual predictors as some studies have pointed out (for example, [2,10,27–29]). Additionally, a restricted least squares weight scheme (see [19], for the original unrestricted Leas Squares approach; or [5] for the restricted version) is considered as well, where the $\beta's$ weights (restricted to sum to one) are estimated by minimizing the sum of squared errors e_{it}.

Our comparison is extended to include the proposals made in recent forecasting literature, where forecasts based on Bayesian model averaging (BMA) has received considerable attention (see [49,50]). In this approach, the weights are determined based on the Bayesian information criterion (BIC) as:

$$\beta_i = \frac{exp\left[-\frac{1}{2}BIC_i\right]}{\sum_{i=1}^{K} exp\left[-\frac{1}{2}BIC_i\right]}; \tag{28}$$

and

$$BIC_i = Tln\left(\hat{\sigma}_i^2\right) + ln(T) \tag{29}$$

where $\hat{\sigma}_i^2$ stands for the LS estimation of σ_i^2.

These techniques for combining the individual predictors x_i will be compared with the estimation of the optimal $\beta's$ weights when the DWP estimator is applied. Consequently, specifying some support for the set of parameters to be estimated and the errors is required. We have fixed the same vector b for all the $\beta's$ parameters. In particular, the proposed DWP estimator assumes as a prior value for each β_i the solution provided by the simple mean of forecasters, where all are equally weighted as $1/K$. More specifically, we have considered that each unknown parameter β_i has $M = 3$ possible realizations with values $b' = (1/K-1, 1/K, 1/K+1)$; in other words, the bounds with the minimum and maximum possible values for the weights are set as the center $1/K \pm 1$.

For the weighting parameters, we have considered a support vector with two possible realizations $N = 2$ and values $b' = (0,1)$. Finally, the supports of the random error terms have been specified by guarantying symmetry around zero and by using the three-sigma rule $(-3s, \ 0, \ 3s)$, with s being the sample standard deviation of the dependent variable.

Tables 1 and 2 summarize the results of comparing the actual target values of our variable of interest (y_t) with the combined individual forecasts $(\hat{y}_t)$ obtained according to the different methods, namely; the simple mean (mean), Least Squares (LS), Bayesian Information Criterion (BIC) and the proposed Data Weighted Prior (DWP), and following two different deviation measures: (i) The mean squared forecast errors (MSFE); and (ii), the mean absolute percentage forecast error (MAPFE), respectively, defined by the two following expressions:

$$MSFE = \sum_{f=1}^{F} \left(y_f - \hat{y}_f\right)^2 \tag{30}$$

$$MAPFE = 100 \sum_{f=1}^{F} \left|y_f - \hat{y}_f\right| \tag{31}$$

Table 1. Mean squared forecasting error (MSFE); 1000 trials.

Mean Squared Forecasting Error (MSFE)					
		Method			
K	G	mean	LS	BIC	DWP
6	5 good	0.0160	0.0136	0.0298	0.0156
	3 good	0.0269	0.0180	0.0379	0.0261
12	10 good	0.0077	0.0099	0.0256	0.0076
	6 good	0.0128	0.0141	0.0288	0.0125
24	20 good	0.0040	0.0147	0.0191	0.0039
	12 good	0.0064	0.0205	0.0243	0.0062

Table 2. Mean absolute percentage forecasting error (MAPFE); 1000 trials.

Mean Absolute Percentage Forecasting Error (MAPFE)					
		Method			
K	G	mean	LS	BIC	DWP
6	5 good	2.0312	1.8454	2.7303	2.0023
	3 good	2.6217	2.1553	3.1300	2.5799
12	10 good	1.4251	1.5797	2.5231	1.4079
	6 good	1.8182	1.8762	2.7280	1.7976
24	20 good	1.0132	1.8501	2.1976	0.99836
	12 good	1.2749	2.2305	2.5106	1.2556

The mean values of these deviation measures are computed from 1000 trials and for a forecast horizon of four periods ahead $(f = 1, \ldots, 4)$, which means that the last four periods in our sample are not included in the estimation of the weights, but taken as reference for evaluating the performance of our combination of predictions.

Error figures in Tables 1 and 2 show how the simple mean outperforms the combining methods based on some regression analysis (LS or BIC) in situations where the number of potential forecasters is large relative to the available sample size. When the predictors considered are 12 or 24, the combination based on LS and BIC presents problems derived from an ill-conditioned dataset (the number of parameters is large relative to the small sample size), whereas the arithmetic mean of predictors is not affected by this problem. The proposed DWP estimator seems to beat the competing combination techniques, given that it takes the weighting scheme as the arithmetic mean and only departs from these weights if the sample contains information providing strong empirical evidence to weigh differently than equal. On the contrary, when the number of predictors is low, an LS-based combination of forecasters performs better than any of the other techniques, given that now the sample size is large enough in relative terms to the number of predictors considered. One important aspect to consider, however, is that the performance of the proposed combined forecast methods has only been evaluated under the criterion of accuracy (measured through some forecast error-based indicators). However, other criteria could be considered (such as forecast error variance or asymmetry) leading to a different relative performance of the combining methods [9].

5. Conclusions

One of the most widespread strategies for combining individual forecasts is to take a simple average of the forecasts. Empirically, many studies have shown that the mean outperforms complex combining strategies. Theoretically, the use of the simple arithmetic mean could be justified when all the forecasters have shown the same forecasting ability or when the available information about their ability seems to be not enough to calibrate the forecasters differently. This paper proposes the use of an entropy-based technique estimator to obtain an affine transformation of the equal weighted forecast combination by using the small available information, a data-weighted prior (DWP) estimator.

We tested the validity of the proposed model by a simulation exercise and compared its ex-ante forecasting performance with other combining methods. The benchmarks for comparing the competing method were the arithmetic mean of the forecasters, a restricted least squares, and weight scheme forecasts based on Bayesian model averaging (where the weights are determined on the basis of the Bayesian information criterion).

We set three different values for the number of individual forecasts to be combined (6, 12, and 24) and we have divided our set of forecasters in two different subsets, which can be classified as "good" or "bad" predictors. The obtained results of the simulation indicate that the proposed DWP estimator seems to beat the competing combination techniques, given that it takes the weighting scheme as the arithmetic mean and only departs from these weights if the sample contains information providing strong enough empirical evidence to weigh differently than equal. The most relevant advantage of this estimator is that, even in situations characterized by a large number of forecasters, the DWP estimator generates a better set of recovered forecasters´ weights than the arithmetic mean which is capable to identify groups of forecasters into groups of "good" and "bad" forecasts. Additionally, the empirical application could be extended by comparing the forecasting performance of the proposed method with other combining methods based on an information-theoretic approach [6].

Author Contributions: Conceptualization, E.F.-V., B.M. and G.J.D.H.; Methodology, E.F.-V.; Validation, E.F.-V. and B.M.; Formal Analysis, E.F.-V.; Resources, B.M.; Writing-Original Draft Preparation, E.F.-V., B.M. and G.J.D.H.; Writing-Review & Editing, E.F.-V., B.M. and G.J.D.H.; Funding Acquisition, E.F.-V. and B.M.

Funding: This research was partially funded by the research project "Integrative mechanisms for addressing spatial justice and territorial inequalities in "Europe (IMAJINE)" in the EU Research Framework Programme H2020.

Acknowledgments: The authors acknowledge the support of the guest editors of this special issues and the comments received by two anonymous reviewers.

Conflicts of Interest: The authors declare no conflict of interest.

References

1. Holden, K.; Peel, D.A. Combining Economic Forecasts. *J. Oper. Res. Soc.* **1988**, *39*, 1005–1010. [CrossRef]
2. Stock, J.H.; Watson, M.W. Combining Forecasts of Output Growth in a Seven-Country Data Set. *J. Forecast.* **2004**, *23*, 405–430. [CrossRef]
3. Marcellino, M. Forecast Pooling for European Macroeconomic Variables. *Oxf. Bull. Econ. Stat.* **2004**, *66*, 91–112. [CrossRef]
4. Greer, M.R. Combination forecasting for directional accuracy: An application to survey interest rate forecasts. *J. Appl. Stat.* **2005**, *32*, 607–615. [CrossRef]

5. Timmermann, A. Forecast Combinations. In *Handbook of Economic Forecasting*; Elliott, G., Granger, C.W.J., Timmermann, A., Eds.; North-Holland: Amsterdam, The Netherlands, 2006; Volume 1, pp. 135–196.

6. Moreno, B.; López, A.J. Combining economic forecasts by using a Maximum Entropy Econometric. *J. Forecast.* **2013**, *32*, 124–136. [CrossRef]

7. Fernandez-Vazquez, E.; Moreno, B. Entropy econometrics for combining regional economic forecasts: A data-weighted prior estimator. *J. Geogr. Syst.* **2017**, *19*, 349–370. [CrossRef]

8. Bates, J.M.; Granger, C.W.J. The Combination of Forecasts. *Oper. Res. Q.* **1969**, *20*, 451–468. [CrossRef]

9. De Menezes, L.M.; Bunn, D.W.; Taylor, J.W. Review of Guidelines for the Use of Combined Forecasts. *Eur. J. Oper. Res.* **2000**, *120*, 190–204. [CrossRef]

10. Genre, V.; Kenny, G.; Meyler, A.; Timmermann, A. Combining expert forecasts: Can anything beat the simple average? *Int. J. Forecast.* **2013**, *29*, 108–121. [CrossRef]

11. Newbold, P.; Granger, C.W.J. Experience with Forecasting Univariate Time Series and the Combination of Forecasts. *J. Royal Stat. Soc. Ser. A* **1974**, *137*, 131–165. [CrossRef]

12. Bunn, D.W. A Bayesian approach to the linear combination of forecasts. *Oper. Res. Q.* **1975**, *26*, 325–329. [CrossRef]

13. Bordley, R.F. The combination of forecast: A Bayesian approach. *J. Oper. Res. Soc.* **1982**, *33*, 171–174. [CrossRef]

14. Winkler, R.L. Combining probability distributions from dependent information sources. *Manag. Sci.* **1981**, *27*, 479–488. [CrossRef]

15. Winkler, R.L.; Makridakis, S. The combination of forecasts. *J. Royal Stat. Soc. Ser. A* **1983**, *146*, 150–157. [CrossRef]

16. Agnew, C.E. Bayesian consensus forecast of macroeconomic variables. *J. Forecast.* **1985**, *4*, 363–376. [CrossRef]

17. Anandalingam, G.; Chen, L. Linear combination of forecasts: A general Bayesian model. *J. Forecast.* **1983**, *8*, 199–214. [CrossRef]

18. Clemen, R.T.; Winkler, R.L. Aggregating point estimates: A flexible modelling approach. *Manag. Sci.* **1999**, *39*, 501–515. [CrossRef]

19. Granger, C.W.J.; Ramanathan, C. Improved Methods of Combining Forecasts. *J. Forecast.* **1984**, *3*, 197–204. [CrossRef]

20. Chan, Y.; Stock, J.; Watson, M.A. Dynamic Factor Model Framework for Forecast Combination. *Span. Eco. Rev.* **1999**, *1*, 91–121. [CrossRef]

21. Stock, J.H.; Watson, M.W. Forecasting Using Principal Components from a Large Number of Predictors. *J. Am. Stat. Association* **2002**, *97*, 147–162. [CrossRef]

22. Fang, Y. Forecasting combination and encompassing tests. *Int. J. Forecast.* **2003**, *19*, 87–94. [CrossRef]

23. Aiolfi, M.; Timmerman, A. Persistence in forecasting performance and conditional combination strategies. *J. Eco.* **2006**, *135*, 31–53. [CrossRef]

24. Guerard, J.B.; Clemen, R.T. Collinearity and the use of latent root regression for combining GNP forecasts. *J. Forecast.* **1989**, *8*, 231–238. [CrossRef]

25. De Mol, C.; Giannone, D.; Reichlin, L. Forecasting using a large number of predictors: Is Bayesian shrinkage a valid alternative to principal components? *J. Eco.* **2008**, *146*, 318–328. [CrossRef]

26. Conflitti, C.; De Mol, C.; Giannone, D. Optimal Combination of Survey Forecasts. 2012. Available online: https://ideas.repec.org/p/eca/wpaper/2013-124527.html (accessed on 6 February 2019).

27. Makridakis, S.A.; Andersen, R.; Carbone, R.; Fildes, M.; Hibon, R.; Lewandowski, J.; Newton, E.; Parsen, E.; Winkler, R. The accuracy of extrapolation (time series) methods: Results of a forecasting competition. *J. Forecast.* **1982**, *1*, 111–153. [CrossRef]

28. Makridakis, S.; Winkler, R.L. Averages of Forecasts: Some empirical results. *Manag. Sci.* **1983**, *29*, 987–996. [CrossRef]

29. Smith, J.; Wallis, K.F. A Simple Explanation of the Forecast Combination Puzzle. *Oxf. Bull. Eco. Stat.* **2009**, *71*, 331–355. [CrossRef]

30. Golan, A. A Simultaneous Estimation and Variable Selection Rule. *J. Eco.* **2001**, *10*, 165–193. [CrossRef]

31. Diebold, F.X.; Pauly, P. Structural change and the combination of forecast. *J. Forecast.* **1987**, *6*, 21–40. [CrossRef]

32. Deutsch, M.; Granger, C.W.; Teräsvirta, T. The combination of forecasts using changing weights. *Int. J. Forecast.* **1994**, *10*, 47–57. [CrossRef]

33. Coulson, N.E.; Robins, R.P. Forecast Combination in a Dynamic Setting. *J. Forecast.* **1993**, *12*, 63–67. [CrossRef]

34. Hallman, J.; Kamstra, M. Combining algorithms based on robust estimation techniques and co-integration restrictions. *J. Forecast.* **1989**, *8*, 189–198. [CrossRef]

35. Miller, C.M.; Clemen, R.T.; Winkler, R.L. The effect of nonstationarity on combined forecasts. *Int. Forecast.* **1992**, *7*, 515–529. [CrossRef]

36. Capistrán, C.; Timmermann, A. Forecast combination with entry and exit of expert. *J. Bus. Eco. Stat.* **2009**, *27*, 428–440. [CrossRef]

37. Golan, A.; Judge, G.; Miller, D. *Maximum Entropy Econometrics: Robust Estimation with Limited Data*; John Wiley & Sons Ltd.: London, UK, 1996.

38. Kapur, J.N.; Kesavan, H.K. *Entropy Optimization Principles with Applications*; Academic Press: New York, NY, USA, 1992.

39. Pukelsheim, F. The three sigma rule. *Am. Stat.* **1994**, *48*, 88–91.

40. Zellner, A. Bayesian Method of Moments/Instrumental Variable (bmom/iv) Analysis of Mean and Regression Models. In *Modeling and Prediction: Honoring Seymour Geisser*; Lee, J.C., Johnson, W.C., Zellner, A., Eds.; Springer: Berlin, Germany, 1996; pp. 61–75.

41. Zellner, A. The Bayesian Method of Moments (BMOM): Theory and Applications. In *Advances in Econometrics*; Fomby, T., Hill, R.C., Eds.; Emerald Group Publishing Limited: Bingley, UK, 1997; Volume 12, pp. 85–106.

42. Bernardini-Papalia, R. A Composite Generalized Cross Entropy formulation in small samples estimation. *Eco. Rev.* **2008**, *27*, 596–609. [CrossRef]

43. Fernandez-Vazquez, E. Recovering matrices of economic flows from incomplete data and a composite Prior. *Entropy* **2012**, *12*, 516–527. [CrossRef]

44. Fernandez-Vazquez, E.; Rubiera-Morollon, F. Estimating Regional Variations of R&D Effects on Productivity Growth by Entropy Econometrics. *Spat. Eco. Anal.* **2013**, *8*, 54–70.

45. Wu, X. A weighted generalized maximum entropy estimator with a data-driven weight. *Entropy* **2009**, *11*, 917–930. [CrossRef]

46. Angelelli, M.; Ciavolino, E. Streaming Generalized Cross Entropy. *arXiv* **2018**, arXiv:1811.09710.

47. Men, B.; Long, R.; Li, Y.; Liu, H.; Tian, W.; Wu, Z. Combined Forecasting of Rainfall Based on Fuzzy Clustering and Cross Entropy. *Entropy* **2017**, *19*, 694. [CrossRef]

48. Carpita, M.; Ciavolino, E. A Generalized Maximum Entropy Estimator to Simple Linear Measurement Error Model with a Composite Indicator. *Adv. Data Anal. Classif.* **2017**, *11*, 139–158. [CrossRef]

49. Buckland, S.T.; Burnham, K.P.; Augustin, N.H. Model selection: An integral part of inference. *Biometrics* **1997**, *53*, 603–618. [CrossRef]

50. Burnham, K.P.; Anderson, D.R. *Model Selection and Multimodel Inference: A Practical Information-Theoretic Approach*; Springer: New York, NY, USA, 2002.

Article

Assessing the Performance of Hierarchical Forecasting Methods on the Retail Sector

José Manuel Oliveira [1,2,*] and Patrícia Ramos [1,3]

1   INESC Technology and Science, Rua Dr. Roberto Frias, 4200-465 Porto, Portugal; patricia@iscap.ipp.pt
2   Faculty of Economics, University of Porto, Rua Dr. Roberto Frias, 4200-464 Porto, Portugal
3   School of Accounting and Administration of Porto, Polytechnic Institute of Porto, Rua Jaime Lopes Amorim, 4465-004 S. Mamede de Infesta, Portugal
*   Correspondence: jmo@fep.up.pt; Tel.: +351-225-571-100

Received: 18 March 2019; Accepted: 22 April 2019; Published: 24 April 2019

Abstract: Retailers need demand forecasts at different levels of aggregation in order to support a variety of decisions along the supply chain. To ensure aligned decision-making across the hierarchy, it is essential that forecasts at the most disaggregated level add up to forecasts at the aggregate levels above. It is not clear if these aggregate forecasts should be generated independently or by using an hierarchical forecasting method that ensures coherent decision-making at the different levels but does not guarantee, at least, the same accuracy. To give guidelines on this issue, our empirical study investigates the relative performance of independent and reconciled forecasting approaches, using real data from a Portuguese retailer. We consider two alternative forecasting model families for generating the base forecasts; namely, state space models and ARIMA. Appropriate models from both families are chosen for each time-series by minimising the bias-corrected Akaike information criteria. The results show significant improvements in forecast accuracy, providing valuable information to support management decisions. It is clear that reconciled forecasts using the Minimum Trace Shrinkage estimator (MinT-Shrink) generally improve on the accuracy of the ARIMA base forecasts for all levels and for the complete hierarchy, across all forecast horizons. The accuracy gains generally increase with the horizon, varying between 1.7% and 3.7% for the complete hierarchy. It is also evident that the gains in forecast accuracy are more substantial at the higher levels of aggregation, which means that the information about the individual dynamics of the series, which was lost due to aggregation, is brought back again from the lower levels of aggregation to the higher levels by the reconciliation process, substantially improving the forecast accuracy over the base forecasts.

Keywords: hierarchical forecasting; information criteria; entropy; model selection; ARIMA; state space models; retail

---

1. Introduction

Retailers need demand forecasts at different levels of aggregation to support decision-making at operational and short-term strategic levels [1]. Consider a retailer warehouse storing inventory that is used to replenish multiple retail stores: Store-level forecasts at different product levels are needed to manage inventory in the store or to allocate shelf space, but aggregate forecasts are also required for the inventory decisions of the retailer warehouse [2]. Understanding whether these aggregate forecasts should be generated independently at each level of the hierarchy, based on the aggregated demand, or obtained using an hierarchical forecasting method, which depends on the aggregation constraints of the hierarchy but ensures coherent decision-making at the different levels, is the gap we seek to address in this paper.

SKUs (Stock Keeping Units) are naturally grouped together in hierarchies, with the individual sales of each product at the bottom level of the hierarchy, sales for groups of related products (such as categories, families, or areas) at increasing aggregation levels, and the total sales at the top level [3]. Generating accurate forecasts for hierarchical time-series can be particularly difficult. Time-series at different levels of the hierarchical structure have different scales and can exhibit very different patterns. The time-series at the most disaggregated level can be very noisy and are often intermittent, being more challenging to model and forecast. Aggregated series at higher levels are usually much smoother and, therefore, easier to forecast. Additionally, in order to ensure coherent decision-making at the different levels of the hierarchy, it is essential that forecasts of each aggregated series be equal to the sum of the forecasts of the corresponding disaggregated series. However, it is very unlikely that these aggregation constraints will be satisfied if the forecasts for each series in the hierarchical structure are generated independently. Finally, hierarchical forecasting methods should take advantage of the interrelations between the series at each level of the hierarchy.

The most traditional approaches to hierarchical forecasting are bottom-up and top-down methods. The bottom-up method involves forecasting each series at the bottom level, and then summing these to obtain forecasts at the higher levels of the hierarchy [4–7]. The main advantage of this approach is that, since forecasts are obtained at the bottom level, no information is lost due to aggregation. However, it ignores the inter-relations between the series and usually performs poorly on highly aggregated data. The top-down method involves forecasting the most aggregated series at the top level, and then disaggregating these, using either historical [8] or forecasted proportions [9], to obtain bottom level forecasts. Top-down approaches based on historical proportions tend to produce less accurate forecasts at lower levels of the hierarchy. The middle-out approach combines both bottom-up and top-down methods. First, forecasts for each series of an intermediate level of the hierarchy chosen previously are obtained. The forecasts for the series above the intermediate level are produced using the bottom-up approach, while the forecasts for the series below the intermediate level are produced using the top-down approach. Empirical studies comparing the performance of bottom-up and top-down methods have mixed results as to a preference for either bottom-up or top-down [4,6,10–12].

Recent work in the area tackles the problem using a two-stage approach: In the first step, forecasts for all series at all the levels of the hierarchy, rather then at a single level, are independently produced (these are called base forecasts). Then, a regression model is used to combine these to give coherent forecasts (these are called reconciled forecasts). Athanasopoulos et al. [9] and Hyndman et al. [13] used the Ordinary Least Squares (OLS) estimator and showed that their approach worked well, compared to most traditional methods. Hyndman et al. [14] suggested the Weighted Least Squares (WLS) estimator, proposing the variances of the base forecast errors as a proxy to the diagonal of the errors covariance matrix, with null off-diagonal elements. They also introduced several algorithms to make the computations involved more efficient under a very large number of series. To extend the work of Hyndman et al. [14], Wickramasuriya et al. [15] proposed a closed-form solution, based on the Generalised Least Squares (GLS) estimator, that minimised the sum of the variances of the reconciled forecast errors incorporating information from a full covariance matrix of the base forecast errors. The authors evaluated the performance of their method, compared to the most commonly-used methods and the results showed that it worked well with both artificial and real data.

Erven and Cugliari [16] proposed a Game-Theoretically OPtimal (GTOP) reconciliation method that selected the set of reconciled predictions, such that the total weighted quadratic loss of the reconciled predictions will never be greater than the total weighted quadratic loss of the base predictions. The authors illustrated the benefits of their approach on both simulated data and real electricity consumption data. This approach required fewer assumptions about the forecasts and forecast errors, but it did not have a closed-form solution and did not scale well for a huge set of time-series.

Mircetic et al. [17] proposed a top-down approach for hierarchical forecasting in a beverage supply chain, based on projecting the ratio of bottom and top level series into the future. Forecast projections were then used to disaggregate the base forecasts of the top level series. The disadvantage of all top-down approaches, including this one, is that they do not produce unbiased coherent forecasts [13].

The remainder of the paper is organized as follows. The next section presents a brief description of the two most widely-used approaches to time-series forecasting: State space models and ARIMA models. The procedure for using information criteria in model selection is also discussed. Section 3 describes the methods more commonly used to forecast hierarchical time-series. Section 4 presents the case study of a Portuguese retailer, explains the evaluation setup implemented and error measures used, and discusses the results obtained. Finally, Section 5 offers the concluding remarks.

2. Pure Forecasting Models

We consider two alternative forecasting methods for generating the base forecasts used by hierarchical forecasting approaches; namely, state space models and ARIMA models. These are briefly described in this section, giving a special focus on the use of information criteria for model selection.

2.1. State Space Models

Forecasts generated by exponential smoothing methods are weighted averages of past observations, where the weights decrease exponentially as the observations get older. The component form representation of these methods comprises the forecast equation and one smoothing equation for each of the components considered, which can be the level, the trend, and the seasonality. The possibilities for each of these components are: Trend $= \{N, A, A_d\}$ and Seasonality $= \{N, A, M\}$, where N, A, A_d and M mean, respectively, none, additive, additive damped, and multiplicative. By considering all combinations of the trend and seasonal components, nine exponential smoothing methods are possible. Each method is usually labelled by a pair of letters, (T,S), specifying the type of trend and seasonal components. Denoting the time-series by y_t, $t = 1, 2, \ldots, n$ and the forecast of y_{t+h}, based on all data up to time t by $\hat{y}_{t+h|t}$, the component form of the additive Holt-Winters' method, (A, A), is

$$\hat{y}_{t+h|t} = l_t + h b_t + s_{t+h-m(k+1)} \tag{1}$$

$$l_t = \alpha \left(y_t - s_{t-m} \right) + (1 - \alpha) \left(l_{t-1} + b_{t-1} \right) \tag{2}$$

$$b_t = \beta^* \left(l_t - l_{t-1} \right) + (1 - \beta^*) b_{t-1} \tag{3}$$

$$s_t = \gamma \left(y_t - l_{t-1} - b_{t-1} \right) + (1 - \gamma) s_{t-m} \tag{4}$$

$$0 \leq \alpha \leq 1, \qquad 0 \leq \beta^* \leq 1, \qquad 0 \leq \gamma \leq 1 - \alpha,$$

where l_t, b_t, and s_t denote, respectively, the estimates of the series level, trend (slope), and seasonality at time t; m denotes the period of seasonality; and k is the integer part of $(h - 1)/m$. The smoothing parameters α, β^*, and γ are constrained, to ensure that the smoothing equations can be interpreted as weighted averages. Fitted values are calculated by setting $h = 1$ with $t = 0, 1, \ldots, n - 1$. H-step ahead forecasts, for $h = 1, 2, \ldots$, can then be obtained using the last estimated values of the level, trend, and seasonality $(t = n)$. Details about all the other methods may be found in Hyndman and Athanasopoulos [18]. To be able to produce forecast intervals and use a model selection criteria, Hyndman et al. [19] (amongst others) developed a statistical framework, where an innovation state space model can be written for each of the exponential smoothing methods. Each state space model comprises a measurement equation, which describes the observed data, and state equations which describe how the unobserved components (level, trend, and seasonality) change with time. For each exponential smoothing method, two possible state space models are considered, one with additive errors and one with multiplicative errors, giving a total of 18 models. To distinguish state space models with additive and multiplicative errors, an extra letter E was added: The triplet (E, T, S) identifies the type of error, trend, and seasonality. The general state space model is

$$y_t = w(x_{t-1}) + r(x_{t-1})\varepsilon_t \tag{5a}$$

$$x_t = f(x_{t-1}) + g(x_{t-1})\varepsilon_t, \tag{5b}$$

where y_t denotes the observation at time t, x_t is the state vector, $\{\varepsilon_t\}$ is a white noise process with variance σ^2 referred to as the innovation (new and unpredictable), $w(.)$ is the measurement function, $r(.)$ is the error term function, $f(.)$ is the transition function, and $g(.)$ is the persistence function. Equation (2a) is the measurement equation and Equation (2b) gives the state equations. The measurement equation shows the relationship between the observations and the unobserved states. The transition equation shows the evolution of the state through time. The equations of the ETS(A, A, A) model (underlying additive Holt-Winters' method with additive errors) are [18]

$$y_t = l_{t-1} + b_{t-1} + s_{t-m} + \varepsilon_t \tag{6a}$$

$$l_t = l_{t-1} + b_{t-1} + \alpha \varepsilon_t \tag{6b}$$

$$b_t = b_{t-1} + \beta \varepsilon_t \tag{6c}$$

$$s_t = s_{t-m} + \gamma \varepsilon_t, \tag{6d}$$

and the equations of the ETS(M, A, A) model (underling additive Holt-Winters' method with multiplicative errors) are [19]

$$y_t = (l_{t-1} + b_{t-1} + s_{t-m})(1 + \varepsilon_t) \tag{7a}$$

$$l_t = l_{t-1} + b_{t-1} + \alpha (l_{t-1} + b_{t-1} + s_{t-m}) \varepsilon_t \tag{7b}$$

$$b_t = b_{t-1} + \beta (l_{t-1} + b_{t-1} + s_{t-m}) \varepsilon_t \tag{7c}$$

$$s_t = s_{t-m} + \gamma (l_{t-1} + b_{t-1} + s_{t-m}) \varepsilon_t. \tag{7d}$$

2.1.1. Estimation of State Space Models

Maximum likelihood estimates of the parameters and initial states of the state space model (2) can be obtained by minimizing its likelihood. The probability density function for $y = (y_1, \ldots, y_n)'$ is given by [19]

$$p(y \mid \theta, x_0, \sigma^2) = \prod_{t=1}^{n} p(y_t \mid x_{t-1}) = \prod_{t=1}^{n} p(\varepsilon_t)/|r(x_{t-1})|, \tag{8}$$

where θ is the parameters vector, x_0 is the initial states vector, and σ^2 is the innovation variance. By assuming that the distribution of $\{\varepsilon_t\}$ is Gaussian, this likelihood has the form

$$\mathcal{L}(\theta, x_0, \sigma^2 \mid y) = (2\pi\sigma^2)^{-n/2} \left| \prod_{t=1}^{n} r(x_{t-1}) \right|^{-1} \exp\left(-\frac{1}{2} \sum_{t=1}^{n} \varepsilon_t^2/\sigma^2 \right), \tag{9}$$

and its logarithm is

$$\log \mathcal{L} = -\frac{n}{2} \log(2\pi\sigma^2) - \sum_{t=1}^{n} \log |r(x_{t-1})| - \frac{1}{2} \sum_{t=1}^{n} \varepsilon_t^2/\sigma^2. \tag{10}$$

The maximum likelihood estimate of σ^2 can be obtained by taking the partial derivative of (10) with respect to σ^2 and setting it to zero:

$$\hat{\sigma}^2 = n^{-1} \sum_{t=1}^{n} \varepsilon_t^2. \tag{11}$$

This estimate can be used to eliminate σ^2 from the likelihood (9), which becomes

$$\mathcal{L}(\theta, x_0 \mid y) = (2\pi e \hat{\sigma}^2)^{-n/2} \left| \prod_{t=1}^{n} r(x_{t-1}) \right|^{-1}. \tag{12}$$

Hence, twice the negative logarithm of this likelihood is

$$-2 \log \mathcal{L}(\boldsymbol{\theta}, \boldsymbol{x}_0 \mid \boldsymbol{y}) = c_n + n \log \left(\sum_{t=1}^{n} \epsilon_t^2 \right) + 2 \sum_{t=1}^{n} \log |r(\boldsymbol{x}_{t-1})|, \tag{13}$$

where $c_n = n \log(2 \pi e) - n \log(n)$. Thus, maximum likelihood estimates for the parameters $\boldsymbol{\theta}$ and the initial states $\boldsymbol{x}_0$ can be obtained by minimizing

$$\mathcal{L}^*(\boldsymbol{\theta}, \boldsymbol{x}_0) = n \log \left(\sum_{t=1}^{n} \epsilon_t^2 \right) + 2 \sum_{t=1}^{n} \log |r(\boldsymbol{x}_{t-1})|. \tag{14}$$

The innovations can be computed recursively, using the relationships

$$\epsilon_t = [y_t - w(\boldsymbol{x}_{t-1})]/r(\boldsymbol{x}_{t-1}) \tag{15}$$
$$\boldsymbol{x}_t = f(\boldsymbol{x}_{t-1}) + g(\boldsymbol{x}_{t-1})\epsilon_t. \tag{16}$$

2.1.2. Information Criteria for Model Selection

Forecast accuracy measures can be used to select a model for a given time-series, as long as the errors are computed from a test set and not from the training set used to estimate the model. However, the errors usually available are not enough to draw reliable conclusions. One possible solution is to use an information criterion (IC), based on the likelihood $\mathcal{L}(\boldsymbol{\theta}, \boldsymbol{x}_0 \mid \boldsymbol{y})$, that would include a regularization term to compensate for potential overfitting. The Akaike Information Criteria (AIC) for state space models is defined as [18]

$$\text{AIC} = -2 \log \mathcal{L}(\boldsymbol{\theta}, \boldsymbol{x}_0 \mid \boldsymbol{y}) + 2k, \tag{17}$$

where $\mathcal{L}(\boldsymbol{\theta}, \boldsymbol{x}_0 \mid \boldsymbol{y})$ is the likelihood and k is the number of parameters and initial states of the estimated model. Akaike based his model selection criteria on the Kullback-Liebler (K-L) discrimination information, also known as negative entropy, defined by

$$I(f, g) = \int f(x) \log \left(\frac{f(x)}{g(x|\theta)} \right) dx, \tag{18}$$

which measures the information lost when the model g is used to approximate the real model f. He found that he could estimate the expectation of K-L information by the maximized log-likelihood corrected for bias. This bias can be approximated by the number of estimated parameters in the approximating model. Thus, the model selection procedure is to choose the model amongst the candidates having the minimum value of the AIC. The Bayesian Information Criteria (BIC) is defined as [20]

$$\text{BIC} = \text{AIC} + k[\log(n) - 2]. \tag{19}$$

The BIC is order-consistent, but is not asymptotically efficient like the AIC. The AIC corrected for small-sample bias, denoted by AIC_c, is defined as [19]

$$\text{AIC}_c = \text{AIC} + \frac{k(k+1)}{n-k-1}. \tag{20}$$

Appropriate models can be selected by minimizing the AIC, the BIC, or the AIC_c.

2.2. ARIMA Models

ARIMA models are generally accepted as one of the most versatile classes of models for forecasting time-series [21,22]. Many different types of stochastic seasonal and non-seasonal time-series can be represented by them. These include pure autoregressive (AR), pure moving average (MA), and mixed AR and MA processes, all requiring stationary data so that they can be applied. Although many time-series are non-stationary, they can be transformed to stationary time-series by taking proper degrees of differencing (regular and/or seasonal). The multiplicative seasonal ARIMA model, denoted as ARIMA$(p, d, q) \times (P, D, Q)_m$, has the following form [23]:

$$\phi_p(B)\Phi_P(B^m)(1 - B)^d(1 - B^m)^D y_t = c + \theta_q(B)\Theta_Q(B^m)\varepsilon_t, \tag{21}$$

where

$$\begin{aligned} \phi_p(B) &= 1 - \phi_1 B - \cdots - \phi_p B^p & \Phi_P(B^m) &= 1 - \Phi_1 B^m - \cdots - \Phi_P B^{Pm}, \\ \theta_q(B) &= 1 + \theta_1 B + \cdots + \theta_q B^q & \Theta_Q(B^m) &= 1 + \Theta_1 B^m + \cdots + \Theta_Q B^{Qm}, \end{aligned}$$

m is the period of seasonality, D is the degree of seasonal differencing, d is the degree of ordinary differencing, B is the backward shift operator, $\phi_p(B)$ and $\theta_q(B)$ are the regular autoregressive and moving average polynomials of orders p and q, respectively, $\Phi_P(B^m)$ and $\Theta_Q(B^m)$ are the seasonal autoregressive and moving average polynomials of orders P and Q, respectively, $c = \mu(1 - \phi_1 - \cdots - \phi_p)(1 - \Phi_1 - \cdots - \Phi_P)$, where μ is the mean of $(1 - B)^d(1 - B^m)^D y_t$, and ε_t is a zero-mean Gaussian white noise process with variance σ^2. To ensure causality and invertibility, the roots of the polynomials $\phi_p(B)$, $\Phi_P(B^m)$, $\theta_q(B)$, and $\Theta_Q(B^m)$ should lie outside the unit circle. One of the main tasks in ARIMA forecasting is selecting the values of p, q, P, Q, d, and D. Usually, the following steps are used [23]: Plot the series, identify outliers, and choose a proper variance-stabilizing transformation. For that purpose, a Box-Cox transformation may be applied [24]:

$$y_t' = \begin{cases} \ln(y_t), & \lambda = 0 \\ (y_t^\lambda - 1)/\lambda, & \lambda \neq 0 \end{cases}, \tag{22}$$

where the parameter λ is a real number, often between -1 and 2. Then, the sample ACF (Auto-Correlation Function) and sample PACF (Partial Auto-Correlation Function) can be computed to decide appropriate degrees of differencing (d and D). Alternatively, unit-root tests may be applied. The Canova–Hansen test [25] can be used to choose D. After D is selected, d can be chosen by applying successive KPSS (Kwiatkowski, Phillips, Schmidt & Shin) tests [26]. Finally, the sample ACF and sample PACF are matched with the theoretical patterns of known models, to identify the orders of p, q, P, and Q.

Information Criteria for Model Selection

As for state space models, the values of p, q, P, and Q may be selected by an information criterion, such as the Akaike Information Criteria [18]:

$$\text{AIC} = -2\log \mathcal{L}(\boldsymbol{\theta}, \sigma^2 | \boldsymbol{y}) + 2(p + q + P + Q + k + 1), \tag{23}$$

where $k = 1$ if $c \neq 0$ and 0 otherwise, and $\log \mathcal{L}(\boldsymbol{\theta}, \sigma^2 | \boldsymbol{y})$ is the log-likelihood of the model fitted to the properly transformed and differenced data, given by [27]

$$\log \mathcal{L}(\boldsymbol{\theta}, \sigma^2 | \boldsymbol{y}) = -\frac{n}{2}\log(2\pi) - \frac{n}{2}\log(\sigma^2) - \sum_{t=1}^{n}\frac{\varepsilon_t^2}{2\sigma^2}, \tag{24}$$

where $\boldsymbol{\theta}$ is the parameter vector of the model and σ^2 is the innovation variance (the last term in parentheses in (23) is the total number of parameters that have been estimated, including the innovation variance). Note that the AIC is defined by considering the same principles of maximum likelihood and negative entropy discussed in Section 2.1. The AIC corrected for small sample sizes, AIC_c, is defined as

$$\text{AIC}_c = \text{AIC} + \frac{2(p + q + P + Q + k + 1)(p + q + P + Q + k + 2)}{n - p - q - P - Q - k - 2}. \tag{25}$$

The Bayesian Information Criterion is defined as

$$\text{BIC} = \text{AIC} + [\log(n) - 2](p + q + P + Q + k + 1). \tag{26}$$

As for the state space models, appropriate ARIMA models may be obtained by minimizing either the AIC, AIC_c, or BIC.

3. Hierarchical Forecasting

3.1. Hierarchical Time-Series

For the purpose of illustration, consider the example of the hierarchical structure shown in Figure 1. At the top of the hierarchy (level 0) is the most aggregated time-series, denoted by *Total*. The observation at time t of the *Total* series is denoted by $y_{Total,t}$. The *Total* series is disaggregated into series A and series B, at level 1. The t-th observation of series A is denoted as $y_{A,t}$ and the t-th observation of series B is denoted as $y_{B,t}$. The series A and B are disaggregated, respectively, into two and three series that are at the bottom level (level 2). For example, $y_{AA,t}$ denotes the t-th observation of series AA. In this case, the total number of series is $n = 8$ and the number of series at the bottom level is $m = 5$. For any time t, the observations at the bottom level will sum to the observations of the series above. Hence, in this case, we have

$$y_{Total,t} = y_{AA,t} + y_{AB,t} + y_{BA,t} + y_{BB,t} + y_{BC,t}, \quad y_{A,t} = y_{AA,t} + y_{AB,t}, \quad y_{B,t} = y_{BA,t} + y_{BB,t} + y_{BC,t}. \tag{27}$$

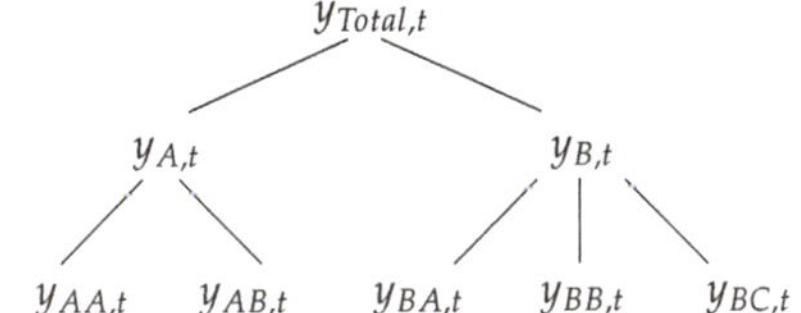

Figure 1. Example of a two-level hierarchical structure.

These aggregation constraints can be easily represented using matrix notation

$$\mathbf{y}_t = S\mathbf{b}_t, \tag{28}$$

where $\mathbf{y}_t = (y_{Total,t}, y_{A,t}, y_{B,t}, y_{AA,t}, y_{AB,t}, y_{BA,t}, y_{BB,t}, y_{BC,t})'$ is an n-dimensional vector, $\mathbf{b}_t = (y_{AA,t}, y_{AB,t}, y_{BA,t}, y_{BB,t}, y_{BC,t})'$ is an m-dimensional vector, and S is the summing matrix of order $n \times m$, given by

$$S = \begin{bmatrix} 1 & 1 & 1 & 1 & 1 \\ 1 & 1 & 0 & 0 & 0 \\ 0 & 0 & 1 & 1 & 1 \\ & & I_5 & & \end{bmatrix}. \tag{29}$$

Note that the first three rows of S correspond, respectively, to the three aggregation constraints in (27). The identity matrix I_5 below guarantees that each bottom level observation on the right-hand side of the equation is equal to itself on the left hand side. These concepts can be applied to an arbitrary set of n time-series that are subject to an aggregation structure, with m series at the bottom level [18]. The goal is to produce coherent forecasts for each series in the hierarchy; that is, forecasts that add up according to the aggregation constraints of the hierarchical structure.

3.2. Hierarchical Forecasting Methods

Let $\hat{\mathbf{y}}_{t+h|t}$ be an n-dimensional vector containing the forecasts of the values of all series in the hierarchy at time $t + h$ (with $h = 1, 2, \ldots$), obtained using observations up to and including time

t, and stacked in the same order as y_t. These are usually called base forecasts. They are calculated independently for each time-series, not taking into account any relationship that might exist between them due to the aggregation constraints. Any forecasting method, such as ETS or ARIMA, can be used to generate these forecasts. The issue is that it is very unlikely that these will be coherent forecasts, hence some reconciliation method should be further applied. All existing reconciliation methods can be expressed as

$$\tilde{y}_{t+h|t} = SP\hat{y}_{t+h|t},\tag{30}$$

where $\tilde{y}_{t+h|t}$ is an n-dimensional vector of reconciled forecasts, which are now coherent, and P is a matrix of dimension $m \times n$, which maps the base forecasts $\hat{y}_{t+h|t}$ into reconciled bottom level forecasts, which are then aggregated by the summing matrix S. If the bottom-up (BU) approach is used, then $P = [0_{m\times(n-m)} \mid I_m]$, where $0_{m\times(n-m)}$ is the null matrix of order $m \times (n-m)$ and I_m is the identity matrix of order m [4–6,9,10,28,29]. For the hierarchy shown in Figure 1, P is given by

$$P = \begin{bmatrix} 0 & 0 & 0 & 1 & 0 & 0 & 0 & 0 \\ 0 & 0 & 0 & 0 & 1 & 0 & 0 & 0 \\ 0 & 0 & 0 & 0 & 0 & 1 & 0 & 0 \\ 0 & 0 & 0 & 0 & 0 & 0 & 1 & 0 \\ 0 & 0 & 0 & 0 & 0 & 0 & 0 & 1 \end{bmatrix}.\tag{31}$$

This approach is computationally very efficient, since it only requires summing the bottom level base forecasts. It also has the advantage of forecasting the series at the most disaggregated level and, although it is more difficult to model, no information about the dynamics of the series is lost due to aggregation. However, it usually provides very poor forecasts for the upper levels in the hierarchy [13]. If a top-down (TD) approach is used, then $P = [p \mid 0_{m\times(n-1)}]$, where $p = [p_1, \ldots, p_m]'$ is an m-dimensional vector containing the disaggregation proportions, which indicate how the top level base forecast at time $t + h$ is to be distributed to obtain forecasts for the bottom level series, which are then summed by S [8,17,30–33]. For the hierarchy shown in Figure 1, P is given by

$$P = \begin{bmatrix} p_1 & 0 & 0 & 0 & 0 & 0 & 0 & 0 \\ p_2 & 0 & 0 & 0 & 0 & 0 & 0 & 0 \\ p_3 & 0 & 0 & 0 & 0 & 0 & 0 & 0 \\ p_4 & 0 & 0 & 0 & 0 & 0 & 0 & 0 \\ p_5 & 0 & 0 & 0 & 0 & 0 & 0 & 0 \end{bmatrix}.\tag{32}$$

The most common top-down methods performed quite well in Gross and Sohl [8]. In method "a" of Gross and Sohl [8] (referred to in the results that follow as TD_{GSa}), each proportion p_i is the average of the historical proportions of bottom level series $y_{i,j}$, relative to top level series $y_{T,j}$, over the time period $j = 1, \ldots, t$:

$$p_i = \frac{1}{t}\sum_{j=1}^{t}\frac{y_{i,j}}{y_{T,j}}, \qquad i = 1, \ldots, m.\tag{33}$$

In method "f" (referred to in the results that follow as TD_{GSf}), each proportion p_i is the average value of the historical data of bottom level series $y_{i,j}$, relative to the average value of the historical data of top level series $y_{T,j}$, over the time period $j = 1, \ldots, t$:

$$p_i = \sum_{j=1}^{t}\frac{y_{i,j}}{t} \Big/ \sum_{j=1}^{t}\frac{y_{T,j}}{t}, \qquad i = 1, \ldots, m.\tag{34}$$

These two methods are very simple to implement, since they only require forecasts for the most aggregated series in the hierarchy. They seem to provide reliable forecasts for the aggregate levels. However, they are not able to capture the individual dynamics of the series that is lost due to aggregation.

Moreover, since they are based on historical proportions, they tend to produce less accurate forecasts than the bottom-up approach at lower levels of the hierarchy, as they do not take into account how these proportions may change over time. To address this issue, Athanasopoulos et al. [9] proposed to obtain proportions based on forecasts rather than historical data:

$$p_i = \prod_{l=0}^{k-1} \frac{\hat{y}_{i,t+h|t}^{(l)}}{\hat{S}_{i,t+h|t}^{(l+1)}}, \qquad i = 1, \ldots, m, \tag{35}$$

where k is the level of the hierarchy, $\hat{y}_{i,t+h|t}^{(l)}$ is the base forecast at the time $t + h$ of the series that corresponds to the node which is l levels above i, and $\hat{S}_{i,t+h|t}^{(l+1)}$ is the sum of the base forecasts at the time $t + h$ of the series that corresponds to the nodes that are below the node that is l levels above node i and are directly connected to it. In the results that follow, this top-down method is referred as TD_{fp}. In the methods discussed so far, no real reconciliation has been performed, because these have been based on base forecasts from a single level of the hierarchy. However, processes that reconcile the base forecasts from the whole hierarchy structure in order to produce coherent forecasts can also be considered. Hyndman et al. [13] proposed an approach based on the regression model

$$\hat{y}_{t+h|t} = S\beta_{t+h|t} + \varepsilon_h, \tag{36}$$

where $\beta_{t+h|t}$ is the unknown conditional mean of the most disaggregated series and ε_h is the coherency error assumed with mean zero and covariance matrix Σ_h. If Σ_h was known, the generalised least squares (GLS) estimator of $\beta_{t+h|t}$ would lead to the following reconciled forecasts

$$\tilde{y}_{t+h|t} = S\hat{\beta}_{t+h|t} = S(S'\Sigma_h^{-1}S)^{-1}S'\Sigma_h^{-1}\hat{y}_{t+h|t} = SP\hat{y}_{t+h|t}, \tag{37}$$

where $P = (S'\Sigma_h^{-1}S)^{-1}S'\Sigma_h^{-1}$. Hyndman et al. [13] also showed that, if the base forecasts $\hat{y}_{t+h|t}$ are unbiased, then the reconciled forecasts $\tilde{y}_{t+h|t}$ will be unbiased, provided that $SPS = S$. This condition is true for this reconciliation approach and also for the bottom-up, but not for top-down methods. So, the top-down approaches will never give unbiased reconciled forecasts, even if the base forecasts are unbiased. Recently, Wickramasuriya et al. [15] showed that, in general, Σ_h is not identifiable. They showed that the covariance matrix of the h-step ahead reconciled forecast errors is given by

$$\text{Var}(y_{t+h} - \tilde{y}_{t+h|t}) = SPW_hP'S', \tag{38}$$

for any P such that $SPS = S$, where $W_h = \text{Var}(y_{t+h} - \hat{y}_{t+h|t}) = \text{E}(\hat{e}_{t+h|t}\hat{e}'_{t+h|t})$ is the covariance matrix of the corresponding h-step ahead base forecast errors. The goal is to find the matrix P that minimises the error variances of the reconciled forecasts, which are on the diagonal of the covariance matrix $\text{Var}(y_{t+h} - \tilde{y}_{t+h|t})$. Wickramasuriya et al. [15] showed that the optimal reconciliation matrix P that minimises the trace of $SPW_hP'S'$, such that $SPS = S$, is

$$P = (S'W_h^{-1}S)^{-1}S'W_h^{-1}. \tag{39}$$

Therefore, the optimal reconciled forecasts are given by

$$\tilde{y}_{t+h|t} = S(S'W_h^{-1}S)^{-1}S'W_h^{-1}\hat{y}_{t+h|t}, \tag{40}$$

which is referred to as the MinT (Minimum Trace) estimator. Note that the MinT and GLS estimators only differ in the covariance matrix. We still need to estimate W_h, which is a matrix of order n that can be quite large. The following simplifying approximations were considered by Wickramasuriya et al. [15]:

(1) $W_h = k_h I_n$ for all h with $k_h > 0$. In this case, the MinT estimator corresponds to the ordinary least squares (OLS) estimator of $\beta_{t+h|t}$. It is the most simplifying approximation considered,

being P-independent of the data (it only depends on S), which means that this method does not account for differences in scale between the levels of the hierarchy (captured by the error variances of the base forecasts), or the relationships between the series (captured by the error covariances of the base forecasts). This is optimal only when the base forecast errors are uncorrelated and equivariant, which are unrealistic assumptions for an hierarchical time-series. In the results that follow, this method is referred to as OLS.

(2) $W_h = k_h \mathrm{diag}(\hat{W}_1)$ for all h with $k_h > 0$, where $\hat{W}_1$ is the sample covariance estimator of the in-sample 1-step ahead base forecast errors. Then, W_h is a diagonal matrix with the diagonal entries of $\hat{W}_1$, which are the variances of the in-sample 1-step ahead base forecast errors, stacked in the same order as y_t. This approximation scales the base forecasts, using the variance of the residuals. In the results that follow, this specification is referred to as MinT-VarScale.

(3) $W_h = k_h \Lambda$ for all h with $k_h > 0$, and $\Lambda = \mathrm{diag}(S1)$ where 1 is a unit vector of dimension n. This method was proposed by Athanasopoulos et al. [34] for temporal hierarchies, and assumes that the bottom level base forecasts errors are uncorrelated between nodes and have variance k_h. Hence, the diagonal entries in Λ are the number of forecast error variances contributing to each node, stacked in the same order as y_t. This estimator only depends on the aggregation constraints, being independent of the data. Therefore, it is usually referred to as structural scaling, and we label it as MinT-StructScale. Notice that this specification only assumes equivariant base forecast errors at the bottom level, which is an advantage over OLS. It is particularly useful when the residuals are not available, which is the case when the base forecasts are generated by judgmental forecasting.

(4) $W_h = k_h \hat{W}_{1,D}^*$ for all h with $k_h > 0$, where $\hat{W}_{1,D}^* = \lambda \hat{W}_{1,D} + (1-\lambda)\hat{W}_1$ is a shrinkage estimator that shrinks the off-diagonal elements of $\hat{W}_1$ towards zero (while the diagonal elements remain unchanged), $\hat{W}_{1,D}$ is a diagonal matrix with the diagonal entries of $\hat{W}_1$, and λ is the shrinkage intensity parameter. By parameterizing the shrinkage in terms of variances and correlations, rather than variances and covariances, and assuming that the variances are constant, Schäfer and Strimmer [35] proposed the following shrinkage intensity parameter

$$\hat{\lambda} = \frac{\sum_{i \neq j} \widehat{\mathrm{Var}}(\hat{r}_{ij})}{\sum_{i \neq j} \hat{r}_{ij}^2},\tag{41}$$

where $\hat{r}_{ij}$ is the ij^{th} element of $\hat{R}_1$, the sample correlation matrix of the in-sample 1-step ahead base forecast errors. In contrast to variance and structure scaling estimators, which are diagonal covariance estimators accommodating only differences in scale between the levels of the hierarchy, this shrinkage estimator, which is a full covariance estimator, also accounts for the relationships between the series, while the shrinkage parameter regulates the complexity of the matrix W_h. In the results that follow, this method is referred to as MinT-Shrink. In all estimators, k_h is a proportionality constant that needs to be estimated only to obtain prediction intervals.

4. Empirical Study

4.1. Case Study Data

The Jerónimo Martins Group is an international company, based in Portugal, with 225 years of accumulated experience in the retail sector. Food distribution is its main business and represents more than 95% of their consolidated sales. In Portugal, it leads the supermarket segment through a supply chain called Pingo Doce. This empirical study was performed using a real database of product sales from one of the largest stores of Pingo Doce. The data were aggregated on a weekly basis and span the period between 3 January 2012 and 27 April 2015, comprising a total of 173 weeks. Only the products that have at least one sale every week were considered, since these are the most challenging for inventory planning. The hierarchical structure of products adopted by the retailer, from the top level to the bottom level, is: Store > Area > Division > Family > Category > Sub-category > SKU.

The total number of time-series considered is 1751 (aggregated and disaggregated) and their split in the six levels of the hierarchy is summarised in Table 1. The most aggregated level, referred to as the top level, comprises the total sales at the store level. Level 1 comprises these sales disaggregated by the six main areas: Grocery, specialized perishables, non-specialized perishables, beverages, detergents and cleaning, and personal care. These are further disaggregated, at level 2, into 21 divisions; at level 3, into 73 families; at level 4, into 203 categories; at level 5, into 459 subcategories; and, at the bottom level, into 988 SKUs (Stock Keeping Units).

Table 1. Number of series in each hierarchical level by area.

Area	Divisions	Families	Categories	Subcategories	SKUs
Specialized perishables	6	19	50	102	193
Non-specialized perishables	4	16	48	117	287
Grocery	3	14	51	144	309
Beverages	4	6	16	32	103
Personal care	2	9	19	37	59
Detergents & cleaning	2	9	19	27	37
Total	21	73	203	459	988

Figure 2 plots the sales at the top level and at level 1 of the hierarchy, aggregating these by the store and by each of the 6 main areas. The scale on the y axis was removed due to confidentiality reasons. The strong peak in sales in 2012, observed in all series, is relative to a promotional event carried out at a national level by Pingo Doce on 1 May (Labour day), after which the company shifted from an Every Day Low Price strategy to a continuous promotional cycle.

All the series show local upward and downward trends, although less prominent in the detergents/cleaning and personal care time-series. The store time-series shows a similar behaviour to the perishables time-series, as the later represent the major proportion of the total sales. These aggregate series do not show any seasonal variation.

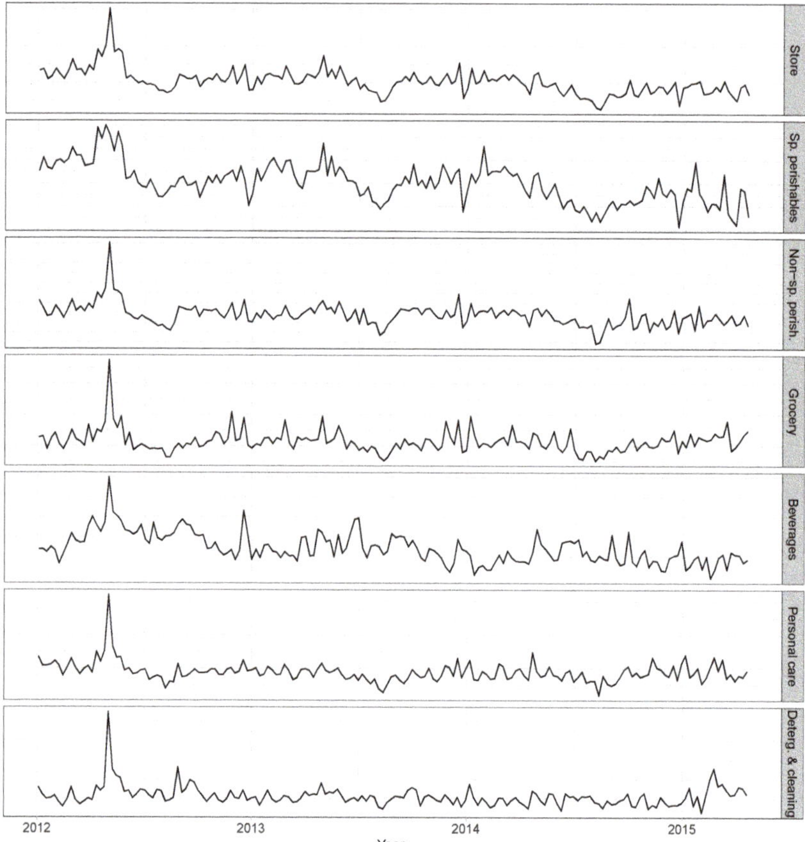

Figure 2. Total sales (top level, or store) and sales aggregated by area (level 1).

For a better understanding of the hierarchical structure of the data, we show, in Table 2, the complete hierarchy for the milk division (level 2). The total sales of the milk division are disaggregated, at level 3, into 2 families: Raw and UHT. The raw family is disaggregated into the Pasteurized category at level 4, which is further disaggregated into the Brik sub-category at level 5, which comprises 5 SKUs. The UHT family is disaggregated into the Current and Special categories. The Current category is disaggregated into the Semi-skimmed and Skimmed sub-categories, which comprise 2 and 3 SKUs, respectively. The Special category is disaggregated into the Semi-skimmed, Skimmed, and Flavored sub-categories, which comprise 10, 10, and 3 SKUs, respectively. The plots in Figure 3 show the sales of the SKUs within each subcategory of the milk division. These help us to visualise the diverse individual dynamics within each sub-category and the relative importance of each SKU. As we move down the hierarchy, the signal-to-noise ratio of the series decreases. Therefore, the series at the bottom level shows a lot more random variation, compared to the higher levels.

Table 2. Hierarchical structure of the milk division.

Area	Division	Families	Categories	Subcategories	SKUs
Non-specialized perishables	Milk	Raw	Pasteurized	Brik	5
		UHT	Current	Semi-skimmed	2
				Skimmed	3
			Special	Semi-skimmed	10
				Skimmed	3
				Flavored	3

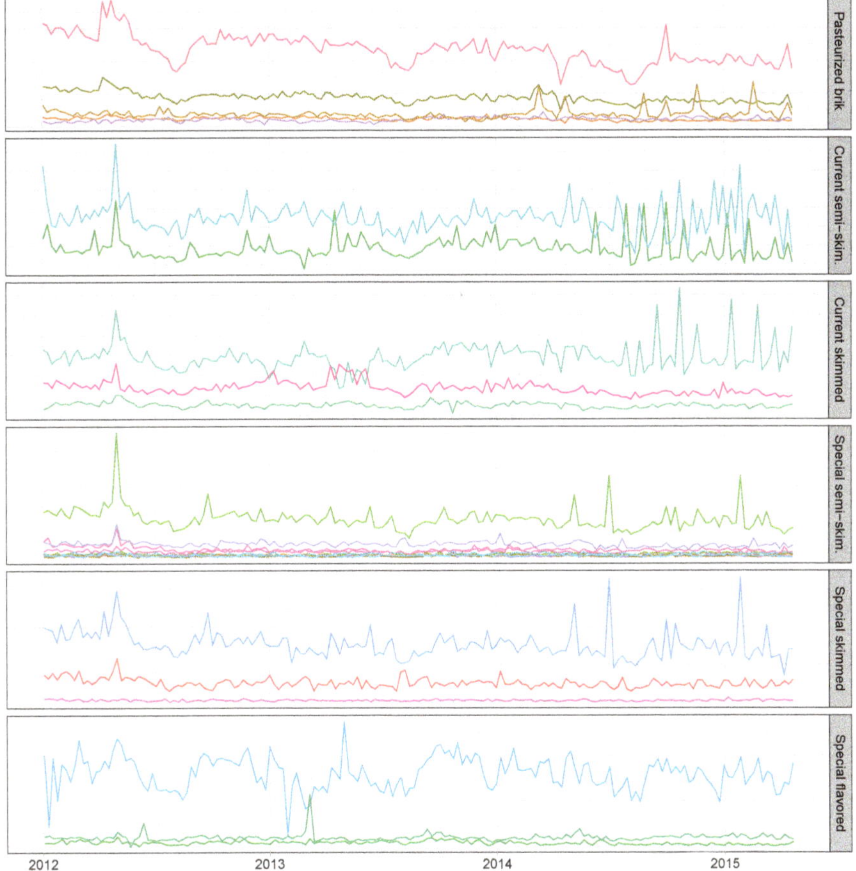

Figure 3. Sales of the SKUs within each sub-category of the milk division.

4.2. Experimental Setup

Generating accurate forecasts for each of the 1751 time-series within the hierarchical structure is crucial for the planning operations of the store. We can always forecast the series at each level of the hierarchy independently (we refer to these as base forecasts), based on forecasting models fitted individually for each series. However, by ignoring the aggregation constraints, it is very unlikely that the resulting forecasts will be coherent. To ensure aligned decision-making across the various levels of management, it is essential that these forecasts are reconciled across all levels of the hierarchy.

We consider two alternative forecasting model families for generating the base forecasts; namely, ETS and ARIMA, as discussed in Section 2. The appropriate ETS model for each time-series is chosen from the 18 potential models by minimising AIC_c, and the smoothing parameters and initial states are estimated by maximising the likelihood $\mathcal{L}$ [19], as implemented in the `forecast` package in the R software [36]. The ARIMA model is chosen following the algorithm proposed by Hyndman and Khandakar [37], also implemented in the `forecast` package. First, the number of seasonal and ordinary differences D and d required for stationarity are selected, and then the orders of $p, q, P,$ and Q are identified, based on AIC_c. ETS and ARIMA models are the two most widely-used approaches to time-series forecasting. They are based on different perspectives to the problem and often, but not always, perform differently, although they share some mathematically equivalent models [21,22,38–40]. ARIMA can potentially capture higher-order time-series dynamics than ETS [34]. Therefore, we use both approaches to generate base forecasts, in order to evaluate how these can influence the performance of each reconciliation process. To make incoherent ETS and ARIMA forecasts coherent, we use the implementations of the hierarchical forecasting approaches, as discussed in Section 3.2, available in the `hts` package [41] for R.

We evaluate the forecasting accuracies of several competing methods using a rolling origin, as illustrated in Figure 4. By increasing the number of forecast errors available, we increase the confidence in our results.

Figure 4. Cross-validation procedure, based on a rolling forecast origin with 1- to 12-week ahead forecasts.

We start with the training set containing the first 139 weeks and generate 1- to 12-week ahead base forecasts for each of the 1751 series using ETS and ARIMA. These base forecasts are then reconciled, using the alternative hierarchical methods. The training set is then expanded by one week, and the process is repeated until week 161. This gives a total of 23 forecast origins for each of the 1751 series. For each forecast origin, new ETS and ARIMA models based on the updated training data are specified, from which we generate new base forecasts which are again reconciled using the corresponding errors for both calculated. The performance of the hierarchical forecasting methods was evaluated by using the Average Relative Mean Squared Error (AvgRelMSE) [42]. As we are comparing forecast accuracy across time-series with different units, it is important to use a scale-independent error measure. For each time-series i, we calculate the Relative Mean Squared Error (RelMSE) [43]

$$\text{RelMSE}_{i,h} = \frac{\text{MSE}_{i,h}}{\text{MSE}_{i,h}^{\text{base}}}, \quad i = 1, \ldots, 1751; \ h = 1, 2, 4, 8, 12, \tag{42}$$

where $\text{MSE}_{i,h}$ is the mean squared error of the forecast of interest averaged across all forecast origins and forecast horizons h, and $\text{MSE}_{i,h}^{\text{base}}$ is the mean squared error of the base forecast averaged across all forecast origins and forecast horizons h, which is used as a benchmark. If the hierarchical forecasting method reconciles with ARIMA (ETS) base forecasts, then the ARIMA (ETS) base forecasts are taken as the benchmark. For each forecast horizon h, we averaged (42) across the time-series of the hierarchy using the following geometric mean

$$\text{AvgRelMSE}_{L,h} = \left(\prod_{i \in L} \text{RelMSE}_{i,h} \right)^{\frac{1}{\#L}}, \quad h = 1, 2, 4, 8, 12. \tag{43}$$

where L is the level (i.e., Top level, Level 1, ..., Level 5, Bottom level, All). The geometric mean should be used for averaging benchmark ratios, since it gives equal weight to reciprocal relative changes [44]. An advantage of AvgRelMSE is its interpretability. When it is smaller than 1, (1-AvgRelMSE)100% is the average percentage of improvement in MSE of the evaluated forecast over the benchmark.

4.3. Results

Table 3 presents the results of AvgRelMSE for the series of each hierarchical level, while Table 4 presents the results of AvgRelMSE for the complete hierarchy. BU refers to bottom-up method, TD_{GSa} refers to top-down "a" method of Gross and Sohl [8], TD_{GSf} refers to top-down "f" method of Gross and Sohl [8], TD_{fp} refers to top-down with forecast proportions, OLS refers to Ordinary Least Squares, MinT-VarScale refers to Minimum Trace Variance Scaling estimator, MinT-StructScale refers to Minimum Trace Structural Scaling estimator, MinT-Shrink refers to Minimum Trace Shrinkage estimator and Base refers to base forecasts. The left side of these tables shows the results using ARIMA base forecasts, while the right side shows the results using ETS base forecasts. As the base forecasts were used to scale the errors, in the rows labelled Base the AvgRelMSE is equal to 1 across all columns. We provide forecast results for 1 week, 2 weeks, 4 weeks (about one month), 8 weeks (about two months), and 12 weeks (about three months). The column labelled Rank provides the mean rank of each method across all forecast horizons. A method with rank of 1 is interpreted as being the best on all the horizons, while that with a rank of 9 it is always the worst. To support the comparisons between the methods that are expected to perform better, Figure 5 visualises the results of AvgRelMSE for the MinT-VarScale, MinT-StructScale, MinT-Shrink, and Base methods, presented in the Tables 3 and 4. The results for the complete hierarchy are highlighted with a light grey background.

It is immediately clear that the MinT-Shrink forecasts improved on the accuracy of the ARIMA base forecasts for all levels and for the complete hierarchy, across all forecast horizons. The only exception was the bottom level for the short-term horizons $h = 1$ and $1 - 2(h = 2)$, albeit with marginal differences. The gains in forecast accuracy were more substantial at the higher levels of aggregation. This was not the case for all other reconciliation methods, attesting to the difficulty of producing reconciled forecasts that were (at least) as accurate as the base forecasts. Furthermore, the MinT-Shrink method using ARIMA base forecasts returned the most accurate coherent forecasts for all levels, the only exceptions being the Store level, for which the MinT-VarScale returned the most accurate forecasts, and the Area level, where the MinT-StructScale performed best. The improvements on the accuracy of MinT-Shrink forecasts, across all forecast horizons, are more pronounced with the ARIMA base forecasts, compared to the ETS base forecasts (with the exception of horizon $h = 1$ at the bottom level), although the former was almost always more accurate than the latter (see Table 5). This could have be due to the limitation of the `ets()` function in the `forecast` package, which restricts seasonality to have a maximum period of 24. Without this limitation, ARIMA can potentially capture seasonalities of a higher order than ETS.

Clearly, the least accurate method was the OLS, for both ETS and ARIMA forecasts and across all forecast horizons. OLS only improved forecast accuracy over the base forecasts at the top level. This was due to ignoring the differences in scale between the levels of the hierarchy and any relationships between the series. A major drawback of the TD_{GSa} and TD_{GSf} methods was that they only considered information from the top level. Interestingly, their forecasts only improved on the accuracy of the ARIMA base forecasts for the Area level, never improving over the ETS base forecasts (the forecasts at the top level are equal to the base forecasts). The TD_{fp} proportions were based on forecasts from all disaggregated levels of the hierarchy, but it performed badly, never improving the forecast accuracy over the ARIMA base forecasts across all forecast horizons. This could be expected, since top-down approaches never give unbiased reconciled forecasts, even if the base forecasts are unbiased. BU provided poor forecasts for all aggregate levels in the hierarchy, showing average increases in the MSE relative to the base forecasts for all levels of aggregation and all forecast horizons (the forecasts at the bottom level are equal to the base forecasts). These losses in forecast accuracy were more substantial at higher levels of aggregation.

Table 3. Average Relative Mean Squared Error (AvgRelMSE) for each level of the hierarchy obtained with ARIMA and ETS base forecasts.

	ARIMA						ETS					
Method	$h=1$	$1-2$	$1-4$	$1-8$	$1-12$	Rank	$h=1$	$1-2$	$1-4$	$1-8$	$1-12$	Rank
Top-level: Store												
BU	2.074	2.179	2.489	2.569	2.237	9	1.748	1.721	1.869	1.990	1.914	9
TD_{GSa}	1	1	1	1	1	6.5	1	1	1	1	1	4.6
TD_{GSf}	1	1	1	1	1	6.5	1	1	1	1	1	4.6
TD_{fp}	1	1	1	1	1	6.5	1	1	1	1	1	4.6
OLS	0.949	0.951	0.959	0.950	0.947	4	0.985	0.990	0.998	1	0.999	2.5
MinT-VarScale	0.736	0.754	0.778	0.762	0.750	1	0.967	0.972	1.021	1.046	1.036	5
MinT-StructScale	0.749	0.777	0.836	0.837	0.796	2.4	1.035	1.031	1.099	1.142	1.123	8
MinT-Shrink	0.737	0.764	0.848	0.877	0.856	2.6	0.915	0.913	0.993	1.011	0.999	2.1
Base	1	1	1	1	1	6.5	1	1	1	1	1	4.6
Level 1: Area												
BU	1.096	1.154	1.242	1.274	1.268	8.6	1.264	1.274	1.314	1.327	1.288	9
TD_{GSa}	0.895	0.899	0.922	0.950	0.972	5	1.077	1.074	1.069	1.092	1.083	7.6
TD_{GSf}	0.886	0.888	0.911	0.938	0.961	4	1.067	1.063	1.057	1.080	1.071	6.6
TD_{fp}	1.020	1.012	1.002	1.009	1.015	7	1.021	1.009	0.998	0.998	0.998	4.5
OLS	1.189	1.186	1.150	1.134	1.125	8.4	1.123	1.079	1.004	0.990	0.977	5.3
MinT-VarScale	0.754	0.763	0.794	0.814	0.827	2.8	0.962	0.965	0.980	0.990	0.985	2.3
MinT-StructScale	0.717	0.734	0.777	0.804	0.817	1	0.980	0.983	0.997	1.008	0.998	3.9
MinT-Shrink	0.733	0.741	0.786	0.812	0.830	2.2	0.906	0.917	0.936	0.946	0.947	1
Base	1	1	1	1	1	6	1	1	1	1	1	4.8
Level 2: Division												
BU	1.082	1.131	1.175	1.212	1.192	7.6	1.278	1.278	1.277	1.256	1.227	8
TD_{GSa}	1.081	1.098	1.130	1.146	1.138	5.8	1.259	1.219	1.190	1.156	1.132	6
TD_{GSf}	1.089	1.104	1.136	1.151	1.142	7	1.269	1.226	1.197	1.162	1.137	7
TD_{fp}	1.091	1.091	1.082	1.068	1.056	5.6	1.026	1.020	1.002	1.004	1.006	3.6
OLS	1.966	1.953	1.994	2.029	2.027	9	1.523	1.495	1.461	1.471	1.457	9
MinT-VarScale	0.848	0.864	0.881	0.887	0.889	2.4	1.009	1.007	1.005	1.006	1.003	3.4
MinT-StructScale	0.842	0.861	0.885	0.903	0.908	2.6	1.036	1.031	1.029	1.029	1.023	5
MinT-Shrink	0.795	0.802	0.819	0.839	0.851	1	0.964	0.969	0.978	0.994	0.998	1
Base	1	1	1	1	1	4	1	1	1	1	1	2
Level 3: Family												
BU	1.016	1.022	1.031	1.040	1.036	5	1.083	1.083	1.073	1.067	1.061	6.4
TD_{GSa}	1.194	1.182	1.174	1.155	1.130	7	1.217	1.176	1.132	1.079	1.043	6.8
TD_{GSf}	1.200	1.188	1.179	1.159	1.134	8	1.223	1.181	1.136	1.083	1.046	7.8
TD_{fp}	1.101	1.094	1.079	1.079	1.075	6	1.024	1.018	1.008	1.005	1.005	4
OLS	2.348	2.314	2.338	2.405	2.399	9	1.567	1.542	1.533	1.524	1.503	9
MinT-VarScale	0.930	0.927	0.924	0.927	0.929	2	0.989	0.988	0.983	0.981	0.981	2
MinT-StructScale	0.982	0.979	0.981	0.991	0.998	3	1.035	1.031	1.026	1.023	1.021	5
MinT-Shrink	0.898	0.888	0.883	0.885	0.890	1	0.961	0.963	0.963	0.970	0.975	1
Base	1	1	1	1	1	4	1	1	1	1	1	3
Level 4: Category												
BU	1.014	1.015	1.019	1.029	1.029	4	1.027	1.028	1.027	1.028	1.027	4.2
TD_{GSa}	1.300	1.290	1.271	1.249	1.233	7	1.295	1.263	1.219	1.159	1.122	7
TD_{GSf}	1.306	1.296	1.276	1.253	1.237	8	1.302	1.269	1.224	1.163	1.125	8
TD_{fp}	1.129	1.121	1.108	1.107	1.103	5.1	1.033	1.031	1.028	1.027	1.030	4.8
OLS	2.463	2.418	2.403	2.398	2.375	9	1.636	1.618	1.602	1.563	1.537	9
MinT-VarScale	0.977	0.973	0.969	0.966	0.966	2	0.988	0.990	0.989	0.988	0.989	1.8
MinT-StructScale	1.129	1.125	1.121	1.115	1.112	5.9	1.076	1.073	1.069	1.063	1.062	6
MinT-Shrink	0.940	0.932	0.926	0.928	0.933	1	0.972	0.976	0.980	0.986	0.992	1.2
Base	1	1	1	1	1	3	1	1	1	1	1	3
Level 5: Subcategory												
BU	1.008	1.009	1.012	1.015	1.014	3.8	1.011	1.009	1.009	1.009	1.009	4
TD_{GSa}	1.326	1.301	1.274	1.231	1.208	6.8	1.314	1.270	1.220	1.155	1.117	7
TD_{GSf}	1.335	1.309	1.282	1.238	1.215	7.8	1.323	1.278	1.228	1.161	1.123	8
TD_{fp}	1.155	1.143	1.131	1.122	1.115	5	1.052	1.046	1.044	1.039	1.039	5
OLS	2.478	2.426	2.408	2.378	2.353	9	1.677	1.651	1.626	1.582	1.558	9
MinT-VarScale	1.009	1.004	1.001	0.994	0.992	2.8	1.000	0.997	0.997	0.995	0.995	1.7
MinT-StructScale	1.260	1.250	1.243	1.225	1.219	6.4	1.135	1.125	1.117	1.106	1.103	6
MinT-Shrink	0.970	0.962	0.955	0.948	0.949	1	0.989	0.992	0.996	1.001	1.005	1.8
Base	1	1	1	1	1	2.4	1	1	1	1	1	2.5
Bottom-level: SKU												
BU	1	1	1	1	1	2.1	1	1	1	1	1	1.5
TD_{GSa}	1.381	1.355	1.321	1.267	1.243	6.2	1.387	1.346	1.293	1.217	1.177	7
TD_{GSf}	1.393	1.366	1.331	1.276	1.251	7.4	1.398	1.357	1.303	1.225	1.184	8
TD_{fp}	1.182	1.166	1.148	1.129	1.126	5	1.080	1.079	1.075	1.068	1.069	5
OLS	2.077	2.038	2.009	1.972	1.959	9	1.506	1.496	1.479	1.448	1.433	9
MinT-VarScale	1.035	1.029	1.022	1.012	1.012	4	1.015	1.016	1.014	1.011	1.012	3.6
MinT-StructScale	1.378	1.364	1.347	1.320	1.315	7.4	1.204	1.200	1.191	1.177	1.172	6
MinT-Shrink	1.011	1.004	0.995	0.987	0.990	1.8	1.004	1.009	1.011	1.015	1.020	3.4
Base	1	1	1	1	1	2.1	1	1	1	1	1	1.5

Like OLS, MinT-StructScale only depended on the structure of the aggregations and not on the actual data, resulting in poor forecasts, especially at the lower levels of aggregation; in our case, at the Category, Sub-category, and SKU levels, which comprised about 94% of the time-series of the complete hierarchy (see Figure 5). On the other hand, by accommodating the differences in scale between the levels of the hierarchy, MinT-VarScale performed well almost always, generally improving the forecast accuracy over the base forecasts. MinT-Shrink also accounted for the inter-relationships between the series in the hierarchy, always performing better than MinT-VarScale, across both ETS and ARIMA forecasts for all forecast horizons; the only exception being at the Store level (which comprised only one time-series).

Table 4. AvgRelMSE for the complete hierarchy obtained with ARIMA and ETS base forecasts.

| | ARIMA | | | | | | | ETS | | | | | |
Method	$h=1$	$1-2$	$1-4$	$1-8$	$1-12$	Rank		$h=1$	$1-2$	$1-4$	$1-8$	$1-12$	Rank
						All							
BU	1.006	1.008	1.010	1.013	1.012	3.7		1.013	1.013	1.013	1.012	1.012	4
TD$_{GSa}$	1.343	1.320	1.292	1.248	1.225	6.8		1.346	1.306	1.256	1.186	1.148	7
TD$_{GSf}$	1.353	1.329	1.301	1.255	1.232	7.8		1.356	1.315	1.264	1.193	1.154	8
TD$_{fp}$	1.163	1.150	1.134	1.121	1.117	5		1.064	1.061	1.057	1.052	1.052	5
OLS	2.223	2.182	2.159	2.132	2.116	9		1.565	1.549	1.530	1.496	1.477	9
MinT-VarScale	1.013	1.008	1.003	0.996	0.995	2.9		1.006	1.006	1.005	1.003	1.003	2.6
MinT-StructScale	1.286	1.276	1.265	1.246	1.242	6.4		1.160	1.154	1.146	1.135	1.131	6
MinT-Shrink	0.983	0.975	0.968	0.963	0.966	1		0.994	0.998	1.001	1.006	1.011	2
Base	1	1	1	1	1	2.4		1	1	1	1	1	1.4

Table 5. AvgRelMSE results of ARIMA base forecasts with ETS base forecasts used as benchmark.

	$h=1$	$1-2$	$1-4$	$1-8$	$1-12$
Top-level	0.592	0.572	0.549	0.563	0.617
Level 1	1.007	0.98	0.958	0.947	0.929
Level 2	1.075	1.01	0.962	0.914	0.913
Level 3	0.986	0.961	0.931	0.902	0.894
Level 4	0.985	0.967	0.95	0.921	0.905
Level 5	0.984	0.969	0.955	0.937	0.925
Bottom-level	1.007	0.998	0.987	0.972	0.961
All	0.998	0.985	0.971	0.953	0.941

To improve on the accuracy of the base forecasts, the reconciliation methods have to take advantage of the combination of informative signals from all levels of aggregation. It is clear that MinT-Shrink was able do this and, hence, improvements in forecast accuracy over the base forecasts were attained. For the complete hierarchy, the accuracy gains generally increased with the forecast horizon varying between 1.7% and 3.7%. It is also evident that the gains in forecast accuracy were more substantial at higher levels of aggregation, which means that information about the individual dynamics of the series which was lost due to aggregation, was brought back again from the lower levels of aggregation to the higher levels by the reconciliation process, substantially improving the forecast accuracy over the base forecasts.

These results are in accordance with those obtained by Kourentzes and Athanasopoulos [45], which compared MinT-Shrink and MinT-VarScale forecasts with base forecasts in the context of generating coherent cross-temporal forecasts for Australian tourism. Both MinT-Shrink and MinT-VarScale improved the forecast accuracy over the base ETS and ARIMA forecasts for the bottom level and the complete hierarchy. MinT-Shrink performed better than MinT-VarScale across both ETS and ARIMA forecasts.

In order to find out if the forecast error differences between the several competing methods were statistically significant or not, we conducted a Nemenyi test [46]. The results of this test are shown in Figure 6. The panels on the left side show the results for the complete hierarchy using ARIMA base forecasts, for each forecast horizon; while the panels on the right side show the respective results using ETS base forecasts. In the vertical axis, the methods are sorted by MSE mean rank. In the horizontal axis, they are ordered as in Tables 3 and 4. In each row, the cell in black represents the method being tested and any blue cell indicates a method with no evidence of statistically significant differences, at a 5% level, while the white cells indicate methods without such evidence. We use the Nemenyi test implementation available in the `tsutils` [47] package for R.

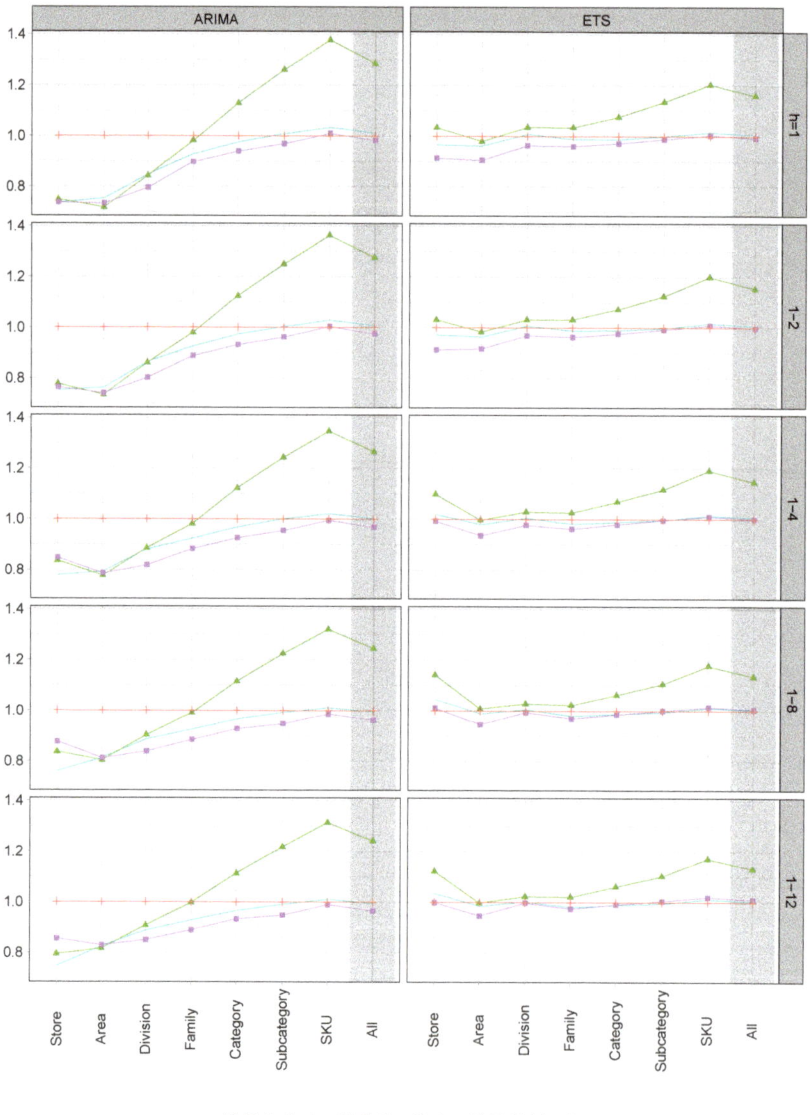

Figure 5. AvgRelMSE for the MinT-VarScale, MinT-StructScale, MinT-Shrink, and Base methods with ARIMA and ETS.

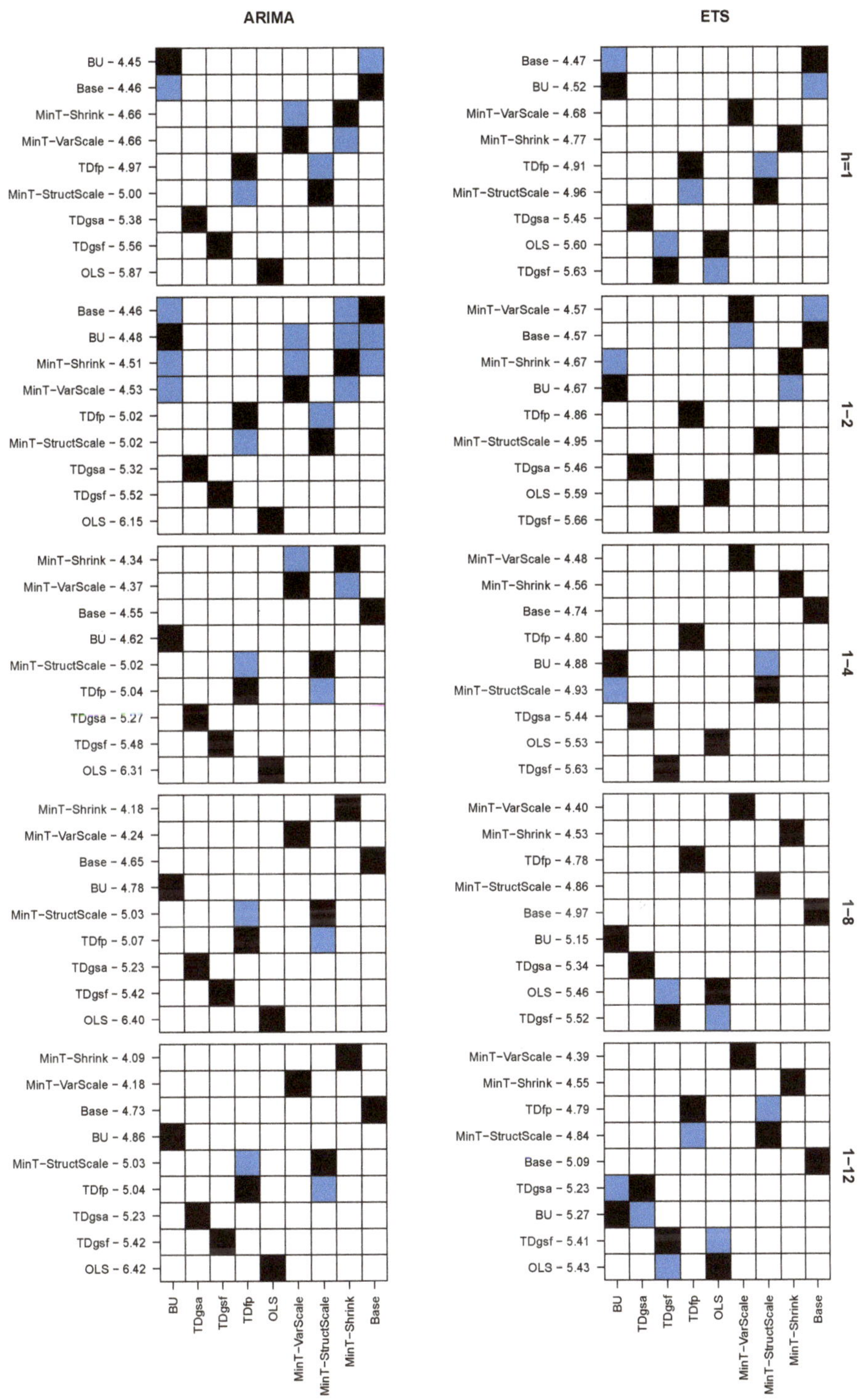

Figure 6. Nemenyi test results, at a 5% significance level, for the complete hierarchy.

Analysing the results for ARIMA presented in the panels on the left side, we observe that, for $h = 1$, BU and Base are grouped together as the top-performing methods. They are immediately followed by MinT-Shrink and MinT-VarScale, which are found to be statistically indifferent. For the forecast horizon $1 - 2$ ($h = 2$), BU, Base, MinT-Shrink, and MinT-VarScale are now grouped together as top-performing methods. For the forecast horizon $1 - 4$ ($h = 4$), MinT-Shrink and MinT-VarScale belong to the top-performing group of forecasts and BU and Base perform significantly worse. For the long-term forecasts, MinT-Shrink performs significantly better than MinT-VarScale, BU, and Base. The TD_{fp} and MinT-StructScale methods perform significantly worse than MinT-Shrink, MinT-VarScale, BU, and Base across all forecast horizons, and are found to be statistically indifferent, outperforming only TD_{GSa}, TD_{GSf}, and OLS.

Analysing the results for ETS presented in the panels on the right side, we observe that, for $h = 1$, BU and Base are again grouped together as top-performing methods, followed by MinT-VarScale and MinT-Shrink. For the forecast horizon $1 - 2$ ($h = 2$), MinT-VarScale and Base are grouped together as top-performing methods, being immediately followed by MinT-Shrink and BU; which are found to be statistically indifferent. For the other forecast horizons, MinT-VarScale performs better, being always followed by MinT-Shrink. Overall, for both ETS and ARIMA, the MinT approach outperforms the other competing methods, with the exception for the short horizon $h = 1$.

5. Conclusions

Retailers need forecasts for a huge number of related time-series which can be organised into an hierarchical structure. Sales at the SKU level can be naturally aggregated into categories, families, areas, stores, and regions. To ensure aligned decision-making across the hierarchy, it is essential that forecasts at the most disaggregated level add up to forecasts at the aggregate levels above. It is not immediately clear if these aggregate forecasts should be generated independently or by using an hierarchical forecasting method that ensures coherent decision-making at the different levels but does not guarantee (at the least) the same accuracy. To give guidelines on this issue, our empirical study investigates the relative performance of independent and reconciled forecasting approaches.

We use weekly data of SKU sales from one big store of a Portuguese retailer, spanning the period between 3 January 2012 and 27 April 2015, and consider the hierarchical structure of products adopted by the company from the top level to the bottom level, comprising six levels overall. We generate the independent forecasts using two alternative forecasting model families; namely, ETS and ARIMA. These are compared to the most commonly-used hierarchical forecasting approaches. We evaluate the forecast accuracies of several competing methods, through the Average Relative Mean Squared Error, by using a cross-validation based on a rolling forecast origin.

It is clear that MinT-Shrink forecasts generally improve on the accuracy of the ARIMA base forecasts for all levels and for the complete hierarchy, across all forecast horizons. The accuracy gains generally increase with the horizon, varying between 1.7% and 3.7% for the complete hierarchy. That is not the case for all other reconciliation methods, attesting to the difficulty of producing reconciled forecasts that are at least as accurate as base forecasts. The improvements on the accuracy of MinT-Shrink forecasts, across all forecast horizons, are more pronounced with the ARIMA base forecasts, compared to the ETS base forecasts (with the exception to horizon $h = 1$ at the bottom level); although, the former is almost always more accurate than the latter.

To improve on the accuracy of the base forecasts, the reconciliation methods have to take advantage of the combination of informative signals from all levels of aggregation. It is clear that MinT-Shrink is able do this and, hence, improvements in forecast accuracy over the base forecasts are attained. It is also evident that the gains in forecast accuracy are more substantial at higher levels of aggregation, which means that the information about the individual dynamics of the series lost when aggregating, is brought back again from the lower levels of aggregation to the higher levels by the reconciliation process, substantially improving the forecast accuracy over the base forecasts.

Author Contributions: Conceptualization, J.M.O. and P.R.; methodology, J.M.O. and P.R.; software, J.M.O. and P.R.; validation, J.M.O. and P.R.; formal analysis, J.M.O. and P.R.; investigation, J.M.O. and P.R.; resources, J.M.O. and P.R.; data curation, J.M.O. and P.R.; writing–original draft preparation, J.M.O. and P.R.; writing–review and editing, J.M.O. and P.R.; visualization, J.M.O. and P.R..

Funding: This research received no external funding.

Conflicts of Interest: The authors declare no conflict of interest.

References

1. Fildes, R.; Ma, S.; Kolassa, S. Retail forecasting: Research and practice. Working paper. Available online: http://eprints.lancs.ac.uk/128587/ (accessed on 24 April 2019).
2. Kremer, M.; Siemsen, E.; Thomas, D.J. The sum and its parts: Judgmental hierarchical forecasting. *Manag. Sci.* **2016**, *62*, 2745–2764. [CrossRef]
3. Pennings, C.L.; van Dalen, J. Integrated hierarchical forecasting. *Eur. J. Oper. Res.* **2017**, *263*, 412–418. [CrossRef]
4. Orcutt, G.H.; Watts, H.W.; Edwards, J.B. Data aggregation and information loss. *Am. Econ. Rev.* **1968**, *58*, 773–787.
5. Dunn, D.M.; Williams, W.H.; Dechaine, T.L. Aggregate versus subaggregate models in local area forecasting. *J. Am. Stat. Assoc.* **1976**, *71*, 68–71. [CrossRef]
6. Shlifer, E.; Wolff, R.W. Aggregation and proration in forecasting. *Manag. Sci.* **1979**, *25*, 594–603. [CrossRef]
7. Kohn, R. When is an aggregate of a time series efficiently forecast by its past? *J. Econom.* **1982**, *18*, 337–349. [CrossRef]
8. Gross, C.W.; Sohl, J.E. Disaggregation methods to expedite product line forecasting. *J. Forecast.* **1990**, *9*, 233–254. [CrossRef]
9. Athanasopoulos, G.; Ahmed, R.A.; Hyndman, R.J. Hierarchical forecasts for Australian domestic tourism. *Int. J. Forecast.* **2009**, *25*, 146–166. [CrossRef]
10. Dangerfield, B.J.; Morris, J.S. Top-down or bottom-up: Aggregate versus disaggregate extrapolations. *Int. J. Forecast.* **1992**, *8*, 233–241. [CrossRef]
11. Widiarta, H.; Viswanathan, S.; Piplani, R. Forecasting aggregate demand: An analytical evaluation of top-down versus bottom-up forecasting in a production planning framework. *Int. J. Prod. Econ.* **2009**, *118*, 87–94. [CrossRef]
12. Syntetos, A.A.; Babai, Z.; Boylan, J.E.; Kolassa, S.; Nikolopoulos, K. Supply chain forecasting: Theory, practice, their gap and the future. *Eur. J. Oper. Res.* **2016**, *252*, 1–26. [CrossRef]
13. Hyndman, R.J.; Ahmed, R.A.; Athanasopoulos, G.; Shang, H.L. Optimal combination forecasts for hierarchical time series. *Comput. Stat. Data Anal.* **2011**, *55*, 2579–2589. [CrossRef]
14. Hyndman, R.J.; Lee, A.; Wang, E. Fast computation of reconciled forecasts for hierarchical and grouped time series. *Comput. Stat. Data Anal.* **2016**, *97*, 16–32. [CrossRef]
15. Wickramasuriya, S.L.; Athanasopoulos, G.; Hyndman, R.J. Optimal forecast reconciliation for hierarchical and grouped time series through trace minimization. *J. Am. Stat. Assoc.* **2018**. [CrossRef]
16. Erven, T.; Cugliari, J. Game-Theoretically Optimal Reconciliation of Contemporaneous Hierarchical Time Series Forecasts. In *Modeling and Stochastic Learning for Forecasting in High Dimensions*; Antoniadis, A., Poggi, J.M., B.X., Eds.; Springer: Cham, Switzerland, 2015; Volume 217, pp. 297–317. [CrossRef]
17. Mircetic, D.; Nikolicic, S.; Đurđica Stojanovic.; Maslaric, M. Modified top down approach for hierarchical forecasting in a beverage supply chain. *Transplant. Res. Procedia* **2017**, *22*, 193–202. [CrossRef]
18. Hyndman, R.J.; Athanasopoulos, G. Forecasting: Principles and Practice; Online Open-access Textbooks, 2018. Available online: https://OTexts.com/fpp2/ (accessed on 24 April 2019).
19. Hyndman, R.J.; Koehler, A.B.; Ord, J.K.; Snyder, R.D. *Forecasting with Exponential Smoothing: The State Space Approach*; Springer: Berlin, Germany, 2008. [CrossRef]
20. Schwarz, G. Estimating the dimension of a model. *Ann. Stat.* **1978**, *6*, 461–464. [CrossRef]
21. Ramos, P.; Santos, N.; Rebelo, R. Performance of state space and ARIMA models for consumer retail sales forecasting. *Robot. Comput. Integr. Manuf.* **2015**, *34*, 151–163. [CrossRef]
22. Ramos, P.; Oliveira, J.M. A procedure for identification of appropriate state space and ARIMA models based on time-series cross-validation. *Algorithms* **2016**, *9*, 76. [CrossRef]

23. Box, G.E.P.; Jenkins, G.M.; Reinsel, G.C.; Ljung, G.M. *Time Series Analysis: Forecasting and Control*, 5th ed.; John Wiley & Sons, Inc.: Hoboken, NJ, USA, 2015.
24. Box, G.E.P.; Cox, D.R. An analysis of transformations. *J. R. Stat. Soc.* **1964**, *26*, 211–252. [CrossRef]
25. Canova, F.; Hansen, B.E. Are seasonal patterns constant over time? A test for seasonal stability. *J. Bus. Econ. Stat.* **1985**, *13*, 237–252. [CrossRef]
26. Kwiatkowski, D.; Phillips, P.C.; Schmidt, P.; Shin, Y. Testing the null hypothesis of stationarity against the alternative of a unit root: How sure are we that economic time series have a unit root? *J. Econom.* **1992**, *54*, 159–178. [CrossRef]
27. Hamilton, J. *Time Series Analysis*; Princeton University Press: Princeton, NJ, USA, 1994.
28. Theil, H. *Linear Aggregation of Economic Relations*; North-Holland: Amsterdam, The Netherlands, 1974.
29. Zellner, A.; Tobias, J. A note on aggregation, disaggregation and forecasting performance. *J. Forecast.* **2000**, *19*, 457–465. [CrossRef]
30. Grunfeld, Y.; Griliches, Z. Is aggregation necessarily bad? *Rev. Econ. Stat.* **1960**, *42*, 1–13. [CrossRef]
31. Lutkepohl, H. Forecasting contemporaneously aggregated vector ARMA processes. *J. Bus. Econ. Stat.* **1984**, *2*, 201–214. [CrossRef]
32. McLeavey, D.W.; Narasimhan, S. *Production Planning and Inventory Control*; Allyn and Bacon Inc.: Boston, MA, USA, 1974.
33. Fliedner, G. An investigation of aggregate variable timesSeries forecast strategies with specific subaggregate time series statistical correlation. *Comput. Oper. Res.* **1999**, *26*, 1133–1149. [CrossRef]
34. Athanasopoulos, G.; Hyndman, R.J.; Kourentzes, N.; Petropoulos, F. Forecasting with temporal hierarchies. *Eur. J. Oper. Res.* **2017**, *262*, 60–74. [CrossRef]
35. Schäfer, J.; Strimmer, K. A shrinkage approach to large-scale covariance matrix estimation and implications for functional genomics. *Stat. Appl. Genet. Mol. Biol.* **2005**, *4*, 151–163. [CrossRef]
36. R Development Core Team. *R: A Language and Environment for Statistical Computing*. R Foundation for Statistical Computing: Vienna, Austria, 2019.
37. Hyndman, R.J.; Khandakar, Y. Automatic time series forecasting: the forecast package for R. *J. Stat. Softw.* **2008**, *26*, 1–22. [CrossRef]
38. Papacharalampous, G.; Tyralis, H.; Koutsoyiannis, D. Predictability of monthly temperature and precipitation using automatic time series forecasting methods. *Acta Geophys.* **2018**, *66*, 807–831. [CrossRef]
39. Papacharalampous, G.; Tyralis, H.; Koutsoyiannis, D. One-step ahead forecasting of geophysical processes within a purely statistical framework. *Geosci. Lett.* **2018**, *5*, 12. [CrossRef]
40. Papacharalampous, G.; Tyralis, H.; Koutsoyiannis, D. Comparison of stochastic and machine learning methods for multi-step ahead forecasting of hydrological processes. *Stoch. Environ. Res. Risk Assess.* **2019**. [CrossRef]
41. Hyndman, R.; Lee, A.; Wang, E.; Wickramasuriya, S. *hts: Hierarchical and Grouped Time Series*, 2018. R package Version 5.1.5. Available online: https://pkg.earo.me/hts/ (accessed on 24 April 2019).
42. Davydenko, A.; Fildes, R. Measuring forecasting accuracy: The case of judgmental adjustments to SKU-level demand forecasts. *Int. J. Forecast.* **2013**, *29*, 510–522. [CrossRef]
43. Fildes, R.; Petropoulos, F. Simple versus complex selection rules for forecasting many time series. *J. Bus. Res.* **2015**, *68*, 1692–1701. [CrossRef]
44. Fleming, P.J.; Wallace, J.J. How not to lie with statistics: The correct way to summarize benchmark results. *Commun. ACM* **1986**, *29*, 218–221. [CrossRef]
45. Kourentzes, N.; Athanasopoulos, G. Cross-temporal coherent forecasts for Australian tourism. *Ann. Tourism Res.* **2019**, *75*, 393–409. [CrossRef]
46. Hollander, M.; Wolfe, D.A.; Chicken, E. *Nonparametric Statistical Methods*; John Wiley & Sons, Inc.: Hoboken, NJ, USA, 2015.
47. Kourentzes, N.; Svetunkov, I.; Schaer, O. *tsutils: Time Series Exploration, Modelling and Forecasting*, 2019. R package Version 0.9.0. Available online: https://rdrr.io/cran/tsutils/ (accessed on 24 April 2019).

Article

A Neutrosophic Forecasting Model for Time Series Based on First-Order State and Information Entropy of High-Order Fluctuation

Hongjun Guan [1], Zongli Dai [1], Shuang Guan [2] and Aiwu Zhao [1,3,*]

[1]   School of Management Science and Engineering, Shandong University of Finance and Economics, Jinan 250014, China; jjxyghj@sdufe.edu.cn (H.G.); studydzl@163.com (Z.D.)
[2]   Courant Institute of Mathematical Sciences, New York University, New York, NY 10012, USA; sg5896@nyu.edu
[3]   School of Management, Jiangsu University, Zhenjiang 212013, China
*   Correspondence: aiwuzh@ujs.edu.cn

Received: 31 March 2019; Accepted: 26 April 2019; Published: 1 May 2019

Abstract: In time series forecasting, information presentation directly affects prediction efficiency. Most existing time series forecasting models follow logical rules according to the relationships between neighboring states, without considering the inconsistency of fluctuations for a related period. In this paper, we propose a new perspective to study the problem of prediction, in which inconsistency is quantified and regarded as a key characteristic of prediction rules. First, a time series is converted to a fluctuation time series by comparing each of the current data with corresponding previous data. Then, the upward trend of each of fluctuation data is mapped to the truth-membership of a neutrosophic set, while a falsity-membership is used for the downward trend. Information entropy of high-order fluctuation time series is introduced to describe the inconsistency of historical fluctuations and is mapped to the indeterminacy-membership of the neutrosophic set. Finally, an existing similarity measurement method for the neutrosophic set is introduced to find similar states during the forecasting stage. Then, a weighted arithmetic averaging (WAA) aggregation operator is introduced to obtain the forecasting result according to the corresponding similarity. Compared to existing forecasting models, the neutrosophic forecasting model based on information entropy (NFM-IE) can represent both fluctuation trend and fluctuation consistency information. In order to test its performance, we used the proposed model to forecast some realistic time series, such as the Taiwan Stock Exchange Capitalization Weighted Stock Index (TAIEX), the Shanghai Stock Exchange Composite Index (SHSECI), and the Hang Seng Index (HSI). The experimental results show that the proposed model can stably predict for different datasets. Simultaneously, comparing the prediction error to other approaches proves that the model has outstanding prediction accuracy and universality.

Keywords: information entropy; aggregation operator; forecasting; neutrosophic set

1. Introduction

Financial markets are a complex system where fluctuation is the result of combined variables. These variables cause frequent market fluctuations with trends exhibiting degrees of ambiguity, inconsistency, and uncertainty. This pattern implies the importance of time series representations, and thus, an urgent demand arises for analyzing time series data in more detail. To some extent, an effective time series representation can be understood from two aspects: traditional time series prediction approaches [1–4]; and the fuzzy time series prediction approaches [5,6]. The former emphasizes the use of a crisp set to represent the time series, while the latter uses the fuzzy set.

Generally speaking, data are not only the source for prediction processes or prediction system inputs. The original data, however, are full of noise, incompleteness, and inconsistency, which limit the function of traditional prediction methods. Therefore, Song and Chissom [7–9] developed a fuzzy time series model to predict real-time scenarios like college admissions. The fuzzification method effectively eliminates part of the noise inside the data, and the prediction performance of the time series is strengthened. Subsequently, with advancing research, the non-determinacy of information has become the main contradiction affecting prediction accuracy. Some studies proposed novel information representation approaches, such as the type 2 fuzzy time series [5], rough set fuzzy time series [10], and intuitionistic fuzzy time series [11].

Although the above work has achieved considerable results for specific problems, certain shortcomings remain that pose a barrier to the accuracy and applicability of predictions. More specifically, complex scenarios and variables in actual situations make it unrealistic to define and classify explicitly the membership and non-membership of elements.

The neutrosophic sets (NSs) method, proposed by Smarandache [12] for the first time, is suitable for the expression of incomplete, indeterminate, and inconsistent information. A neutrosophic set consists of true-, indeterminacy-, and false-memberships. From the perspective of information representation, scholars have proposed two specific concepts based on the neutrosophic set: single-valued NSs [13] and interval-valued NSs [14]. These concepts are intended to seek a more detailed information representation, thereby enabling NSs to quantify uncertain information more accurately. To deal with the above problem, entropy is an important representation of the degree of the complexity and inconsistency. In a nutshell, entropy is more focused on the representation and measure of inconsistency, while NSs tends to describe uncertainty. Zadeh [15] first proposed the entropy of fuzzy events, which measures the uncertainty of fuzzy events by probability. Subsequently, De Luca and Termin [16] proposed the concept of entropy for fuzzy sets (FSs) based on Shannon's information entropy theory and further proposed a method of fuzzy entropy measurement. Since information entropy is an effective measurement in the degree of systematic order, it has been gaining popularity for different applications, such as climate variability [17], uncertainty analysis [18,19], financial analysis [20], image encryption [21], and detection [22]. Specifically, He et al. [23] proposed a collapse hazard forecasting method and applied the information entropy measurement to reduce the influence of collapse activity indices. Bariviera [24] proposed a prediction method based on the maximum entropy principle to predict the market and further monitor market anomalies. In Liang's research [25], information entropy was introduced to analyze trends for capacity assessment of sustainable hydropower development. Zhang et al. [26] proposed a signal recognition theory and algorithm based on information entropy and integrated learning, which applied various types of information entropy including energy entropy and Renyi entropy.

In order to describe the indeterminacy of fluctuations and further measure the inconsistency and uncertainty of dynamic fluctuation trends, we propose a neutrosophic forecasting model based on NSs and information entropy of high-order fuzzy fluctuation time series (NFM-IE). The biggest difference compared to the original models is that the NFM-IE represents both fluctuation trend information and fluctuation consistency information. First of all, a time series is converted to a fluctuation time series by comparing each of the current data and corresponding previous data in the time series. Then, the upward trend of each of the fluctuation data is mapped to the truth-membership of a neutrosophic set and falsity-membership for the downward trend. Information entropy of high-order fluctuation time series is introduced to describe the inconsistency of historical fluctuations and is mapped to the indeterminacy-membership of the neutrosophic set. Finally, an existing similarity measurement method for the neutrosophic set is introduced to find similar states during the forecasting stage, and the weighted arithmetic averaging (WAA) aggregation operator is employed to obtain the forecasting result according to the corresponding similarity. The largest contributions of the proposed model are listed as follows: (1) Introducing information entropy to quantify the inconsistency of fluctuations in related periods and mapping it to the indeterminacy-membership of neutrosophic sets allow NFM-IE to extend traditional forecasting models to a certain level. (2) Employing a similarity measurement method and aggregation operator allows NFM-IE to integrate more possible rules. In order to test its performance, we used the proposed model to forecast some realistic time series, such as the Taiwan Stock Exchange Capitalization Weighted Stock Index (TAIEX), the Shanghai Stock Exchange Composite Index (SHSECI), the Hang Seng Index (HSI), etc. The experimental results show that the model has a stable prediction ability for different datasets. Simultaneously, comparing the prediction error with that from other approaches proves that the model has outstanding prediction accuracy and universality.

The rest of this paper is organized as follows: Section 2 introduces the basic concepts of wave time series and information entropy. Then, the concepts proposed in this paper, such as neutrosophic fluctuation time series (NFTS) and the neutrosophic fluctuation logical relationship, are defined. Section 3 presents the specific modules of the model presented in this paper. Section 4 details the prediction steps and validates the model using TAIEX as the dataset. Section 5 further analyzes the prediction accuracy and universality of the model based on SHSECI and HSI. Finally, the conclusions and prospects are presented in Section 6.

2. Preliminaries

2.1. Fluctuation Time Series

Definition 1. *Let $\{V_t | t = 1, 2, \ldots, T\}$ be a stock time series, where T is the number of observations. Then, $\{U_t | t = 2, 3, \ldots, T\}$ is called a fluctuation time series, where $U_t = V_t - V_{t-1}$ $(t = 2, 3, \ldots, T)$.*

2.2. Information Entropy of the mth-Order Fluctuation in a Time Series

Information entropy (IE) [27] was proposed as a measurement of event uncertainty. The amount of information can be expressed as a function of event occurrence probability. The general formula for information entropy is:

$$E = -\sum_{t=1}^{N} p(x_t) log_2(p(x_t)) \tag{1}$$

where $p(\cdot)$ is the probability function of a set of N events. In addition, the information entropy must satisfy the following conditions: $\sum_{t=1}^{N} p(x_t) = 1$, $0 < p(x_t) < 1$. The information entropy is always positive.

According to the fuzzy set definition by Zadeh [28], each number in a time series can be fuzzified by its membership function of a fuzzy set $L = \{L_1, L_2, \ldots, L_g\}$, which can be regarded as an event in a time series. For example, when $g = 5$, it might represent a set of linguistic event variants as: $L = \{L_1, L_2, L_3, L_4, L_5\} = \{very\ low, low, equal, high, very\ high\}$, etc.

Definition 2. *Let $F(t - 1), F(t - 2), \ldots, F(t - m)$ be fuzzy sets of the mth-order fluctuation time series $\{U_t | t = m + 1, m + 2, \ldots, T\}$. Let $p_{Ut}(L_1), p_{Ut}(L_2), p_{Ut}(L_3), p_{Ut}(L_4)$, and $p_{Ut}(L_5)$ be the probabilities of the occurrence of the linguistic variants L_1, L_2, L_3, L_4, and L_5 for $F(t - 1), F(t - 2), \ldots, F(t - m)$. The information entropy of the mth-order fluctuation is defined as:*

$$E(U_t) = -\sum_{n=1}^{g} p_{Ut}(L_n) \log_2(p_{Ut}(L_n)) \tag{2}$$

where $g = 5$, $E(U_t)$ is the information entropy of the mth-order fluctuation at point t in the fluctuation time series $\{U_t | t = m + 1, m + 2, \ldots, T\}$.

2.3. Neutrosophic Fluctuation Time Series

Definition 3. *(Smarandache [12]) Let W be a space of points (objects), with a generic element in W denoted by w. A neutrosophic set A in W is characterized by a truth-membership function $T_A(w)$, am indeterminacy-membership function $I_A(w)$, and a falsity-membership function $F_A(w)$. The functions $T_A(w)$, $I_A(w)$, and $F_A(w)$ are real* *standard or nonstandard subsets of $]0^-, 1^+[$, where $0^- = 0 - \varepsilon$, $1^+ = 1 + \varepsilon$, $\varepsilon > 0$ is an infinitesimal number. There is no restriction on the sum of $T_A(w)$, $I_A(w)$, and $F_A(w)$.*

Definition 4. *Let $\{U_t | t = 2, 3, \ldots, T\}$ be a fluctuation time series of a stock time series as defined in Definition 1. A number U_t in U is characterized by an upward-trend function $T(U_t)$, a fluctuation-inconsistency function $I(U_t)$, and downward-trend function $F(U_t)$, which can be correspondingly mapped to the truth-membership, indeterminacy-membership, and falsity-membership dimension of a neutrosophic set, respectively. The upward-trend function $T(U_t)$ and downward-function $F(U_t)$ are defined according to the number U_t shown as follows:*

$$T(U_t) = \begin{cases} 0, & U_t \leq m_1 \\ f_1(U_t, m_1, m_2), & m_1 \leq U_t \leq m_2 \\ 1, & \text{otherwise} \end{cases} \qquad F(U_t) = \begin{cases} 1, & U_t \leq o_1 \\ f_2(U_t, o_1, o_2), & o_1 \leq U_t \leq o_2 \\ 1, & \text{otherwise} \end{cases} \qquad (3)$$

where m_j and o_j ($j = 1, 2$) are parameters according to the fluctuation time series.

The fluctuation-inconsistency function $I(U_t)$ can be represented by the information entropy $E(U_t)$ as defined in Equation (2).

Thus, a fluctuation time series $\{U_t | t = 1, 2, 3, \ldots, T\}$ can be represented by a neutrosophic fluctuation time series $\{X_t | t = m + 1, m + 2, \ldots, T\}$, where $X_t = (T(U_t), I(U_t), F(U_t))$ is a neutrosophic set.

2.4. Neutrosophic Logical Relationship

Definition 5. *Let $\{X_t | t = 1, 2, 3, \ldots, T\}$ be a fluctuation time series. If there exists a relation $R(t, t + 1)$, such that:*

$$X_{t+1} = X_t \circ R(t, t + 1) \qquad (4)$$

where $\circ$ is a max–min composition operator, X_{t+1} is said to be derived from X_t, denoted by the neutrosophic logical relationship (NLR) $X_t \rightarrow X_{t+1}$. X_t and X_{t+1} are called the left-hand side (LHS) and the right-hand side (RHS) of the NLR, respectively. X_{t+1} can also represented by D_t. Therefore, $X_t \rightarrow X_{t+1}$ can also be represented by $X_t \rightarrow D_t$.

The Jaccard index, also known as the Jaccard similarity coefficient, is used to compare similarities and differences between finite sample sets [29]. The larger the Jaccard similarity value, the higher the similarity.

Definition 6. *Let X_t, X_j be two NSs. The Jaccard similarity between X_t and X_j in vector space can be expressed as follows:*

$$J(X_t, X_j) = \frac{T_{X_t} T_{X_j} + I_{X_t} I_{X_j} + F_{X_t} F_{X_j}}{(T_{X_t})^2 + (I_{X_t})^2 + (F_{X_t})^2 + (T_{X_j})^2 + (I_{X_j})^2 + (F_{X_j})^2 - (T_{X_t} T_{X_j} + I_{X_t} I_{X_j} + F_{X_t} F_{X_j})} \qquad (5)$$

2.5. Aggregation Operator for NLRs

Definition 7. *Let $X = \{X_1, X_2, \ldots, X_t, \ldots, X_n\}$, $D = \{D_1, D_2, \ldots, D_t, \ldots, D_n\}$ be the LHSs and RHSs of a group of NLRs, respectively. The Jaccard similarities between X_t ($t = 1, 2, \ldots, n$) and X_j are $S_{X_{i,j}}$ ($i = 1, 2, \ldots, n$), respectively. The corresponding D_j can be calculated by an aggregation operator [30] as:*

$$T_{D_j} = \frac{\sum_{t=1}^{n} S_{X_{t,j}} \times T_{D_t}}{\sum_{t=1}^{n} S_{X_{t,j}}}, \quad I_{D_j} = \frac{\sum_{t=1}^{n} S_{X_{t,j}} \times I_{D_t}}{\sum_{t=1}^{n} S_{X_{t,j}}}, \qquad (6)$$

According to the definition of NLR, D_j can be represented by X_{j+1}.

3. Research Methodology

In this section, we will introduce a neutrosophic forecasting model for time series based on first-order state and information entropy of high-order fluctuation. The detailed steps are shown as follow steps and in Figure 1.

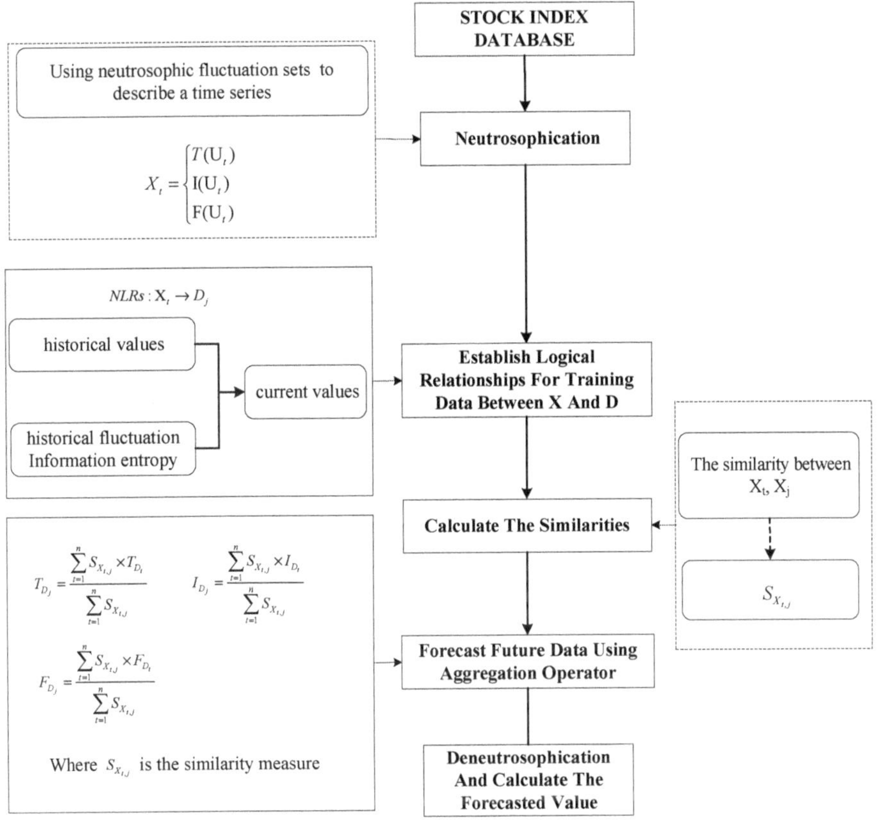

Figure 1. The flowchart of the neutrosophic forecasting model.

3.1. Step 1: Using Neutrosophic Fluctuation Sets to Describe a Time Series

Let $\{V_t | t = 1, 2, 3, \dots, T\}$ be a stock index time series and $\{U_t | t = 2, 3, \dots, T\}$ be its fluctuation time series, where $U_t = V_t - V_{t-1}$ ($t = 2, 3, \dots, T$). Then, we can calculate $len = \frac{\sum_{t=2}^{T} |U_t|}{T-1}$, which is the benchmark for interval division when calculating membership. Let $\{X_t | t = m, m + 1, m + 2, \dots, T\}$ be the mth-order neutrosophic expression of fluctuation time series $\{U_t | t = 2, 3, \dots, T\}$. The conversion rules for the truth-membership T_{X_t} and falsity-membership F_{X_t} of X_t are defined as follows:

$$T_{X_t} = \begin{cases} 0, & U_t \leq -0.5len \\ \frac{U_t}{3/2 \times len} + \frac{1}{3}, & -0.5 \times len \leq U_t \leq len \\ 1, & U_t \geq len \end{cases} \qquad F_{X_t} = \begin{cases} 1, & U_t \leq -len \\ \frac{-U_t}{3/2 \times len} + \frac{1}{3}, & -len \leq U_t \leq 0.5 \times len \\ 0, & U_t \geq 0.5len \end{cases} \quad (7)$$

3.2. Step 2: Using Information Entropy to Represent the Complexity of Historical Fluctuations

$\{U_t | t = 1, 2, 3, \ldots, T\}$ can be fuzzified according to a linguistic set $L = \{l_1, l_2, l_3, l_4, l_5\}$. Specifically, $l_1 = [U_{\min}, -1.5 \times len), l_2 = [-1.5 \times len, -0.5 \times len), l_3 = [-0.5 \times len, 0.5 \times len), l_4 = [0.5 \times len, 1.5 \times len),$ and $l_5 = [1.5 \times len, U_{\max})$. The conversion rule for the indeterminacy-membership I_{X_t} is defined as follows:

$$I_{X_t} = -\sum_{n=1}^{g} p_{X_t}(L_n) \log_2\left(p_{X_t}(L_n)\right) \tag{8}$$

where $g = 5$, $p_{X_t}(L_n)$ indicates the probability of occurrence of the label l_n in the past m days.

3.3. Step 3: Establishing Logical Relationships for Training Data

According to Definition 5, NLRs were established as a training dataset.

3.4. Step 4: Calculating the Similarities between Current Data and Training Data

According to Definition 6, similarities between current data and training data were calculated. Let t be the current data of the point. $S_{X_{t,j}}$ is the similarity of NFTS between the current point t and training data j.

3.5. Step 5: Forecasting Neutrosophic Value Using the Aggregation Operator

According to Definition 7, the future neutrosophic fluctuation number X_{t+1} can be generated based on the training dataset and the similarities with X_t. In order to eliminate very low similarity data, valid NLRs satisfy $S_{X_{t,j}} \geq w'$.

3.6. Step 6: Deneutrosophication for the Neutrosophic Fluctuation Set and Calculating the Forecasted Value

Calculating the expected value of the forecasted neutrosophic set X_{t+1}, the forecasted fluctuation value can be calculated by:

$$V'_{t+1} = \left(T_{X_{t+1}} - F_{X_{t+1}}\right) \times len + V_t \tag{9}$$

4. Empirical Analysis

4.1. Prediction Process

4.1.1. Step 1: After Calculating the Fluctuation Value in Stock Time Series, the Fluctuation Values Will Be Converted to Neutrosophic Time Series

This study needs to select the parameters of the model and estimate its performance. Many studies in the field of fuzzy forecasting have used the data from January–October as the training set and the data from November–December as the test dataset. To facilitate comparison with these existing studies, we also selected data from November–December as the test dataset. Considering the characteristics of time series, traditional cross-validation methods (such as k-fold cross-validation) have poor adaptability. A subset of data after the training subset needs to be retained for validation of model performance. Therefore, we chose a special nested cross-validation, the outer layer of which was used to estimate the model performance and the inner layer of which was used to select the parameters. Specifically, in this paper, we used TAIEX's 1999 data as an example. The closing prices from 1 January–31 October were used as the training dataset. Among them, from January–August was a training subset, and from September–October was for validation. Logical relationships were constructed between each dataset and its closest ninth-order historical values. The closing prices from 1 November–31 December were used as forecast data, and performance was evaluated by comparing forecasting and realistic data.

For example, when the fluctuation value is $U_{12} = 28.7$, the sequence of linguistic variables is l_4, l_5, l_3, l_3, l_2, l_2, l_2, l_5, l_3. $pu_{12}(l_1) = 0$, $pu_{12}(l_2) = 0.3333$, $pu_{12}(l_3) = 0.3333$, $pu_{12}(l_4) = 0.1111$, $pu_{12}(l_5) = 0.2222$. Then, we can calculate the ninth-order fuzzy fluctuation information entropy as follows:

$$E(U_{12}) = E(28.7) = -\sum_{i=1}^{5} pu_{12}(l_i) \log_2(pu_{12}(l_i)) = 1.8911 \tag{10}$$

$$E(U_{13}) = E(-106.5) = -\sum_{i=1}^{5} pu_{13}(l_i) \log_2(pu_{13}(l_i)) = 1.5307 \tag{11}$$

$$E(-33.89) = -\sum_{i=1}^{5} pu_{14}(l_i) \log_2(pu_{14}(l_i)) = 1.3923 \tag{12}$$

$$\cdots$$

The information entropy of fluctuation time proposed in this paper is the intermediate term of NS. In order to maintain the consistency with the other two terms, the above results must be normalized. Normalized information entropy based on the maximum values of information entropy is calculated as follows:

$$E'(U_{12}) = \frac{1.8911}{3.7000} = 0.5111 \tag{13}$$

$$E'(U_{13}) = \frac{1.5307}{3.7000} = 0.4137 \tag{14}$$

$$E'(U_{14}) = \frac{1.3923}{3.7000} = 0.3763 \tag{15}$$

$$\cdots$$

In order to convert the numerical data of stock market fluctuation time series into NS, it is necessary to calculate the elements corresponding to the truth-membership term and the falsity-membership term of NS. According to Equation (7), neutrosophic set membership can be calculated. For example, when the fluctuation value is $U_{12} = 28.7$, then truth-membership $T_{X_{12}}$ of X_{12} is $\frac{28.7}{3/2 \times len} + \frac{1}{3} = 0.5584$ and falsity-membership $F_{X_{12}}$ of X_{12} is $\frac{-28.7}{3/2 \times len} + \frac{1}{3} = 0.1082$. Then, the fluctuation can be represented by the neutrosophic set as follows:

$$X_{12}(28.7) \rightarrow (0.5584, 0.5111, 0.1082) \tag{16}$$

$$X_{13}(-106.5) \rightarrow (0.0000, 0.4137, 1.0000) \tag{17}$$

$$X_{14}(-33.89) \rightarrow (0.0675, 0.3763, 0.5991) \tag{18}$$

$$\cdots$$

$$X_{223}(148.18) \rightarrow (1.0000, 0.3910, 0.0000) \tag{19}$$

$$\cdots$$

4.1.2. Step 2: According to Definition 5, Establishing Mapping Relationships Based on Historical Values, Historical Trends, and Current Values

This step requires establishing neutrosophic logical relationships based on the feature and target sets, where X_{12} is the feature item of X_{13}.

$$X_{12}(x) \rightarrow X_{13}(x) = D_{12}(x) \tag{20}$$

$$X_{13}(x) \rightarrow X_{14}(x) = D_{13}(x) \tag{21}$$

$$\cdots$$

4.1.3. Step 3: Calculating the Jaccard Similarity

Jaccard similarity is usually used to compare similarities and differences of a limited set of samples. The higher the value, the higher the similarity. We used it to compare the current logical group with the logical groups in the training set in order to identify similar groups. $S'_{X_{223,12}}$ indicates the similarity between the 223rd and 12th groups.

$$
\begin{aligned}
S'_{X_{223,12}} &= \frac{0.5584\times1.0000+0.5111\times0.3910+0.1082\times0.0000}{0.5584^2+0.5111^2+0.1082^2+1.0000^2+0.3910^2+0.0000^2-(0.5584\times1.0000+0.5111\times0.3910+0.1082\times0.0000)} \\
&= 0.7742
\end{aligned}
\tag{22}
$$

4.1.4. Step 4: Forecasting the Neutrosophic Fluctuation Point Using the Aggregation Operator

First, we applied the Jaccard similarity measure method to locate similar LHSs of NLRs. We tested different threshold values for the training data. In this example, it was set to 0.89, and we identified 65 groups that met the criteria.

Furthermore, we calculated the forecasting NFTS using the aggregation operator:

$$D_{224} = (0.5005, 0.5067, 0.3401)$$

4.1.5. Step 5: Calculating the Forecasted Value

Then, we calculated the predicted fuzzy fluctuation:

$$Y'(t+1) = 0.5005 - 0.3401 = 0.1604 \tag{23}$$

We also calculated the real number of the fluctuation:

$$U'(t+1) = Y'(t+1) \times len = 0.1604 \times 85 = 13.63 \tag{24}$$

Finally, the predicted value was obtained from the actual value of the previous day and the predicted fluctuation value:

$$V'(t+1) = V(t) + U'(t+1) = 7854.85 + 13.63 = 7868.47 \tag{25}$$

For the sample dataset, the complete prediction result of stock fluctuation trends and the actual values are shown in Table 1 and Figure 2.

Table 1 and Figure 2 show that NFM-IE was able to successfully forecast TAIEX data from 1 November 1999–30 December 1999 based on the logical rules derived from training data.

Table 1. Forecasting results from 1 November 1999–30 December 1999.

Date (MM/DD/YYYY)	Actual	Forecast	(Forecast − Actual)^2	Date (MM/DD/YYYY)	Actual	Forecast	(Forecast − Actual)^2
11/1/1999	7814.89	7868.47	2871.08	12/1/1999	7766.20	7719.40	2190.11
11/2/1999	7721.59	7821.82	10,046.31	12/2/1999	7806.26	7770.62	1270.07
11/3/1999	7580.09	7722.04	20,149.71	12/3/1999	7933.17	7814.75	14,022.27
11/4/1999	7469.23	7577.92	11,813.96	12/4/1999	7964.49	7944.99	380.16
11/5/1999	7488.26	7466.90	456.14	12/6/1999	7894.46	7968.41	5468.57
11/6/1999	7376.56	7489.54	12,764.37	12/7/1999	7827.05	7895.11	4631.50
11/8/1999	7401.49	7374.68	718.73	12/8/1999	7811.02	7826.02	225.13
11/9/1999	7362.69	7399.02	1320.19	12/9/1999	7738.84	7808.59	4864.78
11/10/1999	7401.81	7371.66	909.13	12/10/1999	7733.77	7738.76	24.94
11/11/1999	7532.22	7391.20	19,887.04	12/13/1999	7883.61	7723.92	25,501.56
11/15/1999	7545.03	7543.08	3.82	12/14/1999	7850.14	7897.06	2201.62
11/16/1999	7606.20	7536.55	4851.14	12/15/1999	7859.89	7854.28	31.42
11/17/1999	7645.78	7613.89	1017.07	12/16/1999	7739.76	7860.82	14,654.64
11/18/1999	7718.06	7643.21	5603.26	12/17/1999	7723.22	7738.34	228.50
11/19/1999	7770.81	7729.37	1716.87	12/18/1999	7797.87	7722.01	5754.66
11/20/1999	7900.34	7780.44	14,376.84	12/20/1999	7782.94	7811.00	787.09
11/22/1999	8052.31	7915.24	18,788.73	12/21/1999	7934.26	7782.84	22,929.50
11/23/1999	8046.19	8068.19	483.82	12/22/1999	8002.76	7946.35	3182.30
11/24/1999	7921.85	8046.12	15,443.79	12/23/1999	8083.49	8016.21	4526.63
11/25/1999	7904.53	7919.37	220.29	12/24/1999	8219.45	8096.51	15,113.68
11/26/1999	7595.44	7906.37	96,679.93	12/27/1999	8415.07	8233.25	33,058.13
11/29/1999	7823.90	7592.64	53,479.11	12/28/1999	8448.84	8429.73	365.06
11/30/1999	7720.87	7836.52	13,376.00	Root Mean Square Error (RMSE)			102.02

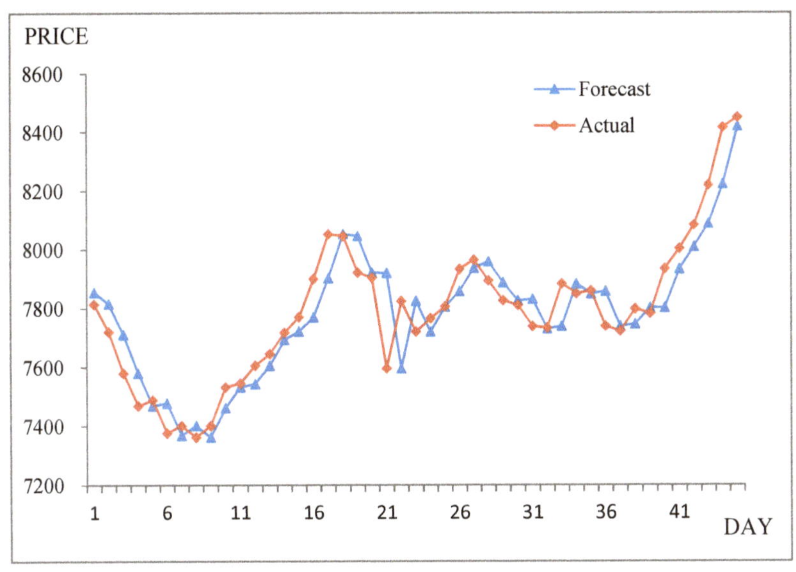

Figure 2. Forecasting results from 1 November 1999–30 December 1999.

4.2. Performance Assessments

During the experimental analysis, some methods were used to measure prediction accuracy in order to quantify model prediction effects. These methods are mainly used in the prediction field, including the mean squared error (MSE), the root mean squared error (RMSE), the mean absolute error (MAE), and the mean absolute percentage error (MAPE).

These expressions are respectively illustrated by Equations (26)–(29):

$$MSE = \frac{\sum_{t=1}^{n}\left(forecast_t - actual_t\right)^2}{n} \tag{26}$$

$$RMSE = \sqrt{\frac{\sum_{t=1}^{n} (forecast_t - actual_t)^2}{n}} \tag{27}$$

$$MAE = \frac{\sum_{t=1}^{n} |(forecast_t - actual_t)|}{n} \tag{28}$$

$$MAPE = \frac{\sum_{t=1}^{n} |(forecast_t - actual_t)|/actual_t}{n} \tag{29}$$

where $forecast_t$ represents the predicted observations and $actual_t$ represents actual observations.

Theil's U index [31] is primarily used to measure the deviation between predicted and actual values. It can get a relative value between zero and one, where zero means that the actual value is equal to the predicted value, that is the prediction model is perfect. At the same time, one indicates that the model prediction effect is not satisfactory. Theil's U index is expressed as follows:

$$U = \frac{\sqrt{\frac{\sum_{t=1}^{n} (forecast_t - actual_t)^2}{n}}}{\sqrt{\frac{\sum_{t=1}^{n} forecast_t^2}{n}} + \sqrt{\frac{\sum_{t=1}^{n} actual_t^2}{n}}} \tag{30}$$

According to Equations (26)–(30), we separately predicted TAIEX data from 1997–2005 and further calculated the error for each year.

From Table 2, the results of different error statistics methods showed that NFM-IE can successfully forecast different time series of TAIEX 1997–2005.

Table 2. Comparing results of different error statistics methods for Taiwan Stock Exchange Capitalization Weighted Stock Index (TAIEX) data collected from 1997–2005.

Year	1997	1998	1999	2000	2001	2002	2003	2004	2005
RMSE	141.42	114.69	102.02	129.94	114.22	66.84	53.88	55.24	53.1
MSE	19,999.62	13,153.80	10,408.08	16,884.40	13,046.21	4467.59	2903.05	3051.46	2819.61
MAE	113.42	96.31	79.38	96.65	92.48	51.65	41.11	38.65	41.27
MAPE	0.0143	0.0138	0.0102	0.0182	0.019	0.0111	0.007	0.0065	0.0067
Theil's U	0.0089	0.0082	0.0065	0.0122	0.0119	0.0072	0.0046	0.0047	0.0043

5. Results Analysis

5.1. Taiwan Stock Exchange Capitalization Weighted Stock Index

In general, TAIEX is a widely-used dataset in stock market forecasting. In order to facilitate comparison with other forecasting models, this paper also uses it as the main dataset to verify the model. Using non-stationary data can lead to spurious regressions, so we first performed a stationarity test based on the unit root test by software Eviews (Eviews10.0 Enterprise Edition, Microsoft, Redmond, WA, USA). It can be concluded that the first-order difference of TAIEX 1997–2005 was stationary data, which indicates that the fluctuation data used in this study were stationary. Further, other datasets in this study were also stationary data.

The model in this paper was based on high order, and thus, different orders may affect the accuracy of the prediction. Hence, the experimental analysis showed that when the order of fuzzy fluctuation information entropy was 9–11, the stability of the model was more ideal. Table 3 shows the experimental errors for different years under different orders.

Table 3. Comparing average RMSEs based on different order fuzzy fluctuation time series from 1997–2005.

Order	8	9	10	11	12	13	14	15	16
1997	141.41	141.42	141.46	141.9	141.53	141.72	141.68	141.8	141.69
1998	114.67	114.69	114.61	114.76	114.63	114.39	114.46	114.29	114.23
1999	101.86	102.02	101.7	101.66	101.55	101.59	101.7	101.26	101.54
2000	129.07	129.94	129.62	129.34	129.87	129.49	128.64	128.6	128.43
2001	113.97	114.22	114.53	114.86	115.37	115.11	115.39	116.06	116.02
2002	67.29	66.84	66.95	66.85	66.76	67.21	66.98	67.02	67.48
2003	53.84	53.88	53.99	53.68	53.74	53.8	53.55	53.48	53.45
2004	54.7	55.24	55.17	55.08	55.07	55.36	55.47	55.1	55.25
2005	53.09	53.1	53.22	53.09	53.14	53.11	53.13	53.04	52.97
average	92.21	92.37	92.36	92.36	92.41	92.42	92.33	92.29	92.34
total	829.9	831.35	831.25	831.22	831.66	831.78	831	830.65	831.06

Not surprisingly, accurate fluctuation trend predictions are very important and needed. Therefore, the performance of different methods must be compared and evaluated, thus verifying the superiority or deficiency of the model. In order to verify the effects of model prediction, this section focuses on comparing this model's experimental results with those from other models. Comparing the errors across model showed that the current model had certain advantages in prediction accuracy. Table 4 shows the prediction errors for the different methods between 1997 and 2005. The NFM-IE hybrid model achieved better prediction accuracy compared to the traditional regression model, autoregressive model, neural network model, and fuzzy model (Table 4). In addition, NFM-IE exhibited better predictive power in some years compared to other hybrid models based on the fuzzy theory.

Table 4. Performance comparison of prediction RMSEs with other models. NFM-IE, neutrosophic forecasting model based on information entropy.

TYPE	Methods	RMSE										
		1997	1998	1999	2000	2001	2002	2003	2004	2005	Average	Total
Regression Model	Univariate conventional regression model (U_R) [32,33]	N/A	N/A	164	420	1070	116	329	146	N/A	374.20	2245
	Bivariate conventional regression model (B_R) [32,33]	N/A	N/A	103	154	120	77	54	85	N/A	98.80	593
Auto-regressive	Autoregressive model for order one (AR_1) [34]	146.22	144.53	116.84	155.12	**112.39**	97.09	91.67	79.94	N/A	117.98	653.05
	Autoregressive model for order two (AR_2) [34]	174.09	135.21	128.15	142.3	129.84	89.8	66.58	60.33	N/A	115.79	617
Neural network	Univariate neural network model (U_NN) [32,33]	N/A	N/A	107	309	259	78	57	60	N/A	145.00	870
	Bivariate neural network mode (B_NN) [32,33]	N/A	N/A	112	274	131	69	**52**	61	N/A	116.40	699
Fuzzy	fuzzy forecasting and fuzzy rule(F-R) [35]	N/A	N/A	123.64	131.1	115.08	73.06	66.36	60.48	N/A	94.95	569.72
	Fuzzy time-series model based on rough set rule (F-RS) [10]	N/A	120.8	110.7	150.6	113.2	66	53.1	58.6	53.5	**90.81**	605.7
	Fuzzy variation groups (F-VG) [36]	140.86	144.13	119.32	129.87	123.12	71.01	65.14	61.94	N/A	106.92	570.4
Fuzzy+	Multi-variable fuzzy and particle swarm optimization (M_F-PSO) [37]	**138.41**	113.88	102.34	131.25	113.62	**65.77**	52.23	56.16	N/A	96.71	521.37
	Univariate fuzzy and particle swarm optimization (U_F-PSO) [38]	143.6	115.34	99.12	**125.7**	115.91	70.43	54.26	57.24	54.68	92.92	577.34
	Autoregressive moving average and fuzzy logical Relationships (ARMA-FR) [39]	141.89	119.85	99.03	128.62	125.64	66.29	53.2	56.11	55.83	94.05	584.72
	Back propagation neural network and high-order fuzzy-fluctuation trends (BPNN-HFT) [40]	142.99	**112.51**	**96.77**	126.85	120.12	66.39	54.87	58.1	54.7	92.59	577.8
	NFM-IE	141.42	114.69	102.02	129.94	114.22	66.84	53.88	**55.24**	**53.1**	92.37	**575.24**

5.2. Forecasting Shanghai Stock Exchange Composite Index

SHSECI is one of the most typical stock indices in China, with certain representativeness. We selected it as an experimental dataset to verify the model's applicability.

Recently, scholars have proposed more comprehensive models based on traditional prediction methods. For example, Guan et al. [39] proposed a two-actor autoregressive moving average model based on the fuzzy logical relationships (ARMA-FR). Guan et al. [40] proposed a model based on back propagation neural network and high-order fuzzy-fluctuation trends (BPNN-HFT). This section compares several typical prediction methods. The results indicated that the model can also effectively predict the stock index. Table 5 and Figure 3 show a comparison of the different prediction methods.

Table 5. RMSEs of forecast errors for the Shanghai Stock Exchange Composite Index SHSECI from 2007–2015.

Year	2007	2008	2009	2010	2011	2012	2013	2014	2015	Average
ARMA-FR (2017) [39]	129.22	79.77	59.96	49.48	29.7	**23.14**	22.13	**44.11**	**58.89**	55.15
BPNN-HFT (2018) [40]	123.89	57.44	**48.92**	47.34	28.37	25.84	21.43	50.59	59.69	51.50
NFM-IE	**112.10**	**51.98**	49.37	**45.58**	**28.22**	24.92	**20.21**	50.44	59.77	**49.17**

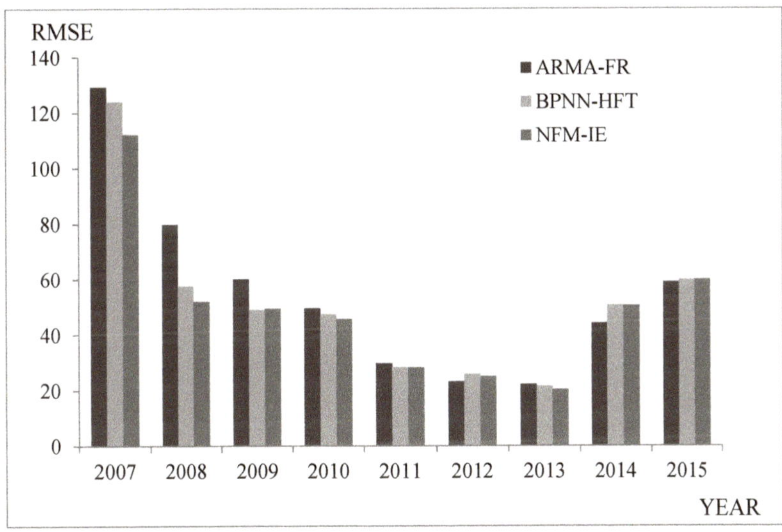

Figure 3. RMSEs of forecast errors for SHSECI from 2007–2015.

The comparison shows that NFM-IE outperformed other methods in predicting SHSECI from 2007–2015.

Comparing the average value of the SHSECI prediction error showed that NFM-IE had better prediction accuracy and stability compared to the neural network-based BPNN-HFT model and the statistical-based ARMA-FR model.

5.3. Forecasting Hong Kong-Hang Seng Index

Finally, the Hong Kong-Hang Seng Index (HSI) was selected as the experimental dataset. Comparing several authoritative prediction methods, we can verify the universality of the model in other stock markets. Table 6 and Figure 4 show a comparison of the different prediction methods from 1998–2012.

Table 6. RMSEs of forecast errors for the Hong Kong-Hang Seng Index (HSI) from 1998–2012.

Method	1998	1999	2000	2001	2002	2003	2004	2005	2006	2007	2008	2009	2010	2011	2012	Average
Yu (2005) [41]	291.4	469.6	297.05	316.85	123.7	186.16	264.34	112.4	252.44	912.67	684.9	442.64	382.06	419.67	239.11	359.66
Wan (2017) [42]	326.62	637.1	356.7	299.43	155.09	226.38	239.63	147.2	466.24	1847.8	2179	437.24	445.41	688.04	477.34	595.26
Ren (2016) [43]	296.67	761.9	356.81	254.07	155.4	199.58	540.19	1127	407.89	1028.7	593.8	435.18	718.33	578.7	442.44	526.46
Cheng (2018) [10]	201.99	231.91	251.7	156.58	106.26	118.74	105.38	103.96	189.2	682.08	460.12	326.65	260.67	346.33	190.13	248.78
NFM-IE	**195.86**	**223.91**	**246.11**	163.49	**105.65**	122.04	**102.23**	105.37	**173.55**	694.89	469.11	**319.7**	274.73	347.2	**181.98**	248.39

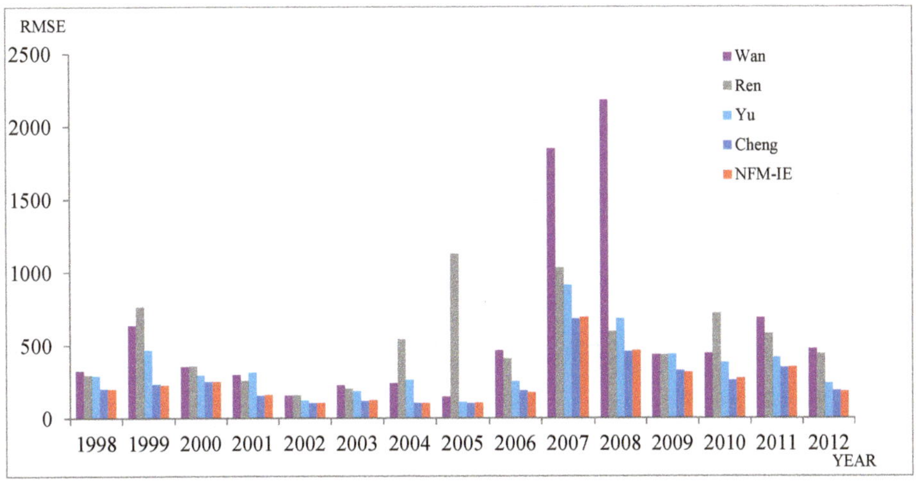

Figure 4. RMSEs of forecast errors for HSI from 1998–2012.

To further evaluate the validity of the proposed model, we used Friedman's test to perform a significance test based on the study of Demšar [44]. For reference, Friedman's test is a parametric statistical test that was proposed by Milton Friedman [45,46]. To further illustrate the significance of the model's predictions compared to other prediction methods, this section will use Friedman's test and the post-hoc test for significance analysis. In the Friedman test phase, SPSS was used for statistical testing, and the post-hoc test phase was based on manual calculations.

In the first stage, Friedman's test requires comparison of the average ranking of different algorithms $R_j = \frac{1}{N}\sum_i r_i^j$, where, r_i^j is the rank of the j-th of k algorithms on the i-th of N datasets. The ranking of each method was based on the analysis of HSI forecast results as shown in Table 7.

Table 7. The rank of the forecasting results of the HSI.

Method	Rank
Yu (2005) [41]	3.40
Wan (2017) [42]	4.40
Ren (2016) [43]	4.20
Cheng (2018) [10]	1.53
NFM-IE	1.47

Through software analysis, we concluded that the method had the highest comprehensive ranking. In addition, according to the Chi-square distribution, there were significant differences between these methods.

$$CD = q_\alpha \sqrt{\frac{k(k+1)}{6N}} \tag{31}$$

In the second stage, in order to further compare the different methods, we used the Nemenyi test [47]. According to Equation (31), $\alpha = 0.05$ and $CD = 1.575$. Upon further comparison, we found that the method proposed in this study had significant advantages over Yu (2005) [41], Wan (2017) [42], Ren (2016) [43], etc. Although it was not significant compared with Cheng's method (2018) [10], the NFM-IE had certain advantages from the perspective of error mean and average level.

5.4. Discussion

The research was mainly focused on two issues. The first was whether the uncertainty of stock market volatility can be used as a key feature of forecasting in a complex environment. The other was whether the prediction method considering uncertainty and trend was effective. We first used the inconsistency of historical fluctuations as a stock forecasting feature and further characterized and quantified it. Then, we applied the neutrosophic set to be the representation of the information and established a neutrosophic logic relationship based on wave inconsistency. Through experimental analysis, the proposed model achieved robustness and stability with relatively few parameters. In addition, it was also proven that predictions that consider inconsistency are meaningful and effective. The advantages were embodied in the following aspects: First, NFM-IE did not need to establish complex assumptions compared to traditional regression-based prediction models. Second, the NFM-IE prediction process was more interpretable than the neural network. Finally, compared with the fuzzy prediction method, NFM-IE effectively utilized data inconsistency as key information. All in all, the model showed satisfactory performance. However, it also showed certain limitations: First, the model used single stock market data as the system input and failed to consider multiple factors fully. Secondly, using information entropy as a key tool for uncertainty measurement requires further optimization in characterizing data.

6. Conclusions

In this paper, we presented the concept of NFTS and proposed a prediction model based on the neutrosophic set and information entropy of high-order fuzzy fluctuation time series. This model had significant performance advantages over existing fuzzy time series models, machine learning prediction models, and traditional economic prediction models. In this paper, we applied three typical test datasets to prove that the model had certain universality and stability. In addition, this paper had a certain degree of scientific contribution in the following aspects: First, the concept of NFTS was proposed. Second, this paper proposed information entropy based on high-order fluctuation time series. Finally, this paper established NLRs based on NFTS and information entropy. This paper discussed the first-order neutrosophic time series to characterize the historical state of uncertainty and high-order information fluctuation entropy to measure the complexity of historical fluctuations. Other types of time series will be tested in the future. Meanwhile, future research should aim to establish detailed high-order neutrosophic time series models indicating the uncertainty of historical trends. In this study, we have considered the Jaccard similarity measure for comparing X_t and X_j. Further work could considered the Jensen–Shannon distance [20], which accomplishes the triangular inequality. Furthermore, in order to verify the robustness of the forecast in longer forecast scenarios, we will extend the model to 2, 3, or 4 periods ahead.

Author Contributions: Data curation, Z.D.; Supervision, H.G.; Validation, H.G., S.G. and A.Z.; Writing—original draft, Z.D.; Writing—review & editing, A.Z.

Funding: This work was supported by the National Natural Science Foundation of China under Grant 71704066.

Acknowledgments: The authors are indebted to the anonymous reviewers for their very insightful comments and constructive suggestions, which helped improve the quality of this paper.

Conflicts of Interest: The authors declare no conflict of interest. The funders had no role in the design of the study; in the collection, analyses, or interpretation of data; in the writing of the manuscript; nor in the decision to publish the results.

References

1. Han, M.; Xu, M. Laplacian Echo State Network for Multivariate Time Series Prediction. *IEEE Trans. Neural Netw. Learn. Syst.* **2018**, *29*, 238–244. [CrossRef] [PubMed]
2. Mishra, N.; Soni, H.K.; Sharma, S.; Upadhyay, A.K. Development and Analysis of Artificial Neural Network Models for Rainfall Prediction by Using Time-Series Data. *Int. J. Intell. Syst. Appl.* **2018**, *10*, 16–23. [CrossRef]

3. Safari, N.; Chung, C.Y.; Price, G.C.D. A Novel Multi-Step Short-Term Wind Power Prediction Framework Based on Chaotic Time Series Analysis and Singular Spectrum Analysis. *IEEE Trans. Power Syst.* **2018**, *33*, 590–601. [CrossRef]

4. Moskowitz, D. Implementing the template method pattern in genetic programming for improved time series prediction. *Genet. Program. Evol. Mach.* **2018**, *19*, 271–299. [CrossRef]

5. Soto, J.; Melin, P.; Castillo, O. *Ensembles of Type 2 Fuzzy Neural Models and Their Optimization with Bio-Inspired Algorithms for Time Series Prediction*; Springer Briefs in Applied Sciences & Technology; Springer: Basel, Switzerland, 2018.

6. Soares, E.; Costa, P., Jr.; Costa, B.; Leite, D. Ensemble of evolving data clouds and fuzzy models for weather time series prediction. *Appl. Soft Comput.* **2018**, *64*, 445–453. [CrossRef]

7. Song, Q.; Chissom, B.S. Forecasting enrollments with fuzzy time series—Part I. *Fuzzy Sets Syst.* **1993**, *54*, 1–9. [CrossRef]

8. Song, Q.; Chissom, B.S. Fuzzy time series and its models. *Fuzzy Sets Syst.* **1993**, *54*, 269–277. [CrossRef]

9. Song, Q.; Chissom, B.S. Forecasting enrollments with fuzzy time series—Part II. *Fuzzy Sets Syst.* **1991**, *62*, 1–8. [CrossRef]

10. Cheng, C.H.; Yang, J.H. Fuzzy Time-Series Model Based on Rough Set Rule Induction For Forecasting Stock Price. *Neurocomputing* **2018**, *302*, 33–45. [CrossRef]

11. Kumar, S.; Gangwar, S. Intuitionistic fuzzy time series: An approach for handling non-determinism in time series forecasting. *IEEE Trans. Fuzzy Syst.* **2016**, *24*, 1270–1281. [CrossRef]

12. Smarandache, F. A unifying field in logics: Neutrosophic logic. *Mult.-Valued Log.* **1999**, *8*, 489–503.

13. Wang, H.; Smarandache, F.; Zhang, Y.Q.; Sunderraman, R. Single valued neutrosophic sets. *Multispace Multistruct* **2010**, *4*, 410–413.

14. Wang, H.; Smarandache, F.; Zhang, Y.Q.; Sunderraman, R. *Interval Neutrosophic Sets and Logic: Theory and Applications in Computing*; Hexis: Phoenix, AZ, USA, 2005.

15. Zadeh, L.A. Probability measure of fuzzy events. *J. Math. Anal. Appl.* **1968**, *23*, 421–427. [CrossRef]

16. DeLuca, A.S.; Termini, S. A definition of nonprobabilistic entropy in the setting of fuzzy set theory. *Inf. Control* **1972**, *20*, 301–312. [CrossRef]

17. Vu, T.M.; Mishra, A.K.; Konapala, G. Information Entropy Suggests Stronger Nonlinear Associations between Hydro-Meteorological Variables and ENSO. *Entropy* **2018**, *20*, 38. [CrossRef]

18. Zeng, X.; Wu, J.; Wang, D.; Zhu, X.; Long, Y. Assessing Bayesian model averaging uncertainty of groundwater modeling based on information entropy method. *J. Hydrol.* **2016**, *538*, 689–704. [CrossRef]

19. Arellano-Valle, R.B.; Contreras-Reyes, J.E.; Stehlík, M. Generalized skew-normal negentropy and its application to fish condition factor time series. *Entropy* **2017**, *19*, 528. [CrossRef]

20. Liu, Z.; Shang, P. Generalized information entropy analysis of financial time series. *Physica A* **2018**, *505*, 1170–1185. [CrossRef]

21. Ye, G.; Pan, C.; Huang, X.; Zhao, Z.; He, J. A Chaotic Image Encryption Algorithm Based on Information Entropy. *Int. J. Bifurcation Chaos* **2018**, *28*, 9. [CrossRef]

22. Tang, Y.; Liu, Z.; Pan, M.; Zhang, Q.; Wan, C.; Guan, F.; Wu, F.; Chen, D. Detection of Magnetic Anomaly Signal Based on Information Entropy of Differential Signal. *IEEE Geosci. Remote Sens. Lett.* **2018**, *15*, 512–516. [CrossRef]

23. He, H.; An, L.; Liu, W.; Zhang, J. Prediction Model of Collapse Risk Based on Information Entropy and Distance Discriminant Analysis Method. *Math. Prob. Eng.* **2017**, *2017*. [CrossRef]

24. Bariviera, A.F.; Martín, M.T.; Plastino, A.; Vampa, V. LIBOR troubles: Anomalous movements detection based on maximum entropy. *Physica A* **2016**, *449*, 401–407. [CrossRef]

25. Liang, X.; Si, D.; Xu, J. Quantitative Evaluation of the Sustainable Development Capacity of Hydropower in China Based on Information Entropy. *Sustainability* **2018**, *10*, 529. [CrossRef]

26. Zhang, Z.; Li, Y.; Jin, S.; Zhang, Z.; Wang, H.; Qi, L.; Zhou, R. Modulation Signal Recognition Based on Information Entropy and Ensemble Learning. *Entropy* **2018**, *20*, 198. [CrossRef]

27. Shannon, C.E. A mathematical theory of communication. *Bell Labs Tech. J.* **1948**, *27*, 379–423. [CrossRef]

28. Zadeh, L.A. The Concept of a Linguistic Variable and its Application to Approximate Reasoning. *Inf. Sci.* **1974**, *8*, 199–249. [CrossRef]

29. Fu, J.; Ye, J. Simplified neutrosophic exponential similarity measures for the initial evaluation/diagnosis of benign prostatic hyperplasia symptoms. *Symmetry* **2017**, *9*, 154. [CrossRef]

30. Ali, M.; Son, L.H.; Thanh, N.D.; Minh, N.V. A neutrosophic recommender system for medical diagnosis based on algebraic neutrosophic measures. *Appl. Soft Comput.* **2017**, *71*, 1054–1071. [CrossRef]

31. Theil, H. *Applied Economic Forecasting*; North-Holland: Amsterdam, The Netherlands, 1966.

32. Yu, T.H.K.; Huarng, K.H. A bivariate fuzzy time series model to forecast the TAIEX. *Expert Syst. Appl.* **2008**, *34*, 2945–2952. [CrossRef]

33. Yu, T.H.K.; Huarng, K.H. Corrigendum to "A bivariate fuzzy time series model to forecast the TAIEX". *Expert Syst. Appl.* **2010**, *37*, 5529. [CrossRef]

34. Sullivan, J.; Woodall, W.H. A comparison of fuzzy forecasting and Markov modeling. *Fuzzy Sets Syst.* **1994**, *64*, 279–293. [CrossRef]

35. Chen, S.M.; Chang, Y.C. Multi-variable fuzzy forecasting based on fuzzy clustering and fuzzy rule interpolation techniques. *Inf. Sci.* **2010**, *180*, 4772–4783. [CrossRef]

36. Chen, S.M.; Chen, C.D. TAIEX Forecasting Based on Fuzzy Time Series and Fuzzy Variation Groups. *IEEE Trans. Fuzzy Syst.* **2011**, *19*, 1–12. [CrossRef]

37. Chen, S.M.; Manalu, G.M.; Pan, J.S.; Liu, H.C. Fuzzy Forecasting Based on Two-Factors Second-Order Fuzzy-Trend Logical Relationship Groups and Particle Swarm Optimization Techniques. *IEEE Trans. Cybern.* **2013**, *43*, 1102–1117. [CrossRef]

38. Jia, J.; Zhao, A.W.; Guan, S. Forecasting Based on High-Order Fuzzy-Fluctuation Trends and Particle Swarm Optimization Machine Learning. *Symmetry* **2017**, *9*, 124. [CrossRef]

39. Guan, S.; Zhao, A. A Two-Factor Autoregressive Moving Average Model Based on Fuzzy Fluctuation Logical Relationships. *Symmetry* **2017**, *9*, 207. [CrossRef]

40. Guan, H.; Dai, Z.; Zhao, A. A novel stock forecasting model based on High-order-fuzzy-fluctuation Trends and Back Propagation Neural Network. *PLoS ONE* **2018**, *13*. [CrossRef]

41. Yu, H.K. A refined fuzzy time-series model for forecasting. *Physica A* **2005**, *346*, 657–681. [CrossRef]

42. Wan, Y.; Si, Y.W. Adaptive neuro fuzzy inference system for chart pattern matching in financial time series. *Appl. Soft Comput.* **2017**, *57*, 1–18. [CrossRef]

43. Ren, Y.; Suganthan, P.N.; Srikanth, N. A Novel Empirical Mode Decomposition With Support Vector Regression for Wind Speed Forecasting. *IEEE Trans. Neural Netw. Learn. Syst.* **2016**, *27*, 1793–1798. [CrossRef] [PubMed]

44. Demšar, J. Statistical comparisons of classifiers over multiple datasets. *J. Mach. Learn. Res.* **2006**. Available online: http://www.jmlr.org/papers/v7/demsar06a.html (accessed on 26 April 2019).

45. Friedman, M. The use of ranks to avoid the assumption of normality implicit in the analysis of variance. *J. Am. Stat. Assoc.* **1937**, *32*, 675–701. [CrossRef]

46. Friedman, M. A comparison of alternative tests of significance for theproblem of m rankings. *Ann. Math. Stat.* **1940**, *11*, 86–92. [CrossRef]

47. Nemenyi, P. Distribution-free Multiple Comparisons. Ph.D. Thesis, Princeton University, Princeton, NJ, USA, 1963.

Article

Evolved-Cooperative Correntropy-Based Extreme Learning Machine for Robust Prediction

Wenjuan Mei [1], Zhen Liu [1], Yuanzhang Su [2,*], Li Du [1] and Jianguo Huang [1]

[1] Department of Instrument Science and Technology, University of Electronic Science and Technology of China, Chengdu 611731, China; meiwenjuan@std.uestc.edu.cn (W.M.); scdliu@uestc.edu.cn (Z.L.); summer_christ@163.com (L.D.); xlhjg@uestc.edu.cn (J.H.)

[2] Department of Applied Linguistics, University of Electronic Science and Technology of China, Chengdu 611731, China

* Correspondence: syz@uestc.edu.cn; Tel.: +86-028-6183-0316

Received: 6 August 2019; Accepted: 12 September 2019; Published: 19 September 2019

Abstract: In recent years, the correntropy instead of the mean squared error has been widely taken as a powerful tool for enhancing the robustness against noise and outliers by forming the local similarity measurements. However, most correntropy-based models either have too simple descriptions of the correntropy or require too many parameters to adjust in advance, which is likely to cause poor performance since the correntropy fails to reflect the probability distributions of the signals. Therefore, in this paper, a novel correntropy-based extreme learning machine (ELM) called ECC-ELM has been proposed to provide a more robust training strategy based on the newly developed multi-kernel correntropy with the parameters that are generated using cooperative evolution. To achieve an accurate description of the correntropy, the method adopts a cooperative evolution which optimizes the bandwidths by switching delayed particle swarm optimization (SDPSO) and generates the corresponding influence coefficients that minimizes the minimum integrated error (MIE) to adaptively provide the best solution. The simulated experiments and real-world applications show that cooperative evolution can achieve the optimal solution which provides an accurate description on the probability distribution of the current error in the model. Therefore, the multi-kernel correntropy that is built with the optimal solution results in more robustness against the noise and outliers when training the model, which increases the accuracy of the predictions compared with other methods.

Keywords: correntropy; information theory extreme learning machine; evolved cooperation

---

1. Introduction

With the rapid development of powerful computing environments and rich data sources, artificial intelligence (AI) technology such as neural networks [1–3], adaptive filtering [4–6] and evolutionary algorithms [7–9] has become increasingly more applicable for forecasting problems in various scenarios, such as medicine [10–12], economy [13–15] and electronic engineering [16–18]. The methods have acquired high reputations due to their great approximation abilities.

Although AI methods perform well when solving real world problems, most corresponding models adapt the mean squared error (MSE) as the criterion for training hidden nodes or building the cost functions, assuming that the data satisfy a Gaussian distribution. Moreover, the MSE is a global similarity measure where all the samples in the joint space have the same contribution [19]. Therefore, the MSE is likely to be badly affected by the noise and outliers that are hiding in the samples and this happens commonly in applications, such as speech signals, images, real-time traffic signals and electronic signals from ill-conditioned devices [20–22]. Therefore, MSE-based models are likely to result in poor performance in real world applications.

To conquer the weaknesses of the least mean squares (LMS), over the past decades, a number of studies have proposed methods to improve the robustness of the model against the noise and outliers that are contained in the data [23–27]. Among the existing technologies, M-estimators have been the focus of many academic studies. By detecting the potential outliers during training procedures, the M-estimator can eliminate the negative influences from the output weights that adversely affect the predictions [28]. Using these advantages, Zhou et al. [29] proposed a novel data-driven standard least-squares support vector regression (LSSVR) applying the M-estimator, which reduces the interference of outliers and enhances the robustness. However, there are difficulties accessing clean learning data without noises so that the application on the M-estimator-based forecasting models based is limited.

Recently, information theoretic learning (ITL) has drawn considerable attention due to its good performance avoiding the effect of the noise and outliers [30–35] and it has become an effective alternative to the MSE criterion. In [36], the authors presented a novel training criterion based on the minimum error entropy (MEE) to replace the MSE. By taking advantages of the higher order description on entropy, MEE has become superior for non-Gaussian signal processing compared with traditional training criteria. Inspired by the entropy and Parzen kernel estimator, Liu et al. [37] proposed an extended definition of the correlation function for random processes using a generalized correlation function, known as correntropy. Although different from global measurements, such as the mean squared error (MSE), the correntropy is regarded as a local similarity measurement where its value is primarily determined by the kernel function along x = y line [38], leading to high robustness against noise and outlies. Moreover, the correntropy has many great properties such as symmetry, nonnegativity and boundness. Most of all, it is easy to form convex cost functions based on the correntropy, which is very convenient for training the models [39–42]. Therefore, the correntropy has been widely used in forming robust models [43–45].

To enhance the forecasting ability of the model, in [46], the correntropy was introduced into the affine projection (AP) algorithm to overcome the degradation of the identification performance with impulsive noise environments. From the simulation results, it is easy to verify that the proposed algorithm has achieved better performance than other methods. Another approach to improve the robustness via the correntropy is enhancing the feature selection efficiencies [47–49]. In [50], the kernel modal regression and gradient-based variable identification were integrated together using the maximum correntropy criterion, which guarantees the robustness of the algorithm. Additionally, in [51], a novel principal component analysis (PCA), based on the correntropy and known as the correntropy-optimized temporal PCA (CTPCA), was adapted to enhance the robustness for rejecting the outlier. The outlier improves the models training in simulation experiments. In addition to providing the extractions of the features in neural networks and filtering methods, the correntropy turns out to be a powerful tool for developing robust training methods that generate and adjust the weights in the model. In [52], Wang et al. introduced a feedback mechanism using the kernel recursive maximum correntropy to provide a novel kernel adaptive filters known as the kernel recursive maximum correntropy with multiple feedback (KRMC-MF). The experiments show that the generated filters have high robustness against outliers. In [53], Ahmad et al. proposed the correntropy based conjugate gradient backpropagation (CCG-BP), which can achieve high robustness in environments with both impulsive noise and heavy-tailed noise distributions. Unfortunately, most of the neural networks have to adjust the weights of each node during each training iterations which wastes time during the training process.

Recently, forecasting models with parameters that are free from adjustments have gained increasingly more attention due to their fast training speeds for the models [54–56]. Combined with the correntropy, these algorithms have shown great potential in real-world applications. For example, Guo et al. [57] developed a novel training method for echo state networks (ESNs) based on a correntropy induced loss function (CLF), which provides robust predictions for time-series signals. Similar to ESNs, extreme learning machines (ELMs) have received great attention on fast learning due to the random assignments of the hidden layer and being equipped with simpler structures, such as single layer feedback networks (SLFNs) [58–60]. It has been proven that the hidden nodes can be assigned with any continuous probability distribution, while the model satisfies the universal approximation and classification capacity [61]. In particular, the extreme learning machine has been applied and received a high reputation for predicting production processes [62,63], system anomalies [64], etc. [65]. In [66], the authors first developed the correntropy-based ELM that uses the regularized correntropy criterion in place of the MSE with half quadratic (HQ) optimization which is called the regularized correntropy criterion for an extreme learning machine (RCC-ELM). Later, Chen et al. [67] extended the dimensions of the correntropy by combining two kinds of correntropy together to enhance the flexibility of the model to generate more robust ELM called ELM by maximum mixture correntropy criterion (MMCC-ELM). The experimental results show that the learning method performs better than the conventional maximum correntropy method. Although the RCC-ELM and MMCC-ELM possess high robustness compared with other ELM methods, the corresponding correntropy is constrained by no more than two kernels. The kernel bandwidth required for the assignments by users in advance is likely to degrade the model due to the improper description on the probability distribution of the signal with the correntropy.

To conquer the weakness of the existing correntropy-based ELMs, this paper focuses on providing a more robust predicting model with adaptive generation based on multi-kernel correntropy which can bring an accurate description of the current errors of ELM. This study developed a more flexible and robust forecasting ELM based on a newly developed adaptive multi-dimension correntropy using evolving cooperation. In the proposed method, the output weights of the ELM are trained based on the maximum multi-dimension correntropy with no constraints on the dimensions of the kernels. To achieve the most appropriate assignment of the parameters of each kernel in the correntropy, a novel evolving cooperation method is developed to concurrently optimize the bandwidths and the corresponding influence coefficients to achieve the best estimations of the residual errors of the model. Furthermore, the training approach has been developed based on the properties of the multi-dimension correntropy. The main contribution of the paper can be summarized as follows.

- The proposed method develops a novel correntropy criterion with multiple kernels to improve the flexibility for depicting the probability distribution of the current error of the predicting model. Then, a convex cost function has been developed based on the multiple kernel correntropy, which can provide a more robust training strategy for ELMs, resulting in high performance on the predictions against noise and outliers.
- To accurately describe the probability distribution of the current error, the proposed method develops a cooperating evolution strategy to adaptively generate proper bandwidths and coefficients to suit the error distribution which enhances the accuracy on the approximation for the correntropy, leading to more robust training.

The experiments compare the performance of the proposed method and several state-of-art methods using both simulated data and real-world data, which show that the proposed method obtains more the robust predictions than other methods. Finally, the proposed method is incorporated into the forecasting model for the current transfer ratio (CTR) signals for the optical couplers, and it achieves high accuracies and robustness.

The rest of the paper is as follows. The next section introduces the framework of the proposed method and multi-dimension correntropy. Section 3 describes the evolved cooperation for the kernels with multi-dimension correntropy and Section 4 provides the training procedures of the forecasting model. Then, Section 5 estimates the performance of the proposed method using both simulation data and real-world applications. Finally, the conclusion is drawn in Section 6.

2. The Framework of the Proposed Method

The structure of the prediction model that is built using the proposed method is similar to those of other ELM-based methods. Figure 1 shows the basic structure of the method. Generally, the network includes one input layer, one hidden layer and one output layer. The hidden output is calculated using the given input vectors and the weights and the biases of the hidden nodes which are randomly assigned [54]:

$$h = f(wx + b) \tag{1}$$

where $f(.)$ is the activation function and (w,b) are the weights and bias of the hidden nodes.

With the hidden layer, the network can simulate any kind of function by generating the output weights with the least mean squares (LMS) The cost function is calculated as follows [58]:

$$J_{LS} = \|\mathbf{Y} - \mathbf{T}\| \tag{2}$$

where $\mathbf{T}$ is the expected output and $\mathbf{Y}$ is the predicted output of the model. $\mathbf{Y}$ calculated with the hidden outputs h and the output weights β as follows:

$$\mathbf{Y} = \beta h \tag{3}$$

Therefore, the output layer is calculated as follows:

$$\beta = (H^T H)^{-1} H^T T \tag{4}$$

Further, to constrain the output weights, the output layer is calculated as follows:

$$\beta = (H^T H + \lambda I)^{-1} H^T T \tag{5}$$

where λ is the constraining coefficient.

Although the output weights that are calculated by Equation (4) or Equation (5) can provide good predictions using the training data, the model has suffered with the outliers and noises in the data which negatively affect the predictions. To overcome the problem, the correntropy, as a high order similarity measurement, has been used in some recently developed methods.

In [62], the cost function built using the correntropy as follows:

$$J_{RCC} = max_\beta \left[\sum_{p=1}^{N} G(t_p - h\beta) - \lambda \|\beta\| \right] \tag{6}$$

where $G(t_p - h\beta)$ is the Gussian kernel calculated as follows:

$$G(t_p - h\beta) = \exp\left(-\frac{(t_p - h\beta)}{2\sigma^2}\right) \tag{7}$$

where σ is the bandwidth of the kernel.

Therefore, the output layer is calculated as follows:

$$\beta = (H^T \Lambda H - \lambda I)^{-1} H^T \Lambda T \tag{8}$$

where Λ is the diagonal matrix of the local optimal solution. It is calculated as follows:

$$\alpha_p^{\tau+1} = -G(t_p - h\beta) \tag{9}$$

To further improve the flexibility of the correntropy, the cost function with a mixed correntropy is defined in [67] as follows:

$$J_{MMCC} = 1 - \frac{1}{N} \sum_{i=1}^{N} [\alpha G_{\sigma_1}(e_i) + (1 - \alpha) G_{\sigma_2}(e_i)] + \lambda \|\beta\| \tag{10}$$

Therefore, the output is calculated as follows:

$$\beta = (H^{\mathrm{T}} \Lambda H + \lambda' I)^{-1} H^{\mathrm{T}} \Lambda T \tag{11}$$

where the $\lambda' = 2N\lambda$ and Λ is the diagonal matrix with elements calculated as follows:

$$\Lambda_{ii} = \alpha / \sigma_1 G_{\sigma_1}(e_i) + (1 - \alpha) / \sigma_2 G_{\sigma_2}(e_i) \tag{12}$$

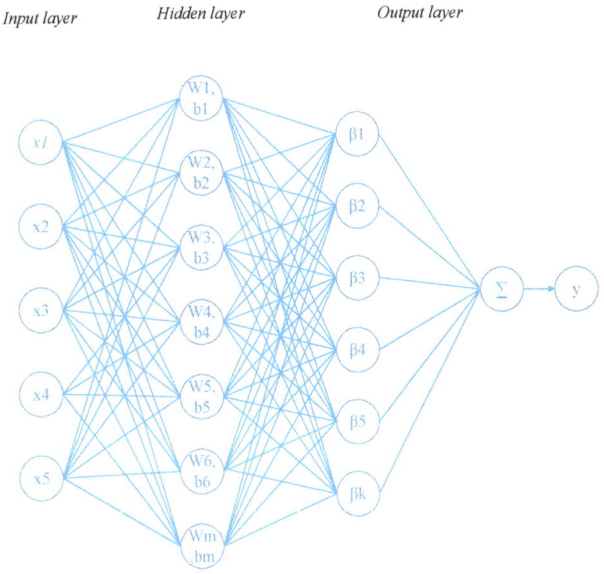

Figure 1. The structure of the prediction model.

With two coefficients, Equation (9) gives a more accurate estimation of the costs of the output layer, leading to a higher robustness of the model. Although Equations (7) and (9) can acquire better local similarity measurements compared with Equation (5), both criterions limit the correntropy into two kernels, leading to an inappropriate description on the probability distribution of the data. Additionally, the bandwidths and the coefficients must be assigned by users, thus limiting the performance of the corresponding model in real world applications which can be badly affected since the bandwidths are not suitable for the estimation of the correntropy. To provide a more flexible criterion for the training strategy with a more appropriate description of the probability distribution of the data, the proposed method develops a multi-kernel correntropy criterion that is calculated as follows:

$$k(\mathbf{T} - \beta \mathbf{H}) = \sum_{i=1}^{K} \alpha_i G_{\sigma_i}(\mathbf{T} - \beta \mathbf{H}) \tag{13}$$

where α_i is the influence coefficients controlling the weight of each kernel. By using multiple kernels to construct the correntropy, the proposed method brings a more accurate approximation on the probability distribution of the samples, leading to a high prediction performance of the model. Based on the corretropy using Equation (13), the proposed method built a convex cost function for training the output weights, which has been analyzed in Section 4. For the suitable assignments of the parameters in Equation (13), a novel generation strategy using an evolved cooperating process based on SDPSO with the MIE to generate the parameters adaptively has been developed. Therefore, the framework of the proposed method can be summarized in Figure 2. The proposed method developed an evolved-cooperation strategy to generate the optimized solution of the influence coefficients and the bandwidths which suits the distribution of the prediction errors. To achieve an accurate estimation, the bandwidth was generated based on switching delayed particle swarm optimization (SDPSO) [68] and the influence coefficients were calculated based on the cost function for estimating the probability distribution function of errors.

The basic procedures of the method are as follows. Supposing that the input vector of the samples is represented as $x = \{x_1, x_2, \ldots, x_N\}$, calculate the output of hidden nodes with randomly assigned weights and biases as Equation (1). Then, adapt the cooperating evolution technology for training the output weights. For each iterations of the evolution, the output of the predicting model can be generated using Equation (3). Compared with the actual outputs, the predicted outputs result in current error e with the model. Based on the current error e, the proposed method makes the best assignments of the bandwidths in the correntropy with SDPSO and accesses the optimal coefficients based on MIE. This is shown in the next section. Using the generated correntropy, a list of diagnostic kernels can be calculated which effects the updating of the output layer to reach higher accuracy. This is presented in Section 4. The processes stop when the cost function of the model is stable.

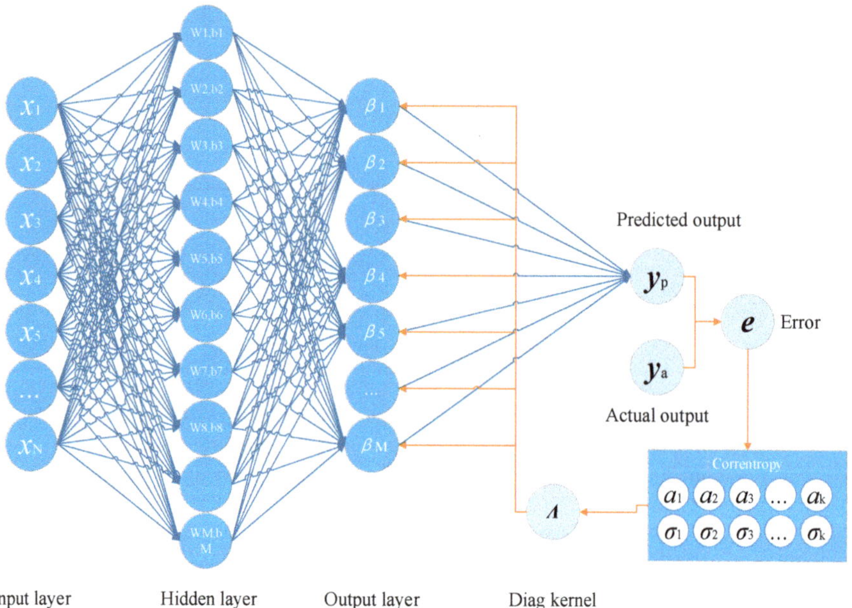

Figure 2. The framework of the proposed method.

More details are presented in the next section.

3. The Cooperating Evolution Process for the Bandwidth and Influence Coefficients of the Kernel

For the correntropy that is defined by Equation (12), the bandwidth and the influence coefficients are for the similarity measurements since the bandwidths act as the zoom lens for the measurements and the coefficients determine the effect that each kernel has on the estimation of the correntropy according to the assigned bandwidth. They are defined as follows:

$$\sigma = \{\sigma_1, \sigma_2, \sigma_3, \ldots, \sigma_M\} \tag{14}$$

$$A = \{\alpha_1, \alpha_2, \alpha_3, \ldots, \alpha_M\} \tag{15}$$

Therefore, the bandwidth and the influence coefficients should be carefully assigned to match the probability distribution of the samples to achieve the best effect of the correntropy on generating the output weights of the prediction model. Since the correntropy depicts the probability distribution of the distance between the actual output and the model response, the bandwidth and the coefficients are able to form the probability distribution (pdf) function as follows:

$$\hat{f}(e) = \sum_{i=1}^{N} \alpha_1 G_{\sigma_1}(e_i) + \alpha_2 G_{\sigma_2}(e_i) + \cdots + \alpha_N G_{\sigma_n}(e_i) \tag{16}$$

In applications, the real joint probability distribution for the cases are unknown. Therefore, the joint pdf can only be estimated for a finite number of samples$\{(t_i, y_i)\}$, where i = 1, 2, ... , N:

$$f(e) = \frac{1}{N} g(\{(t_k, y_k) \mid \|t_k - y_k\| = e\}) \tag{17}$$

where $g(\mathbf{S})$ is the cardinal number of the set $\mathbf{S}$.

Using the kernel contrasts between the pdf estimated with the assigned parameters and the pdf estimated using the data, the least mean integrated error (MIE) can be calculated as follows:

$$\text{MIE} = E\left(\int (\hat{f}(e) - f(e))^2 de \right) \tag{18}$$

Based on the MIE, the performance of the bandwidth and coefficients can be estimated using the contrasts with the pdf from the data. Therefore, the optimization of these parameters can be transformed to finding the solution with the minimum MIE.

In the proposed method, the switching delay particle swarm optimization is adapted to search for the best bandwidth. To achieve this, the particles are initialized with a list of potential bandwidth setting $\sigma c = \{\sigma_{c,1}, \sigma_{c,2}, \ldots, \sigma_{c,N}\}$. With respect to each bandwidth of the particle, the velocities for the evolution of the particles are defined as follows:

$$v\sigma_c = \{v\sigma_{c,1}, v\sigma_{c,2}, \ldots, v\sigma_{c,N}\} \tag{19}$$

Meanwhile, the influence coefficient is denoted as vector A:

$$A_c = \{\alpha_{c,1}, \alpha_{c,2}, \alpha_{c,3}, \ldots, \alpha_{c,M}\} \tag{20}$$

where α_i is the influence coefficient according to $\sigma_{c,i}$.

Since the samples provide disperse values of the outputs, the pdf from the data is estimated using the discrete version of Equation (16):

$$F = \{f(m_1), f(m_2), \ldots, f(m_k)\} \tag{21}$$

$$f(m) = \frac{1}{N} g(\{(t_k, y_k) \mid m - \varepsilon \leq |t_k - y_k| \leq m + \varepsilon\}) \tag{22}$$

where the vector $\mathbf{m} = \{m_1, m_2, \ldots, m_k\}$ is a list of values that satisfy $m_1 < m_2 < \ldots < m_k$ and $|m_i - m_{i-1}| = \varepsilon$. ε is the step length of the estimation.

Accordingly, the values from Equation (15) with respect to $\mathbf{m}$ are equivalent to the following set:

$$\hat{\mathbf{f}} = \{\hat{f}(m_1), \hat{f}(m_2), \cdots, \hat{f}(m_k)\} \tag{23}$$

They can be calculated as:

$$\hat{\mathbf{f}} = \mathbf{AK} \tag{24}$$

where K is the kernel matrix, which is as follows:

$$K = \begin{bmatrix} G_{\sigma_1}(e_1) & G_{\sigma_1}(e_2) & \cdots & G_{\sigma_1}(e_N) \\ G_{\sigma_2}(e_1) & G_{\sigma_2}(e_2) & \cdots & G_{\sigma_2}(e_N) \\ \vdots & \vdots & \ddots & \vdots \\ G_{\sigma_M}(e_1) & G_{\sigma_M}(e_2) & \cdots & G_{\sigma_M}(e_N) \end{bmatrix} \tag{25}$$

By inserting Equations (20) and (22) into Equation (17), the following cost function can be obtained:

$$\text{MIE} = (\mathbf{AK} - \mathbf{F})(\mathbf{AK} - \mathbf{F})^{\text{T}} \tag{26}$$

Then, the following differential equations with respect to $\mathbf{A}$ are calculated:

$$2(\mathbf{AK} - \mathbf{F}) = 0 \tag{27}$$

Therefore, the coefficient can be calculated using the assigned bandwidth as follows:

$$\mathbf{A} = \mathbf{FK}^T(\mathbf{KK}^T)^{-1} \tag{28}$$

Since each particle contains one solution for the kernels' parameters, the personal best solution $p\sigma$ and the global best solution $g\sigma$ is updated by minimizing the costs. Then, the particles are updated as follows:

$$v\sigma_c(k+1) = wv\sigma_c + c_1(k) \times r_1(p\sigma(k-(k)) - \sigma_c(k))) + c_2(k) \times r_2(g\sigma(k - \tau_2(k) - \sigma_c(k))) \tag{29}$$

$$\sigma_c(k+1) = \sigma_c(k) + v\sigma_c(k+1) \tag{30}$$

where $c_1(k)$ and $c_2(k)$ are the acceleration coefficients and $\tau_1(k)$ and $\tau_2(k)$ are the time delays. All the parameters are adjusted based on the evolution factor, Ef, which determines the evolutionary states, and it is calculated as follows:

$$\text{Ef} = (d_g - d_{min})/(d_{max} - d_{min}) \tag{31}$$

where d_g is the global best particle among the mean distance. It is calculated as:

$$d_g = \frac{1}{N} \sum_{i=1}^{N} \|\sigma_{c,i} - g\sigma\| \tag{32}$$

With the estimate on Ef, the parameters can be selected as shown in Table 1.

Table 1. The strategies for selecting the parameters.

State	Range of Ef	c_1	c_2	$p\sigma$	$g\sigma$	τ_1	τ_2
Convergence	$0 \le \text{Ef} < 0.25$	2	2	$p\sigma(k)$	$g\sigma(k)$	0	0
Exploitation	$0.25 \le \text{Ef} < 0.5$	2.1	1.9	$p\sigma(k - \tau_1(k))$	$g\sigma(k)$	$[k \cdot rand_1]$	0
Exploration	$0.5 \le \text{Ef} < 0.75$	2.2	1.8	$p\sigma(k)$	$g\sigma(k - \tau_2(k))$	0	$[k \cdot rand_2]$
Jumping out	$\text{Ef} > 0.75$	1.8	2.2	$p\sigma(k - \tau_1(k))$	$g\sigma(k - \tau_2(k))$	$[k \cdot rand_1]$	$[k \cdot rand_2]$

The final solution of the bandwidth and the influence coefficients are determined as the solution that minimizes the costs during the evolution procedures.

In summary, the cooperative evolution process is shown in Algorithm 1. First, the bandwidth and the corresponding velocity of each particle are randomly assigned. Then, for each iteration of the process, the influence coefficients are evolved using the bandwidth based on the MIE and the particles are updated using the cost function. Finally, the algorithm finds the best solutions for the bandwidth and the influence coefficients, from which the kernel depicts the pdf from the data. Based on the generated kernel, the correntropy can lead to a model with good robustness.

Algorithm 1 Evolved cooperation for the kernel parameters

Input: the samples $\{x_i, t_i\}, i = 1, 2, \ldots, N$
Output: the vector of bandwidth σ and the vector of influence coefficients A
  Parameters: the step length and the number of iterations L
  Initialization: Set the cost function of the best solution MIE_{best} to ∞ and randomly assign the bandwidth of the kernels $\sigma_c = \{\sigma_{c,1}, \sigma_{c,2}, \ldots, \sigma_{c,N}\}$ and the corresponding velocity $v\sigma_c = \{v\sigma_{c,1}, v\sigma_{c,2}, \ldots, v\sigma_{c,N}\}$.
1: **for** $k = 1, 2, \ldots$ L **do**
2:    Generate the best influence coefficients Ac using Equation (26) for each particles.
3:    Calculate value of cost function for each particle $MIEc$ based on Equation (24)
4:    Update the personal best solution $p\sigma$ and the global best solution $g\sigma$ based on minimizing the cost function.
5:    Calculate the Ef of the iteration with Equation (29)
6:    Access the parameters for evolution based on Table 1
7:    Update the swarm with Equations (27) and (28)
8: **end for**
9: Return the global best bandwidth $g\sigma$ and the corresponding influence coefficients

---

4. Training the Extreme Learning Machine Using the Multi-Dimension Correntropy

To improve the robustness of the extreme learning machine, in the proposed method, the training procedure of the output layer as Equation (5), is replaced by the developed calculation using the mixture correntropy that is generated using the evolved kernel from Section 3. The loss function for the output layer is developed according to the following properties.

Property 1. *$K(Y,T)$ is symmetric, which means the following: $K(Y,T) = K(T,Y)$.*

Property 2. *$K(T,Y)$ is positive and bounded, which means the following: $0 < K(Y,T) <\,= 1$ and $K(T,Y) = 1$ if and only if $T = Y$.*

Property 3. *$K(T,Y)$ involves all the even moments of e, which means the following:*

$$K(T, Y) = E[e^{2n}] \sum_{n=0}^{\infty} \frac{(-1)^n \sum_{i=1}^{M} \alpha_i \sigma_i^{2n}}{2^n \prod_{i=1}^{M} (\sigma_i)^{2n} n!} \tag{33}$$

Property 4. *When the first bandwidth is large enough, it satisfies the following:*

$$K(T, Y) \approx \sum_{i=1}^{M} \alpha_i - \frac{\sum_{i=1}^{M} \alpha_i \sigma_i^2}{2 \prod_{i=1}^{M} \sigma_i^2} E[e^2] \tag{34}$$

Proof. For $\lim\limits_{x \to 0} \exp(x) \approx 1 + x$, suppose that σ_1 is large enough, $K(T,Y)$ can be approximated as follows:

$$\begin{aligned}
K(T, Y) &= \alpha_1 G_{\sigma_1}(e) + \alpha_2 G_{\sigma_2}(e) + \ldots + \alpha_m G_{\sigma_m}(e) \\
&= \alpha_1 \left(1 - \frac{e^2}{2\sigma_1^2}\right) + \alpha_2 \left(1 - \frac{e^2}{2\sigma_2^2}\right) + \ldots + \alpha_m \left(1 - \frac{e^2}{2\sigma_m^2}\right) \\
&= \sum_{i=1}^{m} \alpha_i - \frac{\sum_{i=1}^{M} \alpha_i \sigma_i^2}{2 \prod_{i=1}^{M} \sigma_i^2} E[e^2]
\end{aligned} \tag{35}$$

that completes the proof. $\square$

Remark 1. *Based on Property 4, the mixed C-loss is defined as L(**T**,**Y**) = 1 − K(**T**,**Y**), which is approximately equivalent to the mean square error (MSE) with a large enough bandwidth.*

Property 5. *The empirical mixed C-loss L(e) that is a function of e is convex at any point satisfying* $\|e\|_\infty = max|e_i| \le \sigma_1$.

Proof. Build the Hessian matrix of the C-loss function L(e) with respect to e as follows:

$$H_{\mathbf{L(e)}} = \left[\frac{\partial \mathbf{L(e)}}{\partial \mathbf{e_i} \partial \mathbf{e_j}}\right] = \mathbf{diag}(\xi_1, \xi_2, \dots, \xi_N) \tag{36}$$

The elements of matrix ξ is calculated as follows:

$$\xi_i = \sum_{i=1}^{m} \alpha_i \frac{\sigma_i^4 - e_i^4}{N\sigma_i^4} G_{\sigma_i}(e_i) \tag{37}$$

It is obvious that ξ_i is positive. Therefore, L(e) is convex. □

Remark 2. *Using Property 4 and Property 5, the loss function of the output weights is based on the empirical mixed C-loss L(e) from the data observations, which can be defined as follows:*

$$J = L(\mathbf{T}, \mathbf{Y}) + \Lambda\|\boldsymbol{\beta}\|^2 = 1 - \frac{1}{N} \sum_{i=1}^{N} \sum_{j=1}^{M} \alpha_j G_{\sigma_j}(e_i) + \Lambda\|\boldsymbol{\beta}\|^2 \tag{38}$$

Based on Equation (38), the training criterion is generated for improvement on the robustness of the model.

Taking the differential of the loss function, it is easy to get the following:

$$
\begin{aligned}
&\frac{\partial J(\boldsymbol{\beta})}{\partial \boldsymbol{\beta}} = 0 \\
&-\sum_{i=1}^{N} \{[\sum_{j=1}^{M} \frac{\alpha_j}{\sigma_j^2} G_{\sigma_j}(e_i)]e_i \boldsymbol{h}_i^T\} + 2N\Lambda\boldsymbol{\beta} = 0 \\
&\sum_{i=1}^{N} (\varphi(e_i)\boldsymbol{h}_i^T \boldsymbol{h}_i \boldsymbol{\beta} - \varphi(e_i)t_i \boldsymbol{h}_i^T) + \Lambda'\boldsymbol{\beta} = 0 \\
&\sum_{i=1}^{N} (\varphi(e_i)\boldsymbol{h}_i^T \boldsymbol{h}_i \boldsymbol{\beta} + \Lambda'\boldsymbol{\beta} = \sum_{i=1}^{N} (\varphi(e_i)t_i \boldsymbol{h}_i^T) \\
&\boldsymbol{\beta} = [\boldsymbol{H}^T \Lambda \boldsymbol{H} + \Lambda' \boldsymbol{I}]^{-1} \boldsymbol{H}^T \Lambda \boldsymbol{T}
\end{aligned} \tag{39}
$$

where $\Lambda' = 2N\Lambda$, $\varphi(e_i) = \sum_{j=1}^{M} \frac{\alpha_j}{\sigma_j^2} G_{\sigma_j}(e_i)$ and Λ is a diagonal matrix with diagonal elements $\Lambda_{ii} = \varphi(e_i)$, which provides the local similarity measurements between the predicted output and the actual outputs. When the training data contain large noise or many outliers, the corresponding diagonal elements are relatively low which induce the effects of such samples. Therefore, the algorithm can achieve high robustness against noises and outliers in the signals.

Since Equation (37) is a fixed-point equation because the diagonal matrix depends on the weight vector, the optimal solution should be solved by applying the evolved cooperation using Equation (37).

Therefore, combined with the kernel optimization in Section 3, the whole training process can be summarized in Algorithm 2, which is referred to as the ECC-ELM algorithm in this paper.

Algorithm 2 ECC-ELM

Input: the samples $\{x_i, t_i\}$, $i = 1, 2, \ldots, N$
Output: output weights
   Parameters: the number of hidden nodes N, the number of iterations L, the iterations T and termination tolerance ε
   Initialization: Randomly set the weights and bias of the hidden nodes and initialize the output weights β using Equation (5)
1: **for** $t = 1, 2, \ldots, T$ **do**
2:   Calculate the residual error: $ei = t_i - h_i\beta$, $i = 1, 2, \ldots, N$
3:   Calculate the kernel parameters $\{\sigma, A\}$ using Algorithm 1
4:   Calculate the diagonal matrix Λ: $\Lambda_{ii} = \varphi(e_i) = \sum_{j=1}^{M} \frac{\alpha_i}{} G_{\sigma_j}(e_i)$
5:   Update the output weight using Equation (37)
6:   **Until** $|J_k(\beta) - J_{k-1}(\beta)| < \varepsilon$
7: **end for**

---

5. Analysis on Time Complexity and Space Complexity of ECC-ELM

In this section, the time complexity of the proposed method is analyzed and compared with the other algorithms. The main time complexity of the ECCELM comes from the cooperating evolution process and the training process of the model. The cooperative evolution contains the calculations of the influence coefficients and the particles updating with the time complexity of $O(I_t NK^2)$, where I_t is the number of iterations, N is the number of particles and K is the number of disperse values of the outputs. To train the ELM, the procedures share the same time complexity as the RCC-ELM and MMCC-ELM, which is $O(I_h N_l(5M+M^2))$, where I_h is the amount of iterations for training and N_l is the number of training data. Additionally, M is the number of hidden nodes. Therefore, the time complexity of ECC-ELM is $O(I_h N_l(5M+M^2+I_t NK^2))$, which is slightly higher than those of the RCC-ELM and MMCC-ELM but it satisfies the requirements in most applications.

With respect to the spatial complexity, the ECC-ELM has the same complexity as the prediction models using the RCC-ELM, which is $O(N+(N+2)M+N_l^2)$. Additionally, the space complexity consumed by evolving process is $O(2N+K)$. Therefore, the space complexity of ECC-ELM is $O(N+(N+2)M+N_l^2+2N+K)$, which has the same order as RCC-ELM and MMCC-ELM.

In summary, the time complexity and spatial complexity are practical for most applications.

6. Experiments

6.1. The Simulation of the Sinc Function with Sas noises

In this section, the simulation experiments using the Sinc function with random noises are presented. They compare between serval state-of-art algorithms with the proposed method, which are the R-ELM, the RCC-ELM, the MMCC-ELM and our method. The training and test samples were randomly assigned according to the Sinc function and random noises were added with respect to alpha-stable distribution. This is represented as follows:

$$y = \alpha \text{Sinc}(x) + \rho \tag{40}$$

where α is the scale of the function which is set to 8.0 and $\text{Sinc}(x)$ is the Sinc function. The Sinc function is represented as follows:

$$\text{Sinc}(x) = \begin{cases} \sin(x)/x & x \neq 0 \\ 1 & x = 0 \end{cases} \tag{41}$$

Moreover, ρ is the noise that satisfies the following characteristic function [69]:

$$\rho = \begin{cases} \exp\left(-\delta^{\alpha}|\theta|^{\alpha}(1 - j\beta sign(\theta)\tan\left(\frac{\pi\alpha}{2}\right))\right) + j\mu\theta & \alpha \neq 1 \\ \exp\left(-\delta^{1}|\theta|^{1}(1 - j\beta(\pi/2)sign(\theta)\log\left(\frac{\pi\alpha}{2}\right))\right) & \alpha = 1 \end{cases} \tag{42}$$

The parameters α, β, γ and μ are real and characterize the distribution of the random variable X. Here, the alpha-stable probability distribution function is denoted as $S(\alpha,\beta,\gamma,\mu)$. In these experiments, the four parameters were assigned to three different conditions to provide three types of noises. The assignment of the parameters in each sample is presented in Table 2.

Table 2. The assignments of the parameters in each sample.

Sample #	α	β	γ	μ
Sample 1	1	0	0.001	0
Sample 2	0.7	0	0.0001	0
Sample 3	1.2	0	0.001	0

Each sample contained 200 data, with half of the data being used for training and another half for testing. To get a proper estimation of the performances of each method, the experiments were operated with the best optimization of parameters. This is presented in Table 3.

Table 3. The assignment of the parameters for each algorithms.

Algorithm	Parameter	Sample 1	Sample 2	Sample 3
R-ELM	N	100	100	100
	λ	0.00001	0.0001	0.0001
RCC-ELM	N	100	100	100
	λ	0.00001	0.00001	0.00001
	I_{hq}	30	30	30
	ε	0.0001	0.0001	0.0001
	σ	1	1.2	1.2
MMCC-ELM	N	100	100	100
	λ	0.00001	0.00001	0.00001
	I_{hq}	30	30	30
	ε	0.0001	0.0001	0.0001
	Σ_1	2	2.2	4.3
	Σ_2	0.8	0.8	8.5
	α	0.8	0.8	0.9
ECC-ELM	N	100	100	100
	λ	0.00001	0.00001	0.00001
	I_{hq}	30	30	30
	ε	0.0001	0.0001	0.0001

Each experiment was conducted 30 times and the averages were taken. The comparison of the accuracies of these algorithms is presented in Table 4. Compared with other algorithms, the R-ELM and ECC-ELM achieve lower mean square errors due to the advantages of the correntropy. The performance of R-ELM is relatively poor due to the effect of noises. The performance of MMCC-ELM also improved by the correntropy. However, since the fixed dimension of the correntropy, the accuracy can be badly influenced by unnecessary assignments on the second order of the bandwidth. Furthermore, it is clear that the proposed algorithm achieves the lowest training MSE, which means that it is the most accurate method for simulation of the Sinc function.

Table 4. The comparison of the accuracies of the four algorithms.

Samples	ELM		RCC-ELM		MMCC-ELM		ECCC-ELM	
	Training MSE	Testing MSE	Training MSE	Testing MSE	Training MSE	Testing MSE	Training MSE	Testing MSE
Sample 1	0.336	0.6601	0.1339	0.3505	0.7225	1.1085	0.1415	0.3595
Sample 2	0.0828	0.11	0.0507	0.0892	1.363	2.189	0.0257	0.0576
Sample 3	0.2219	0.2572	0.2076	0.2339	0.868	0.7583	0.2046	0.2237

To further analyze the predictive abilities of these four algorithms, Figure 3 depicts the differences between the actual function and the predicted function for each algorithm. It is clear that all the algorithms achieve relatively good prediction on the Sinc function. However, the prediction results of the ELM have been badly influenced by the noises in all three samples. Additionally, the MMCC-ELM performance is poor on sample 2 and sample 3, which is probably due to the assignments with high dimension parameters. The RCC-ELM and ECC-ELM provide good predictions, which are almost identical to the actual functions in all three samples. The ECCELM has the closet predicted function with the Sinc function, which also proves that the method has high reliability against noise.

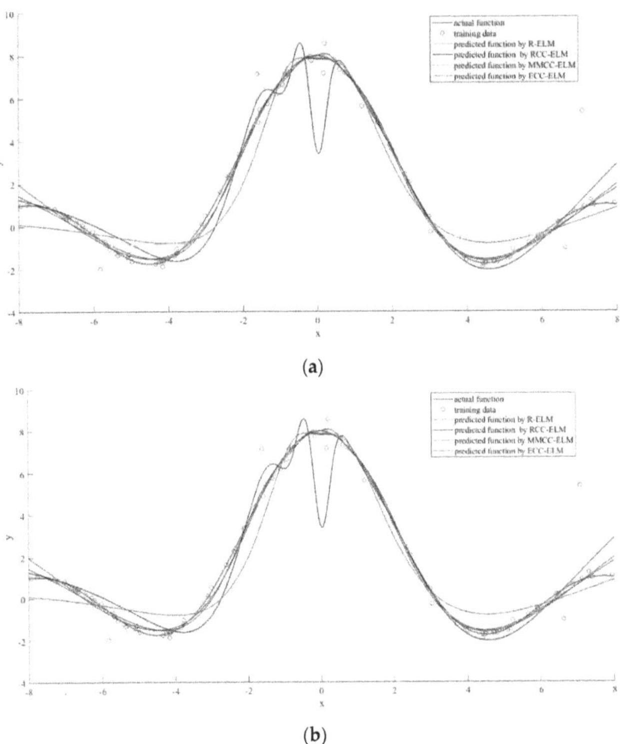

(a)

(b)

Figure 3. *Cont.*

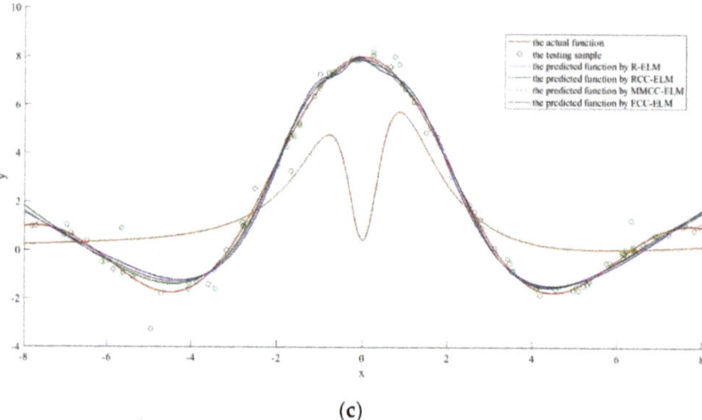

(c)

Figure 3. The performance comparison of each algorithm (**a**) comparison with sample 1; (**b**) comparison with sample 2 and (**c**) comparison with sample 3.

Furthermore, an experiment on sample 1 was conducted to compare the cost function for the output weights with the MMCC-ELM and ECC-ELM since they share similar cost functions. The results are shown in Figure 4, which show that the cost function of ECC-ELM is quite lower than the cost of MMCC-ELM. Additionally, the costs of the ECC-ELM become stable for less than 25 iterations for all three examples than MMCC-ELM. This shows the improvements on training the model with ECC-ELM taking the cooperating evolution technique. Since both algorithms finish the generation of the model when the cost function becomes stable, it can be concluded that the proposed model has faster convergence on training the prediction model.

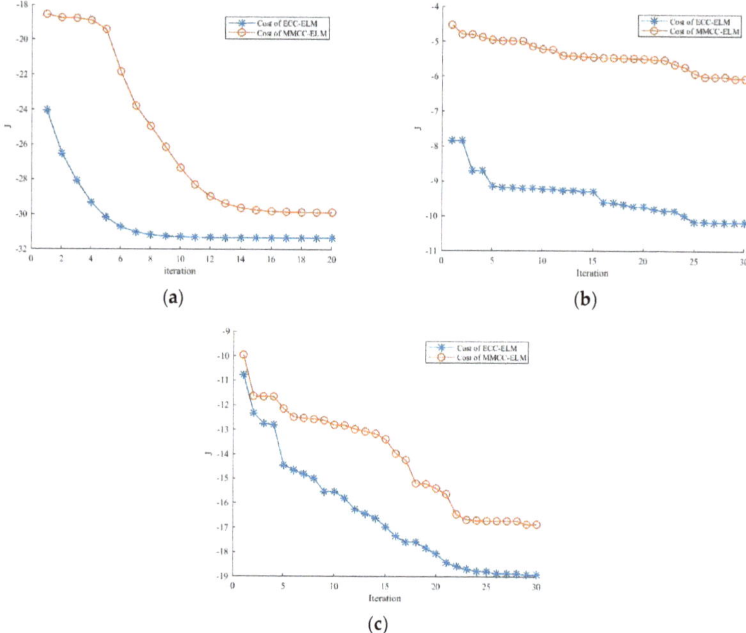

Figure 4. The comparison on the cost function values of the extreme learning machine by maximum mixture correntropy criterion (MMCC-ELM) and ECC-ELM (**a**) comparison with sample 1; (**b**) comparison with sample 2; (**c**) comparison with sample 3.

Figure 5 illustrates the effects of the evolutionary process on the optimization of the kernel bandwidth and influence coefficients. From Figure 5, it can be seen that the cost function for the kernel bandwidth quickly drops during the evolution process. Moreover, Ef continuously decreases during the process, which means that the particle swarm become stable and the best solution occurs. Figure 6 compares the actual pdf function and the estimated pdf function. It can be seen that the algorithm achieves a comparatively accurate estimation of the distribution of the errors.

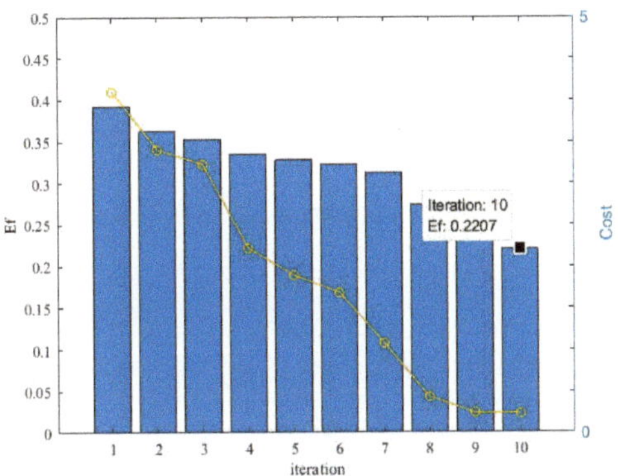

Figure 5. The dynamic changes of the evolution factor (Ef) and costs during the cooperative evolution.

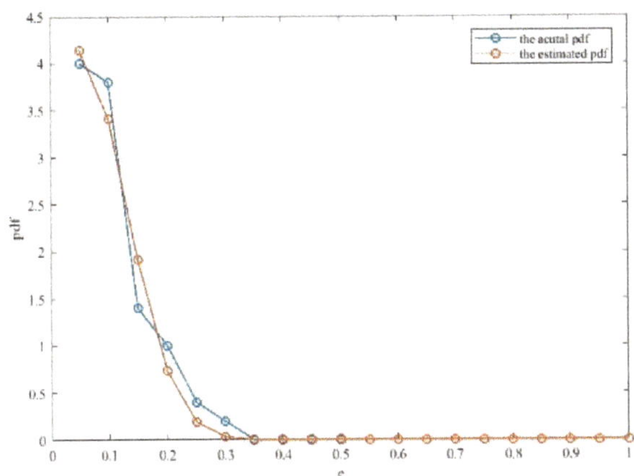

Figure 6. Comparison between the estimated pdf and actual pdf.

6.2. The Performance Comparison on Benchmark datasets

To further assess the proposed algorithm, the performance of the ECC-ELM and other methods were compared using the data set from the UCI machine learning repository [70], awesome public dataset [71] and the United Nations development program [72], which are listed in Table 5. The assignments of the parameters are shown in Table 6, all of which refer to the best performance of each algorithm. Each experiment was conducted 30 times and the average performance was reported.

The performance is compared in Table 7, which shows that the proposed algorithm is able to achieve better prediction accuracies than other methods. Additionally, the performance of the proposed method is relatively stable compared with other correntropy-based extreme learning machines.

Figure 7 compares the actual output value and the predicted value for the Servo data set. It is clear that the predicted values are basically identical to the actual output values, and it has not been influenced by the outliers in the data.

To illustrate the evolutionary processes for optimizing the bandwidth, Figure 8 depicts the distributions of the particles and the evolution of the optimal solutions. It can be seen that the distribution of the particles dynamically changes based on the state of the PSO process. The optimal solution is adjusted and stabilizes during the process, which allows the optimal solution of the bandwidth assignments to generate a more accurate model.

Table 5. The information on the data sets.

Data Set	Features	Observations	
		Training Numbers	Testing Numbers
Servo	5	83	83
Slump	10	52	51
Concrete	9	515	515
Housing	14	253	253
Yacht	6	154	154
Airfoil	5	751	751
Soil moisture	124	340	340
HDI	12	93	93
HIV	10	65	65

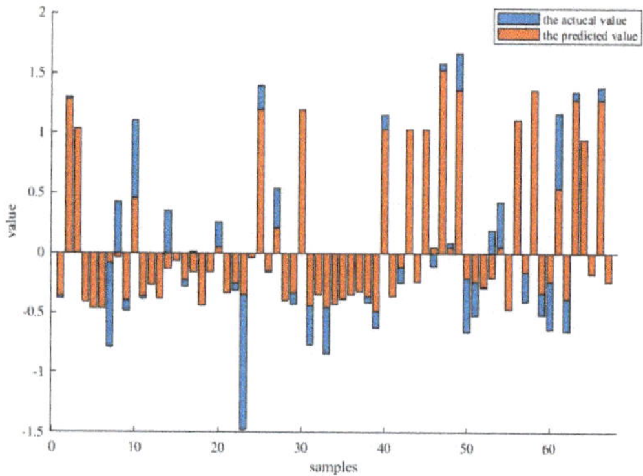

Figure 7. The comparison between the actual values and the predicted values under the data set, Servo.

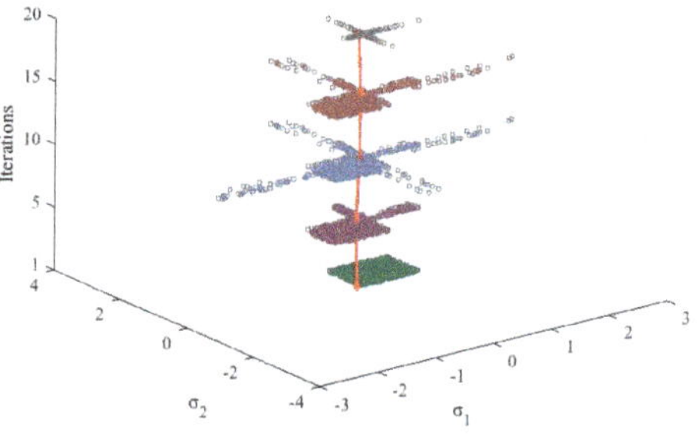

Figure 8. The evolutionary process of the particles.

Table 6. Parameter settings of each algorithm.

Algorithm	Parameter	Servo	Slump	Concrete	Housing	Yacht	Airfoil	Soil Moisture	HDI	HIV
R-ELM	N	90	190	185	180	185	200	200	100	100
	λ	0.00010000	0.00050000	0.00020000	0.00020000	0.00002000	0.00002000	0.00002000	0.00001000	0.00001000
RCC-ELM	N	120	100	200	200	200	180	180	150	120
	λ	0.00001000	0.00010000	0.00000100	0.00010000	0.00000001	0.00000001	0.00000001	0.00000001	0.00000001
	I_{hq}	30	30	30	30	30	30	30	30	30
	ε	0.00010000	0.00010000	0.00010000	0.00010000	0.00010000	0.00010000	0.00010000	0.00010000	0.00010000
	σ	0.00100000	0.00001000	0.00005000	0.01000000	0.00000100	0.00000100	0.00000100	0.00000100	0.00000130
MMCC-ELM	N	90	165	200	200	195	150	150	150	150
	λ	0.00100000	0.00001000	0.00005000	0.01000000	0.00000100	0.00000100	0.00000100	0.00000100	0.00000100
	I_{hq}	30	30	30	30	30	30	30	30	30
	ε	0.00010000	0.00010000	0.00010000	0.00010000	0.00010000	0.00010000	0.00010000	0.00010000	0.00010000
	Σ_1	0.2	0.5	0.5	0.5	0.5	0.2	1.0	1.2	0.7
	Σ_2	2.8	1.6	2.6	2	2	2.7	0.7	0.8	0.3
	α	0.8	0.3	0.5	0.3	0.8	0.5	0.6	0.7	0.6
ECC-ELM	N	90	180	180	180	180	200	200	200	200
	λ	0.00100000	0.00001000	0.00005000	0.01000000	0.00000100	0.00000100	0.00000100	0.00000100	0.00000100
	I_{hq}	30	30	30	30	30	30	30	30	30
	ε	0.00010000	0.00010000	0.00010000	0.00010000	0.00010000	0.00010000	0.00010000	0.00010000	0.00010000

Table 7. The performance comparison.

Data Set	R-ELM		RCC-ELM		MMCC-ELM		ECCC-ELM	
	Training RMSE	Testing RMSE	Training RMSE	Testing RMSE	Training RMSE	Testing RMSE	Training RMSE	Testing RMSE
Servo	0.0590 ± 0.009	0.1039 ± 0.0164	0.0740 ± 0.0106	0.1031 ± 0.0148	0.0839 ± 0.0174	0.0989 ± 0.0187	0.1047 ± 0.0181	0.8742 ± 0.0131
Slump	0.0081 ± 0.0011	0.0461 ± 0.0095	0.0000 ± 0.0000	0.0422 ± 0.0094	0.0001 ± 0.0000	0.0408 ± 0.0101	0.0001 ± 0.0001	0.354 ± 0.1890
Concrete	0.0738 ± 0.0021	0.0917 ± 0.0045	0.0561 ± 0.0018	0.0872 ± 0.0066	0.0560 ± 0.0021	0.0867 ± 0.0064	0.0561 ± 0.0018	0.0852 ± 0.0053
Housing	0.0439 ± 0.0043	0.0896 ± 0.0124	0.0495 ± 0.0045	0.0830 ± 0.0110	0.0554 ± 0.0045	0.0821 ± 0.0101	0.0352 ± 0.0013	0.0791 ± 0.0110
Yacht	0.0366 ± 0.0093	0.0529 ± 0.0090	0.0125 ± 0.0008	0.0349 ± 0.0113	0.0125 ± 0.0008	0.0328 ± 0.0074	0.0172 ± 0.0027	0.0268 ± 0.0031
Airfoil	0.0974 ± 0.0074	0.1031 ± 0.0077	0.0736 ± 0.0022	0.0906 ± 0.0054	0.0736 ± 0.0025	0.0898 ± 0.0051	0.0736 ± 0.0023	0.0889 ± 0.0046
Soil moisture	0.0032 ± 0.0011	0.0095 ± 0.0013	0.0007 ± 0.0001	0.0015 ± 0.0003	0.0006 ± 0.0000	0.0012 ± 0.0002	0.0006 ± 0.0000	0.0009 ± 0.0001
HDI	0.0004 ± 0.0001	0.0006 ± 0.0002	0.0001 ± 0.0000	0.0003 ± 0.0001	0.0001 ± 0.0000	0.0003 ± 0.0001	0.0001 ± 0.0000	0.0003 ± 0.0001
HIV	0.0376 ± 0.0220	0.0599 ± 0.0130	0.0050 ± 0.0017	0.0079 ± 0.0009	0.0047 ± 0.0006	0.0065 ± 0.0004	0.0059 ± 0.0007	0.0059 ± 0.0006

6.3. The Performance Estimations for Forecasting the CTR of Optical Couplers

Finally, to estimate the performance of a real application, the proposed method has been used to predict the current transfer ratio for optical couplers. This is one type of transmission device for electric signals and optical signals with wide applications to the isolation transfer of signals, A/D transmission, D/A transmission, digital communications and high-pressure control. For optical couplers, the CTR is an essential factor for estimating the operating status of optical couplers. In this section, the proposed method was used to give the predictions of CTR for the optical couplers to predict the health condition of the devices.

For the experiments, the degenerating signals of four optical couplers were recorded and transformed into the samples historical CTR value as input vectors and the CTR value of the next time as the expected output. The training data was the samples that were generated from the optical couplers' records over the first ten years and the testing data were the samples that were generated from the last ten years.

Figure 9 depicts the evolutionary process of the PSO procedure. It shows that the Ef value quickly decreases during the evolutionary process and stabilizes within 17 iterations, resulting in the optimal solution that is provided by the swarm.

Finally, the predicted results of the four optical couplers are shown in Figure 10. It is clear that the generated ELM network accurately predicts the CTR value of each optical coupler and is robust with the noises of the signals. Therefore, the proposed method is able to achieve good performance for the optical couplers.

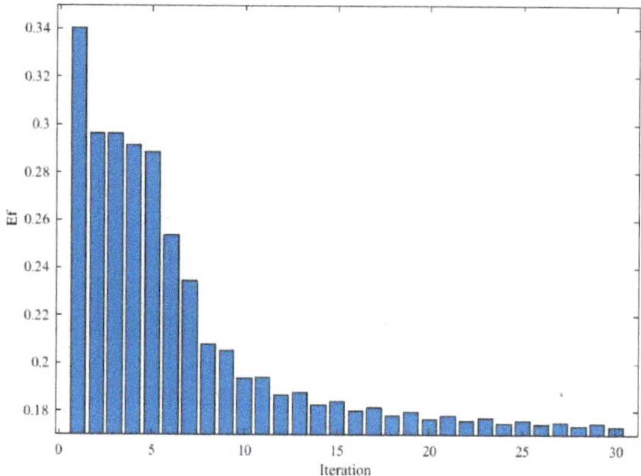

Figure 9. Dynamic changes on Ef and costs during the cooperative evolution.

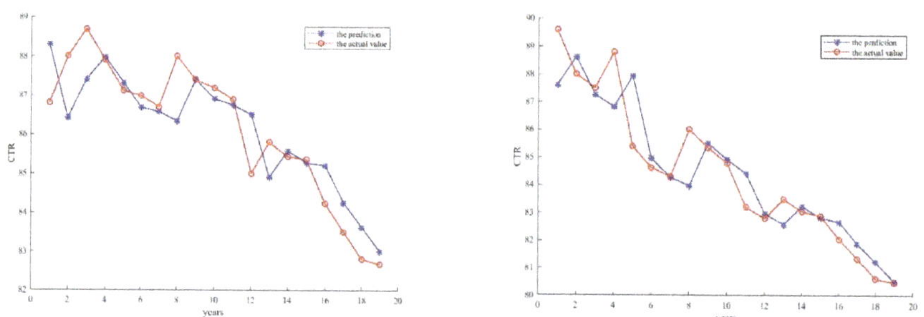

Figure 10. *Cont.*

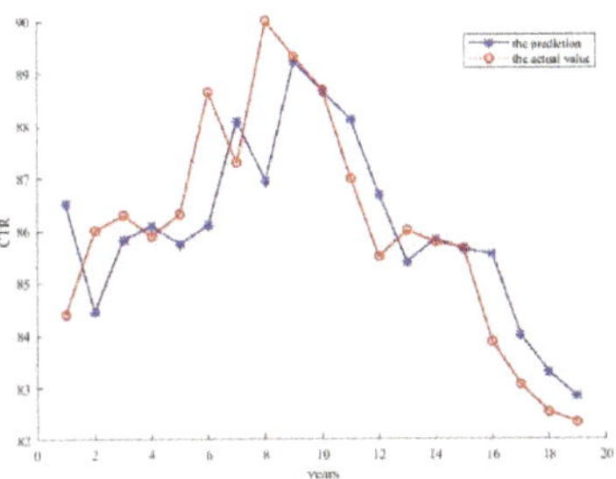 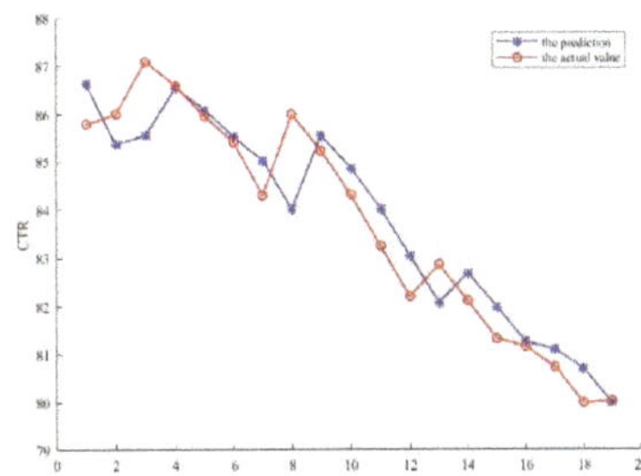

Figure 10. The comparison with the predicted current transfer ratio (CTR) and actual CTR.

Table 8 presents the numerical results of the CTR prediction, which compares the actual CTR and the predicted CTR. It is clear that the proposed method can very accurately provide the prediction on the state of Optical Couplers (OCs). Additionally, the time consumption is presented in Table 8 which shows that the proposed method is able to obtain high accuracy on the prediction of the future CTR of the OC and the predicting time is quite low within 5 ms. Therefore, the proposed method can achieve high performance on real applications.

Table 8. The performance of the predicted model that is generated using the ECCELM.

Time (year)	Actual CTR	Predicted CTR	Normalized Error	Predicting Time (ms)
1	87.90	88.03	0.0037	2.98
2	87.70	88.01	0.0068	4.02
3	87.40	87.94	0.0274	3.92
4	85.50	87.15	0.0095	4.98
5	86.30	87.02	0.0122	2.26
6	85.93	86.61	0.0084	3.22
7	85.86	85.40	0.0188	5.74
8	84.73	85.30	0.0145	4.48
9	84.01	84.33	0.0115	5.85
10	83.31	83.38	0.0023	4.87

7. Conclusions

To improve the robustness of the forecasting model, the paper provides a novel correntropy-based ELM called the ECC-ELM. It uses a multi-dimension correntropy criterion and the evolved cooperation method to adaptively generate the parameters for kernels. In the proposed algorithm, SDPSO is integrated by minimizing the MIE to determine the proper bandwidths and their corresponding influence coefficients to estimate the probability distributions of the residual error of the model. A novel training process was developed based on the properties of the multi-dimension correntropy and it was able to build the convex cost function to calculate the output weights for the ELM. The experiments on the simulated data and real-world application were conducted to estimate the accuracy of the probability distribution of the signal and robustness on predicting the samples. The simulation results with the Sinc function proved that the proposed method can generate the multi-kernel correntropy with high accuracy on describing the probability distribution of the signals and fast converge on the evolution process. This leads to high robustness of the proposed method compared with the other methods. The performance comparisons on the benchmark datasets show that the proposed method can achieve higher accuracy and more stability than the other methods. Finally, the CTR prediction experiments show the proposed method can achieve high accuracy within acceptable time consumption on real world applications. Although the proposed algorithm has predictive advantages, there are still several limitations on the study. One limitation is the proposed method is only applicable for an ELM with one hidden layer, which requires extensions on multi-layer networks. The other limitation is that the proposed method only provides an offline training model. Therefore, how to update the online prediction model becomes another interesting topic for future research. The codes and data of the research are available at https://github.com/mwj1997/ECC-ELM.

Author Contributions: Conceptualization, W.M.; Data curation, L.D.; Funding acquisition, Z.L.; Investigation, J.H.; Methodology, W.M. and Y.S.; Project administration, Z.L.; Resources, J.H.; Software, W.M.; Supervision, Z.L.; Validation, Y.S.; Writing—original draft, W.M.; Writing—review & editing, L.D.

Funding: This work was supported by the National Natural Science Foundation of China under Grants No. U1830133 (NSAF) and No. 61271035.

Conflicts of Interest: The authors declare no conflict of interest.

References

1. Heddam, S.; Keshtegar, B.; Kisi, O. Predicting total dissolved gas concentration on a daily scale using kriging interpolation, response surface method and artificial neural network: Case study of Columbia river Basin Dams, USA. *Nat. Resour. Res.* **2019**, *2*, 1–18. [CrossRef]
2. Ahmadi, N.; Nilashi, M.; Samad, S.; Rashid, T.A.; Admadi, H. An intelligent method for iris recognition using supervised machine learning techniques. *Opt. Laser Thechnol.* **2019**, *120*, 105701. [CrossRef]
3. Aeukumar, R.; Karthigaikumar, P. Multi-retinal disease classification by reduced deep learning features. *Neural Comput. Appl.* **2017**, *28*, 329–334.
4. Pentapati, H.K.; Teneti, M. Robust speaker recognition systems with adaptive filter algorithms in real time under noisy conditions. *Adv. Decis. Sci. Image Process. Secur. Comput. Vis.* **2020**, *4*, 1–18.
5. Eweda, E. Stability bound of the initial mean-square division of high-order stochastic gradient adaptive filtering algorithms. *IEEE Trans. Signal Process.* **2019**, *6*, 4168–4176. [CrossRef]
6. Huang, X.; Wen, G.; Liangm, L.; Zhang, Z.; Tan, Y. Frequency phase space empirical wavelet transform for rolling bearing fault diagnosis. *IEEE Access.* **2019**, *7*, 86306–86318. [CrossRef]
7. Yang, J.; Zhu, H.; Liu, T. Secure and economical multi-cloud storage policy with NSGA-II-C. *Appl. Soft Comput.* **2019**, *83*, 105649. [CrossRef]
8. Albasri, A.; Abdali-Mohammadi, F.; Fathi, A. EEG electrode selection for person identification thru a genetic-algorithm method. *J. Med. Syst.* **2019**, *43*, 297. [CrossRef]
9. Dermanaki Farahani, Z.; Ahmadi, M.; Sharifi, M. History matching and uncertainty quantification for velocity dependent relative permeability parameters in a gas condensate reservoir. *Arab. J. Geosci.* **2019**, *12*, 454. [CrossRef]
10. Shah, P.; Kendall, F.; Khozin, S.; Goosen, R.; Hu, J.; Laramine, J.; Ringel, M.; Schork, N. Artificial intelligence and machine learning in clinical development: A translational perspective. *Nature* **2019**, *2*, 1–5. [CrossRef]
11. Shirwaikar, R.D.; Dinesh, A.U.; Makkithaya, K.; Suruliverlrajan, M.; Srivastava, S.; Leslie, E.S.; Lewis, U. Optimizing neural network for medical data sets: A case study on neonatal apnea prediction. *Artif. Intell. Med.* **2019**, *98*, 59–76. [CrossRef] [PubMed]
12. Lucena, O.; Souza, R.; Rittner, L.; Frayne, R.; Lotufo, R. Convolutional neural network for skull-stripping in brain MR imaging using silver standard masks. *Artif. Intell. Med.* **2019**, *98*, 48–58. [CrossRef] [PubMed]
13. Guan, H.; Dai, Z.; Guan, S.; Zhao, A. A neutrosophic forecasting model for time series based on first-order state and information entropy of high-order fluctuation. *Entropy* **2019**, *21*, 455. [CrossRef]
14. Tymoshchuk, O.; Kirik, O.; Dorundiak, K. Comparative analysis of the methods for assessing the probability of bankruptcy for Ukrainian enterprises. In *Lecture Notes in Computational Intelligence and Decision Making*; Springer: Basel, Switzerland, 2019; pp. 281–293.
15. Yang, T.; Jia, S. Research on artificial intelligence technology in computer network technology, International conference on artificial intelligence and security. In Proceedings of the 5th International Conference on Artificial Intelligence and Security (ICAIS 2019), New York, NY, USA, 26–28 July 2019; pp. 488–496.
16. Senguta, E.; Jain, N.; Garg, D.; Choudhury, T. A review of payment card fraud detection methods using artificial intelligence. In Proceedings of the International Conference on Computational Techniques, Electronics and Mechanical Systems (CTEMS), Belagavi, India, 21–23 December 2018; pp. 494–499.
17. Ampatzidis, Y.; Partel, V.; Meyering, B.; Albercht, U. Citrus rootstock evaluation utilizing UAV-based remote sensing and artificial intelligence. *Comput. Electron. Agric.* **2019**, *164*, 104900. [CrossRef]

18. Yue, D.; Han, Q. Guest editorial special issue on new trends in energy internet: Artificial intelligence-based control, network security and management. *IEEE Trans. Syst. Man Cybern. Syst.* **2019**, *49*, 1551–1553. [CrossRef]

19. Liu, W.; Pokharel, P.P.; Principe, J.C. The kernel least mean square algorithm. *IEEE Trans. Signal Process.* **2008**, *56*, 543–554. [CrossRef]

20. Vega, L.R.; Rey, H.; Benesty, J.; Tressens, S. A new robust variable step-size NLMS algorithm. *IEEE Trans. Signal Process.* **2008**, *56*, 1878–1893. [CrossRef]

21. Vega, L.R.; Rey, H.; Benesty, J.; Tressens, S. A fast robust recursive least-squares algorithm. *IEEE Trans. Signal Process.* **2008**, *57*, 1209–1216. [CrossRef]

22. Ekpenyong, U.E.; Zhang, J.; Xia, X. An improved robust model for generator maintenance scheduling. *Electr. Power Syst. Res.* **2012**, *92*, 29–36. [CrossRef]

23. Huang, Y.; Lee, M.-C.; Tseng, V.S.; Hsiao, C.; Huang, C. Robust sensor-based human activity recognition with snippet consensus neural networks. In Proceedings of the IEEE 16th International Conference on Wearable and Implantable Body Sensor Networks (BSN), Chicago, IL, USA, 19–22 May 2019.

24. Ning, C.; You, F. Deciphering latent uncertainty sources with principal component analysis for adaptive robust optimization. *Comput. Aided Chem. Eng.* **2019**, *46*, 1189–1194.

25. He, C.; Zhang, Q.; Tang, Y.; Liu, S.; Liu, H. Network embedding using semi-supervised kernel nonnegative matrix factorization. *IEEE Access.* **2019**, *7*, 92732–92744. [CrossRef]

26. Bravo-Moncayo, L.; Lucio-Naranjo, J.; Chavez, M.; Pavon-Garcia, I.; Garzon, C. A machine learning approach for traffic-noise annoyance assessment. *Appl. Acoust.* **2019**, *156*, 262–270. [CrossRef]

27. Santos, J.D.A.; Barreto, G.A. An outlier-robust kernel RLS algorithm for nonlinear system identification. *Nonlinear Dyn.* **2017**, *90*, 1707–1726. [CrossRef]

28. Guo, W.; Xu, T.; Tang, K. M-estimator-based online sequential extreme learning machine for predicting chaotic time series with outliers. *Neural Comput. Appl.* **2017**, *28*, 4093–4110. [CrossRef]

29. Zhou, P.; Guo, D.; Wang, H.; Chai, T. Data-driven robust M-LS-SVR-based NARX modeling for estimation and control of molten iron quality indices in blast furnace ironmaking. *IEEE Trans. Neural Netw. Learn.* **2018**, *29*, 4007–4021. [CrossRef]

30. Ma, W.; Qiu, J.; Liu, X.; Xiao, G.; Duan, J.; Chen, B. Unscented Kalman filter with generalized correntropy loss for robust power system forecasting-aided state estimation. *IEEE Trans. Ind. Inf.* **2019**. [CrossRef]

31. Safarian, C.; Ogunfunmi, T. The quaternion minimum error entropy algorithm with fiducial point for nonlinear adaptive systems. *Signal Process.* **2019**, *163*, 188–200. [CrossRef]

32. Dighe, P.; Asaei, A.; Bourlard, H. Low-rank and sparse subspace modeling of speech for DNN based acoustic modeling. *Speech Commun.* **2019**, *109*, 34–45. [CrossRef]

33. Li, L.-Q.; Wang, X.-L.; Xie, W.-X.; Liu, Z.-X. A novel recursive T-S fuzzy semantic modeling approach for discrete state-space systems. *Neurocomputing* **2019**, *340*, 222–232. [CrossRef]

34. Hajiabadi, M.; Hodtani, G.A.; Khoshbin, H. Robust learning over multi task adaptive networks with wireless communication links. *IEEE Trans. Comput. Aided Des.* **2019**, *66*, 1083–1087.

35. Kutz, N.J. Neurosensory network functionality and data-driven control. *Curr. Opin. Syst. Biol.* **2019**, *3*, 31–36. [CrossRef]

36. Chen, B.; Xing, L.; Zheng, N.; Principe, J.C. Quantized minimum error Entropy criterion. *IEEE Trans. Neural Netw. Learn. Syst.* **2019**, *30*, 1370–1380. [CrossRef]

37. Liu, W.; Pokharel, P.P.; Principe, J.C. Correntropy: Properties and applications in non-guassian signal processing. *IEEE Trans. Signal Process.* **2007**, *55*, 5286–5298. [CrossRef]

38. Kuliova, M.V. Factor-form Kalman-like implementations under maximum correntropy criterion. *Signal Process.* **2019**, *160*, 328–338. [CrossRef]

39. Ou, W.; Gou, J.; Zhou, Q.; Ge, S.; Long, F. Discriminative Multiview nonnegative matrix factorization for classification. *IEEE Access.* **2019**, *7*, 60947–60956. [CrossRef]

40. Wang, Y.; Yang, L.; Ren, Q. A robust classification framework with mixture correntropy. *Inform. Sci.* **2019**, *491*, 306–318. [CrossRef]

41. Moustafa, N.; Turnbull, B.; Raymond, K. An ensemble intrusion detection technique based on proposed statical flow features for protecting network traffic of internet of things. *IEEE Internet Things J.* **2019**, *6*, 4815–4830. [CrossRef]

42. Wang, G.; Zhang, Y.; Wang, X. Iterated maximum correntropy unscented Kalman filters for non-Gaussian systems. *Signal Process.* **2019**, *163*, 87–94. [CrossRef]

43. Peng, J.; Li, L.; Tang, Y.Y. Maximum likelihood estimation-based joint sparse representation for the classification of hyperspectral remote sensing images. *IEEE Trans. Neural Netw. Learn. Syst.* **2019**, *30*, 1790–1802. [CrossRef]

44. Masuyama, N.; Loo, C.K.; Wermter, S. A kernel Bayesian adaptive resonance theory with a topological structure. *Int. J. Neural Syst.* **2019**, *29*, 1850052. [CrossRef]

45. Shi, W.; Li, Y.; Wang, Y. Noise-free maximum correntropy criterion algorithm in non-Gaussian environment. *IEEE Trans. Circuits Syst. II Express Briefs* **2019**. [CrossRef]

46. Jiang, Z.; Li, Y.; Hunag, X. A correntropy-based proportionate affine projection algorithm for estimating sparse channels with impulsive noise. *Entropy* **2019**, *21*, 555. [CrossRef]

47. He, R.; Zheng, W.-S.; Hu, B.-G. Maximum correntropy criterion for robust face recognition. *IEEE Trans. Patt. Anal. Mach. Intell.* **2019**, *33*, 1561–1576.

48. Macheshwari, S.; Pachori, R.B.; Rajendra, U. Automated diagnosis of glaucoma using empirical wavelet transform and correntropy features extracted from fundus images. *IEEE J. Biol. Health Inf.* **2017**, *21*, 803–813. [CrossRef]

49. Moharmmadi, M.; Noghabi, H.S.; Hodtani, G.A.; Mashhadi, H.R. Robust and stable gene selection via maximum minimum correntropy criterion. *Geomics* **2016**, *107*, 83–87.

50. Guo, C.; Song, B.; Wang, Y.; Chen, H.; Xiong, H. Robust variable selection and estimation based on modal regression. *Entropy* **2019**, *21*, 403. [CrossRef]

51. Luo, X.; Xu, Y.; Wang, W.; Yuan, M.; Ban, X.; Zhu, Y.; Zhao, W. Towards enhancing stacked extreme learning machine with sparse autoencoder by correntropy. *J. Frankl. Inst.* **2018**, *355*, 1945–1966. [CrossRef]

52. Wang, S.; Dang, L.; Wang, W.; Qian, G.; Chi, K.T.S.E. Kernel adaptive filters with feedback based on maximum correntropy. *IEEE Access.* **2018**, *6*, 10540–10552. [CrossRef]

53. Heravi, A.R.; Hodtani, G.A. A new correntropy-based conjugate gradient backpropagation algorithm for improving training in neural networks. *IEEE Trans. Neural Netw. Learn. Syst.* **2018**, *29*, 6252–6263. [CrossRef]

54. Jaeger, H.; Lukosevicious, M.; Popovivi, D.; Siewert, U. Optimization and applications of echo state networks with leaky integrator neurons. *Neural Netw.* **2007**, *20*, 335–352. [CrossRef]

55. Tanaka, G.; Yamane, T.; Heroux, J.B.; Nakane, R.; Kanazawa, N.; Takeda, S.; Numata, H.; Nakano, D.; Hirose, A. Recent advances in physical reservoir computing: A review. *Neural Netw.* **2019**, *115*, 100–123. [CrossRef]

56. Obst, O.; Trinchi, A.; Hardin, S.G.; Chawick, M.; Cole, I.; Muster, T.H.; Hoschke, N.; Ostry, D.; Price, D.; Pham, K.N. Nano-scale reservoir computing. *Nano Commun. Netw.* **2013**, *4*, 189–196. [CrossRef]

57. Guo, Y.; Wang, F.; Chen, B.; Xin, J. Robust echo state network based on correntropy induced loss function. *Neurocomputing* **2017**, *267*, 295–303. [CrossRef]

58. Huang, G.; Chen, L. Convex incremental extreme learning machine. *Neurocomputing* **2007**, *70*, 3056–3062. [CrossRef]

59. Huang, G.; Zhou, H.; Ding, X.; Zhang, R. Extreme learning machine for regression and multiclass classification. *IEEE Trans. Syst. Man Cybern. Part B* **2012**, *42*, 513–529. [CrossRef]

60. Tang, J.; Deng, C.; Huang, G. Extreme learning machine for multilayer perceptron. *IEEE Trans. Neural Netw. Learn. Syst.* **2016**, *27*, 809–821. [CrossRef]

61. Huang, G.; Bai, Z.; Lekamalage, L.; Vong, C.M. Local receptive fields based extreme learning machine. *IEEE Comput. Intell. Mag.* **2015**, *10*, 18–29. [CrossRef]

62. Arabilli, S.F.; Najafi, B.; Alizamir, M.; Mosavi, A.; Shamshirband, S.; Rabczuk, T. Using SVM-RSM and ELM-RSM Approaches for optimizing the production process of Methyl and Ethyl Esters. *Energies* **2018**, *11*, 2889.

63. Ghazvinei, P.T.; Darvishi, H.H.; Mosavi, A.; Yusof, K.b.W.; Alizamir, M.; Shamshirband, S.; Chau, K.-W. Sugarcane growrh prediction based on meteorological parameter using extreme learning machine and artificial neural network. *Eng. Appl. Comp. Fluid.* **2018**, *12*, 738–749.

64. Shamshirband, S.; Chronopoulos, A.T. A new malware delectation system using a high performance ELM method. In Proceedings of the 23rd international database applications & engineering symposium, Athens, Greece, 10–12 June 2019; p. 33.

65. Bin, G.; Yan, X.; Yang, X.; Gary, W.; Shuyong, L. An intelligent time-adaptive data-driven method for sensor fault diagnosis in induction motor drive system. *IEEE Trans. Ind. Electr.* **2019**, *66*, 9817–9827.

66. Xing, H.; Wang, X. Training extreme learning machine via regularized correntropy criterion. *Neural Comput. Appl.* **2013**, *23*, 1977–1986. [CrossRef]

67. Chen, B.; Wang, X.; Lu, N.; Wang, S.; Cao, J.; Qin, J. Mixture correntropy for robust learning. *Pattern Recognit.* **2018**, *79*, 318–327. [CrossRef]

68. Zeng, N.; Zhang, H.; Liu, W.; Liang, J.; Alsaadi, F.E. A switching delayed PSO optimized extreme learning machine for short-term load forecasting. *Neurocomputing* **2017**, *240*, 175–182. [CrossRef]

69. Weron, A.; Weron, R. Computer simularion of Levy alpha-stable variables and processes. In *Lecture Notes in Pihysics*; Springer: Berlin/Heidelberg, Germany, 1995; pp. 379–392.

70. Frank, A.; Asuncion, A. *UCI Machine Learning Repository*; University of California, School of Information and Computer Science: Irvine, CA, USA, 2010.

71. Awesome Data. Available online: http://www.awesomedata.com/ (accessed on 16 September 2015).

72. Human Development Reports. Available online: http://hdr.undp.org/en/data# (accessed on 15 September 2019).

Article

A Statistical Method for Estimating Activity Uncertainty Parameters to Improve Project Forecasting

Mario Vanhoucke * and Jordy Batselier

Faculty of Economics and Business Administration, Ghent University, Tweekerkenstraat 2, 9000 Gent, Belgium; jordy.batselier@ugent.be
* Correspondence: mario.vanhoucke@ugent.be

Received: 21 August 2019; Accepted: 26 September 2019; Published: 28 September 2019

Abstract: Just like any physical system, projects have entropy that must be managed by spending energy. The entropy is the project's tendency to move to a state of disorder (schedule delays, cost overruns), and the energy process is an inherent part of any project management methodology. In order to manage the inherent uncertainty of these projects, accurate estimates (for durations, costs, resources, . . .) are crucial to make informed decisions. Without these estimates, managers have to fall back to their own intuition and experience, which are undoubtedly crucial for making decisions, but are are often subject to biases and hard to quantify. This paper builds further on two published calibration methods that aim to extract data from real projects and calibrate them to better estimate the parameters for the probability distributions of activity durations. Both methods rely on the lognormal distribution model to estimate uncertainty in activity durations and perform a sequence of statistical hypothesis tests that take the possible presence of two human biases into account. Based on these two existing methods, a new so-called statistical partitioning heuristic is presented that integrates the best elements of the two methods to further improve the accuracy of estimating the distribution of activity duration uncertainty. A computational experiment has been carried out on an empirical database of 83 empirical projects. The experiment shows that the new statistical partitioning method performs at least as good as, and often better than, the two existing calibration methods. The improvement will allow a better quantification of the activity duration uncertainty, which will eventually lead to a better prediction of the project schedule and more realistic expectations about the project outcomes. Consequently, the project manager will be able to better cope with the inherent uncertainty (entropy) of projects with a minimum managerial effort (energy).

Keywords: project management; entropy; managerial effort; distribution fitting; lognormal distribution

1. Introduction

Project Management is the discipline to manage, monitor and control the uncertainty inherent to projects. Project management processes are used to monitor and control the progress of projects in order to reduce the uncertainty, and each such process requires effort from the project manager and her team. The academic literature has been overwhelmed by research studies in project management and control, and many of them focus on the construction of the project baseline schedule to assess the project risk and to monitor the performance of a project in progress. The combination of these three dimensions—schedule, risk and control—is often referred to in the literature as *dynamic scheduling* [1,2] or *integrated project management and control* [3].

This paper starts with the observation that the relation between managerial effort and the ability to reduce the project uncertainty lies at the heart of many research studies, although this relation is often not explicitly mentioned. Especially in some research papers that rely on the concept of *entropy* as a way to express that projects have the natural tendency to move to a state of disorder, authors have referred to the relation between entropy (uncertainty) and energy (effort). They have proposed different entropy measures to enable the project manager to better predict the project uncertainty and eventually reduce it by taking better actions. This concept of entropy is—to the best of the authors' knowledge—not widely used in the previously mentioned dynamic scheduling studies; however, it is believed that it sheds an interesting light on the project management domain and opens ways to look at the dynamic scheduling literature (schedule/risk/control) in a fundamentally different way.

The current study reviews the research on entropy in project management and proposes a new way to accurately estimate project uncertainty to improve project forecasting and decision-making. This paper first elaborates on the link between the traditional dynamic scheduling literature and the much less investigated concept of entropy in project management and argues that entropy is an ideal concept to measure project uncertainty. Then, it will be shown that, in order to reduce a project's entropy, forecasting and estimates are crucial for a project manager and her team to make well-informed decisions. Then, finally, a new so-called *calibration method* is proposed that should help project managers to better quantify the project uncertainty by providing better estimates for the activity durations. Such calibration procedures are relatively new in the literature, since they rely on a combination of statistical data analysis and the correction for human biases.

The paper is organized in the following sections. Section 2 reviews the most important studies on entropy for managing projects that have been used as an inspiration for the current research study. Based on this, the section also explains the basic idea of calibrating project data to better estimate project uncertainty, which constitutes the main theme of our study. In Section 3, two currently known data calibration methods from literature are then briefly reviewed, as they will be used as foundations for a new third calibration method taking the shortcomings of the existing methods into account. This new so-called *statistical partitioning heuristic* is discussed in Section 4. Section 5 presents the results of a computational experiment on a set of 83 empirical projects (mainly construction projects) from a known database. The section shows that the statistical partitioning heuristic outperforms the two other procedures, but also discusses some limitations that can be used as guidelines for future research. Finally, Section 6 draws conclusions and highlights some potentially promising future research avenues.

2. Managing Projects

2.1. Entropy in Project Management

Project Management is the discipline to manage, monitor and control the uncertainty inherent to projects. Whatever specific project management process is used to monitor and control the project progress to reduce the uncertainty, it always requires effort from the project manager and her team. In several studies in the literature, this managerial effort of project management to reduce the project's uncertainty is studied from an entropy point-of-view. In this view, the entropy is the natural tendency of projects to move to a state of disorder, often quantified as schedule delays, cost overruns and/or quality problems, and the managerial effort to monitor and control such projects in progress is then the *energy* of the entropy concept to reduce the uncertainty. The general idea of entropy is proposed by [4] who stated that the uncertainty of a system decreases by receiving information about the possible outcome of the system. From this point of view, *project management* requires energy to cope with the inherent entropy of projects. Note that the term energy cannot be interpreted in a very strict sense here, since energy itself is of course not sufficient for dealing with entropy. Project management is much more than just using energy, and instead requires the right people at the right place to solve problems. Hence, effective project management requires "competences" and "skills" which are composed by many components, and not only the amount of energy by its people. Consequently, the term *energy* is used to refer to all the effort done by people with the right competences to bring projects in danger back on track.

Most project management studies do not explicitly take the concept of entropy into account, but nevertheless all aim at developing new methodologies for project managers to better measure, predict and control the inevitable problems of a project (uncertainty) in the easiest possible way (effort). Consequently, while many excellent studies indirectly deal with the issue of managing project uncertainty, to the best of our knowledge, only three studies explicitly quantified the relation between managerial effort (*energy*) and uncertainty reduction (*entropy*). First, the study of [5] investigated whether the use of *schedule risk analysis* can improve the time performance of projects in progress. In a large simulation study with artificial project data, the author varied the degree of management attention—which is a proxy for the effort of control—and measured whether this has an impact on the quality of the corrective action decision-making process to bring projects in trouble back on track (uncertainty reduction). The study of [6] extended this approach and relied on the same concept of effort (of a project manager) and quality of actions (to cope with uncertainty) and compared two alternative project control approaches. The bottom-up control approach is similar to the previously mentioned schedule risk analysis study and aims at reducing the project uncertainty by focusing on the activities with the highest risk in the project schedule. The second so-called top-down method makes use of the well-known earned value management methodology to monitor the project's performance, which is used as an early warning signal for taking corrective actions. The authors compared these two alternative project control methods, and proposed the so-called *control efficiency* concept which aims at finding the right balance between minimizing effort and maximizing quality of actions. Finally, Ref. [7] measured the impact of managerial effort to reduce the activity variability on the project time and cost performance. Without mentioning the concept of entropy, they defined a so-called effort-uncertainty reduction function to quantify the relation between the managerial effort (energy) and the reduced uncertainty (entropy). Despite the explicit quantification of both *effort* and *uncertainty reduction*, these three studies never have made any attempt to use empirical project data to measure uncertainty. Instead, all results have been obtained using simulation studies on artificial project data using statistical probability distributions with randomly selected values for their parameters to quantify project uncertainty. Hence, since the authors had no idea whether the chosen values correspond with possible real-life values, they have relied on a huge set of simulation runs, varying these values as much as possible to assure that their results provide enough managerial insights relevant for practice. Moreover, none of these studies have explicitly referred to the concept of entropy as a possible way to model project uncertainty.

However, the use of entropy sheds an interesting light on the project management domain. In a study of two decades ago by [8], the authors proposed an entropy model for estimating and management the uncertainty of projects, and argued that controlling projects comes with a certain degree of managerial effort, since:

> "With the aid of the entropy one can estimate the amount of *managerial effort* required to overcome the *uncertainty* of a particular project."

Or course, not all project management studies took the relation between effort and uncertainty so explicitly into account, but nevertheless made use of the entropy concept in project management. Ref. [9] proposed an uncertainty index as a quantitative measure for evaluating the inherent uncertainty of a project, and analysed their approach on a real turbojet engine developing project. In a recent study, Ref. [10] measure the uncertainty related to the evolution of a resource-constrained project scheduling problem with uncertain activity durations using the entropy concept. Ref. [11] proposed a new risk analysis and project control methodology, and used entropy functions for a project's completion time and critical path. In addition, [12] proposed an entropy-based approach for measuring project uncertainty, and argued that management's inability to address uncertainty is one of the major reasons for project failures. According to these authors, the managerial effort to deal with uncertainty in projects should consist of three parts:

Step 1. Identifying sources of project uncertainty,
Step 2. Quantifying project uncertainty,
Step 3. Using the uncertainty metrics for improving decision-making.

The previously mentioned studies have been an inspiration to develop and propose the model of the current paper. However, it should be noted that the literature contains many studies dealing with the three-step process discussed earlier, and an overview of these is outside the scope of this paper. The reader is referred to summary papers about project risk [13] and project control [14] to find interesting references. The current study elaborates on the second part of the required managerial effort (*quantifying uncertainty*) and proposes a new way of quantifying probability distributions for activity duration by making use of empirical project data rather than simply by relying on statistical probability functions with randomly chosen values for the averages and variances (with no known link to practice). Ref. [8] argue that such a study for better quantifying activity duration uncertainty is necessary since "usually in practice we can only estimate the possible duration range of activities and very rarely we have information about the probability distribution curve". Moreover, in the previously mentioned paper by [12], the authors conclude that "a better prediction of project costs, schedule and potential benefits leads to more realistic expectations about project outcomes and lower failures", and, hence, implicitly argue that a more accurate way of estimating probability distributions for project uncertainty is key for making better project management decisions.

As a conclusion, the previous studies have shown that, just like any physical system, projects have entropy that must be managed by spending energy. This energy process—defined as all the effort done by people with the right competences—is a very important aspect of any project management methodology. In order to manage the inherent uncertainty of these projects, accurate estimates (for durations, costs, resources, ...) are crucial to make informed decisions. Without these estimates, managers have to fall back to their own intuition and experience, which—although valuable—are often subject to biases and hard to quantify. The next section discusses the specific approach of the current study to accurately estimate distributions for activity duration, and it is shown that this specific approach—which we refer to as data calibration—is an extended version of an existing methodology of three recently published studies.

2.2. Calibrating Data

In the previous section, it has been shown that forecasting is important for good decision-making in project management, and that such an approach requires the presence of accurate estimates for the activity durations and costs of the project activities. While many studies have investigated the project management domain from different angles, they all—implicitly or explicitly—agree that good forecasting is a necessary requirement for coping with the *entropy of projects*, but this requires *energy*, which is the managerial effort of the project manager and her team.

Hence, accurate estimates should ideally be based on a mix of data for similar past projects and human judgement (the expertise often so readily available in the project team). Many of the simulation studies in the literature clearly opt for using well-known statistical distributions to model activity uncertainty, and randomly vary the parameters for the average duration and standard deviation without really knowing what realistic values are. Despite the relevance of such studies, they do not take any human judgement into account when estimating the distribution parameters, and hardly make use of data of past projects. Instead, they simply rely an arbitrarily chosen numbers for the distribution parameters without a link to real projects. The idea of calibrating data is to overcome the shortcomings of these simulation studies by relying on data of past projects to fit probability distributions, without ignoring the observation that these data are prone to human biases and possible misjudgements.

Figure 1 gives a graphical summary of the central idea of calibrating project data for activity duration distributions. A calibration method is a method to filter data of empirical projects (inputs) by removing parts (calibration) that cannot be used further in the analysis, and to identify the distribution parameters for activity duration that appears the most appropriate in a real-life context. The goal is to classify the project activities in clusters that have identical values for the parameters (average and variance) of a predefined probability distribution (outputs). The three parts (input–calibration–output) are briefly summarized along the following lines, followed by some details about the existing calibration methods.

- **Input:** The input data should exist of a set of empirical projects that are finished and for which the outcome is known. More specifically, the empirical project data should consist of a set of planned activity durations (estimates made during the schedule construction) and a set of known real activity durations (that are collected after the project is finished).
- **Calibration:** The calibration phase makes use of the input data (planned and real activity durations) and performs a sequence of hypothesis tests to split the set of activities into clusters (partitions) with similar characteristics. Throughout these hypothesis tests, it is assumed that the activity durations follow a predefined probability distribution, but a calibration method differs from an ordinary statistical test since it recognizes that the reported values in the empirical data might contain some biases. More precisely, the data might be biased due to the presence of the Parkinson effect (activities that finish early are reported to be on time (hidden earliness)) as well as rounding errors (real activity durations are rounded up or down when reported). In order to overcome these potential biases, the calibration method starts with a sequence of hypothesis tests (for which the null hypothesis is that all activity durations follow the predefined distribution), and, if the hypothesis cannot be accepted, a portion of the activities of the project has to be removed from the set to correct for the previously mentioned biases. This approach continues until the remaining set of project activities follows the predefined distribution (i.e., the test is accepted), and then the value for the average and variance of this distribution can be accurately estimated.
- **Output:** The ultimate goal of the data calibration phase is to define one or multiple clusters of activities with similar and known values for the parameters for the predefined probability distribution (i.e., average durations and standard deviation). These values can be used to better predict the project outcome, and since the activity uncertainty is then no longer set as randomly chosen values (as is often the case in simulation studies) but based on realistic values, it should enable the project manager to better predict the project outcome and reduce the project uncertainty more efficiently. Hence, calibration methods aim at better estimating the activity and project uncertainty (i) based on real project data, (ii) by taking human input biases into account, and (iii) by recognizing that not all activities should have the same values but can be clustered in smaller groups with similar values within each group, but different values between groups.

To the best of our knowledge, only two calibration methods have been proposed in literature that explicitly take the presence of the two human biases—the Parkinson's effect and the effect of rounding errors—into account, and the current study will extend these methods to a third method. The two existing calibration procedures rely on a pre-defined distribution for the activity durations of the project, as outlined in the *calibration* step. More specifically, the *lognormal distribution* is chosen as the distribution for modelling activity duration uncertainty, which means that the null hypothesis for all calibration tests (step 2 in Figure 1) is that the division of real activity duration with the estimated activity duration from the schedule follows a lognormal distribution. While some arguments were given in previous studies why the lognormal distribution is a good candidate distribution for modelling activity duration uncertainty (see, e.g., the study by [15] who advocated the use of this distribution based on theoretical arguments and empirical evidence), this choice obviously restricts the two current and the newly presented calibration methods. Indeed, many other distributions have been used in literature to model activity duration uncertainty, such as the beta distribution (e.g., [16]), the generalised beta distribution (e.g., [17]) or the triangular distributions (e.g., [18]), but a detailed discussion on the choice of distribution and a comparison of these distributions for modelling activity uncertainty is not within the scope of our study. However, this does not mean that our study has no practical or academic value. The main goal of the calibration methods used in the study is that, although they assume that the core distribution of an activity duration is the *lognormal distribution*, it is still true that the parameters for this given distribution (such as the values for the average and standard deviation) cannot be readily seen from empirical data due to distorting human factors such as hidden earliness or rounded data. Consequently, since the calibration methods test whether activity durations follow a lognormal distribution after correcting for the Parkinson effect and rounding errors, we will refer—in line with the previous studies—to the assumed distribution for activity duration as the *Parkinson distribution with lognormal core* (PDLC).

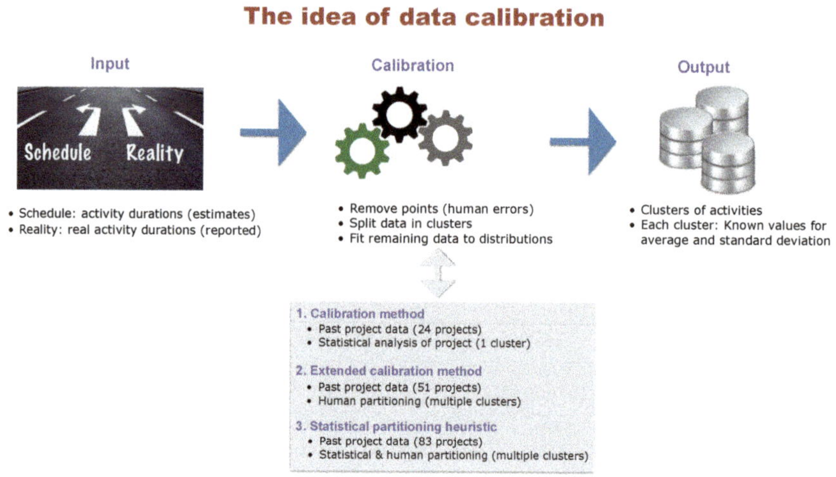

Figure 1. The idea of calibrating project data.

The current paper focuses on extending the two currently existing methods to a third method, taking the weaknesses and shortcomings of the existing methods into account. A summary of the three methods is given below the *calibration* step of Figure 1. The first calibration method has been proposed by [15] and has been validated on only 24 projects by [19]. The procedure consists of a sequence of tests that removes data from the empirical database until the lognormality test is accepted for the project as a whole (no clustering). More recently, this calibration method has been extended by [20] and includes human partitioning as an initialisation step before the calibration actually starts its sequence of hypothesis tests. The underlying idea is that humans can better divide activities into clusters based on their knowledge about the project, and only afterwards, the calibration phase processes the data of each cluster to test the lognormallity of the calibration phase. A summary of both calibration methods (i.e., the original calibration method and its extension to human clustering) is given in Section 3. It is important to review both procedures since they form the foundation of the newly developed statistical partitioning heuristic discussed in the current paper. In the remainder of this paper, we will refer to the two calibrating procedures as the *calibration procedure* and the *extended calibration method*. Since both procedures contain strong similarities, they will sometimes be referred to as the two calibration procedures. The new method that will be presented in the current study—which will be referred to as a *statistical partitioning heuristic*—builds further on two currently known calibration methods in literature. The new method still relies on this basic lognormal core assumption but now extends the current calibration procedures with an automatic partitioning phase to define clusters of activities that each has the same parameters values (average and standard deviation) for their lognormal distribution.

This method will be discussed in Section 4. In the computational experiment of Section 5, the three procedures will be tested on a set of 125 empirical projects (for which 83 could eventually be used for the analysis), and their performance will be compared. It will be shown that the new statistical partitioning heuristic outperforms the two other procedures but still contains some limitations that can be used as guidelines for future research.

We believe that the contribution and relevance of the current calibration study are threefold. First, and foremost, the current study presents an extended calibration method that allows the project manager to test whether clusters of activities follow a lognormal distribution for their duration. When this hypothesis is accepted, the procedure returns the values for the parameters of this distribution (average duration and standard deviation) such that they can be used for forecasting the future progress of a new project using Monte Carlo simulations. Such simulations can then be done using data from the past rather than arbitrarily chosen numbers, which is often criticised in simulation studies in the literature. Secondly, the calibration method is an extension of two previously published methods that take the same two human biases (rounding and Parkinson) into account. The extensions consist of mixing human expertise with automatic statistical testing, as well as allowing partitioning during testing rather than treating the whole project as one cluster of identical activities. Finally, to the best of our knowledge, this is the first study that calibrates data on such a large empirical dataset of 83 projects collected over several years.

Of course, our approach is only one possible approach of improving the accuracy of duration estimates, and all results should be interpreted within this limitation. Moreover, implementing such a procedure in practice requires a certain level of maturity for the project manager as it assumes that historical data are readily available. Consequently, using the new calibration method might require some additional effort as an initial investment to design a data collection methodology for past projects. Finally, even when project data are available, our approach is only beneficial if past projects are representative for future projects, which implies that some project characteristics are general and typical for the company. Consequently, in case every project is unique and totally different from the previous portfolio of projects, calibrating data would be of no use and relevance.

Of course, other studies in the academic literature have also aimed at estimating distribution parameters. However, we believe our calibration method is the first approach that does this by taking the two biases into account, and we therefore compare the new calibration method only with the two other calibration procedures using the same two biases. We believe that, thanks to the automatic nature of statistical testing in the new calibration method, our calibration method will contribute to a better forecasting of new projects, and hence to reducing the inherent uncertainty of a project with a minimum effort.

3. Calibration Procedures

This section gives a short summary of the two versions for calibrating data—the *calibration procedure* and the *extended calibration method*—as discussed earlier. Both calibration procedures form the foundation for the current paper, which is the reason why their main steps are repeated in Section 3.1. After this summary, the main shortcomings and areas for improvements of the extended calibration method are given in Section 3.2, and these limitations are then used to present the newly developed *statistical partitioning procedure* in Section 4.

3.1. Summary of Procedure

The extended calibration method consists of five main building blocks which are graphically summarised in Figure 2. Steps S1 to S4 are identical to the four steps of the original calibration method, apart from some small technical modifications. The extended calibration method added a fifth initialisation step S0 to these four steps to cluster data into so-called human partitions. As said, these five steps (S0 to S4) are used as foundations for the new statistical partitioning heuristic discussed later, which is the reason why they are reviewed here.

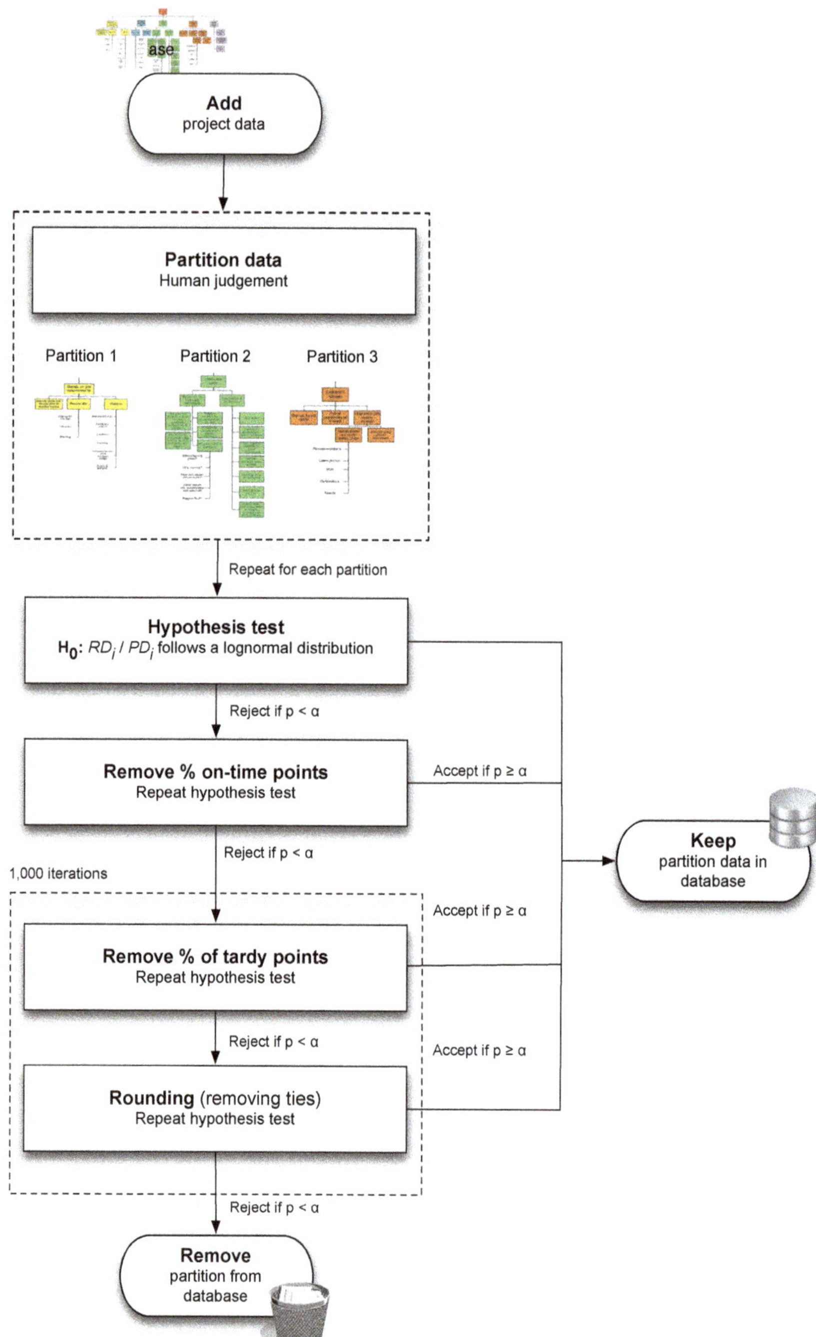

Figure 2. Extended calibration method.

Step 0 (S0). Human Partitioning

The starting point for developing the extended calibration method was inspired by the saying that *"data cannot replace human intuition"*, and that human judgement and experience of the project manager should be taken into account when evaluating data of past projects. Indeed, the original calibration method was merely a sequence of statistical tests to calibrate data, and no human input whatsoever about the project was taken into account. It is, however, well justified to state that the wonders of the human brain, although not always very reliable and subject to biases, cannot simply be replaced by a statistical data analysis, and the extension therefore mainly focused on taking this "human expertise" into account. Consequently, in order to avoid potential users of the calibration method from complaining that their human intuition would be completely ignored and replaced by a black-box statistical analysis, the gap between the dark secrets of statistical testing and the human expertise was narrowed by adding a human initialisation phase (S0) that must be executed prior to the four remaining steps of the calibration method (S1 to S4).

This initialisation phase consists of a so-called *managerial partitioning* step that splits the project data into different clusters (called *partitions*). The general idea is that the human expertise (the project manager's knowledge about the project data) should come before any statistical analysis to create clusters of project data with identical characteristics. Treating these clusters separately in the remaining steps S1 to S4, rather than analysing the project data as a whole, should give the statistical calibration method more power to accept some of the project clusters, and reject others for the same project (rather than simply accepting the project data or not). Consequently, the black box analysis of the statistical calibration method is now preceded by a human input phase, and recognizes that activities of a project do not always adhere to one and the same probability distributions. Hence, the main contribution of the extension is that it assumes that computing probability distributions for activities is best done by comparing clusters of completed activities in a project rather than treating the project data as one big homogeneous dataset.

As mentioned earlier, the four remaining steps (S1 to S4) are copied from the original calibration method, only slightly extended with some minor technical adaptations to increase the acceptance rate. The only difference is that these four steps are now carried out on the different partitions separately, instead of using the project data as a whole. Each of these partitions can now pass the lognormality test (accepted partitions are assumed to contain activities with lognormal distribution, and are therefore added to the project database) or not (rejected partitions are thrown away).

In a set of computational experiments, the authors have shown that the managerial partitioning is a promising additional feature for calibrating data. Three managerial criteria have been taken into account to split the project data into partitions. More precisely, the project data were split up based on the *work packages* (WP) the activities belong to, the *risk profile* (RP) defined by the project manager as well as the estimate for the *planned duration* (PD) of each activity. The extended calibration method has been tested on 83 empirical projects taken from [21] (mainly construction projects) and results show that the additional human partitioning step increased the acceptance rate to 97% of the total created partitions.

The four remaining steps of the calibration method are now briefly summarized along the following lines.

Step 1 (S1). Hypothesis Testing (Lognormal Core)

Testing clusters (or partitions) of data using the four-phased statistical calibration method aims at creating a database of past project data (divided in clusters) in order to better understand and analyse the behaviour of new projects. For each cluster of past project data, it is assumed that the planned and real duration of its activities are known, and it is tested whether the durations of these activities follow a certain predefined probability distribution. Indeed, if the distribution of activity durations is known, its parameters can be estimated and used for analysis of a new project with similar characteristics. The hypothesis test of S1 will be repeated in each of the following steps (S2 to S4) until a final acceptance or rejection is reached. A detailed outline of the hypothesis test is given in the previously mentioned sources for the (extended) calibration method, and its main features are now briefly repeated below.

Testing variable: The ratio between the real duration RD_i and the planned duration PD_i for each activity i is used as the test variable in each cluster. Obviously, when $RD_i/PD_i < 1$, activity i was completed early, $RD_i/PD_i = 1$ signals on-time activities while, for $RD_i/PD_i > 1$, the activity i suffered from a delay (these will be referred to as tardy activities).

Hypothesis test: The hypothesis is now that the testing variable RD_i/PD_i follows a lognormal distribution for each activity i in the partition under study. This corresponds to testing whether $ln(RD_i/PD_i)$ follows a normal distribution or not.

Goodness-of-fit: To assess whether the hypothesis can be accepted or not, a three-phased approach is followed. First, Pearson's linear correlation coefficient R is calculated by performing a linear regression of the test variable on the corresponding Blom scores [22]. The calculated R value can then be compared to the values tabulated by Looney and Gulledge [23] to obtain a p-value. Finally, the hypothesis is accepted when $p \geq \alpha$ with α the significance level equal to e.g., 5%. Each cluster that passes the test is added immediately to the database, while the remaining clusters will be subject to a calibration procedure.

Calibration: If the hypothesis is not accepted ($p < \alpha$), the project data of the cluster is not immediately thrown away. Instead, the data will be calibrated, then put under the same hypothesis test again, and only then a final evaluation and decision will be made. The term *calibration* is used since it adapts/calibrates the data of a cluster by removing some of the data points. It assumes that certain data points in the cluster are subject to human biases and mistakes, and should therefore not be kept in the cluster, while the remaining points should be tested again in a similar way as explained here in S1. Two biases are taken into account, one known as the *Parkinson effect* (S2 and S3) and another to account for *rounding errors* (S4).

Steps 2/3 (S2 & S3). Parkinson's Law

The (clusters of) project data consist of activity durations of past projects, and since the data are collected by humans, they are likely to contain mistakes. Most of the project data used in the previously mentioned studies are collected using the so-called *project card approach* of [24], which prescribes a formal method to collect data of projects in progress, exactly to avoid these human input mistakes. Nevertheless, people are and will continue to be prone to make errors when reporting numbers, and possible mistakes due to optimism bias and strategic misinterpretations will continue to exist.

For this very reason, the (extended) calibration method takes the *Parkinson effect* into account which states that work fills the allocated time. It recognizes that the reported RD_i values are not always accurate or trustful, and they might bias the analysis and the acceptance rate of the lognormality hypothesis (S1). In order to overcome these biases, all on-time data points (S2) and a portion of the tardy data points (S3) are removed from the cluster before a new hypothesis test can be performed.

Remove on-time points (S2): The procedure assumes that *all* on-time points are hidden earliness points and should therefore be removed from the cluster. More precisely, all points that are falsely reported as being completed on time, i.e., each activity with $RD_i/PD_i = 1$ in a cluster that did not pass S1, are removed from the analysis. By taking this Parkinson effect into account, the cluster now only contains early and tardy points. Before a new hypothesis test can be performed, the proportion of tardy points should be brought back to the original proportion, as suggested in S3.

Remove tardy points (S3): The removal of these on-time points—that actually were assumed to be early points—distort the real proportion of early versus tardy points in the data cluster, and this distortion should be corrected first. Consequently, an equal *portion* of the tardy points must be removed from the cluster too to bring the data back to the original proportion of early and tardy activities. Note that the calculation of a proportion of tardy points to remove only defines *how many* tardy activities should be removed from the cluster but does not specify which of these tardy points to remove. In the implementation of the original calibration procedure of [19], the tardy points were selected at random, while in the extended calibration method of [20], the number of tardy points were selected randomly for 1000 iterations and further analyses were carried out on these 1000 iterations to have more stable results.

After the removal of all on-time points, and a portion of the tardy points, the hypothesis test of S1 is executed again on the remaining data in the cluster, now containing a reduced amount of activities. The same goodness-of-fit criteria are applied as discussed in S1 and only when the hypothesis can not be accepted does the procedure continue with S4. Obviously, the data points of accepted clusters are added—as always—to the database.

Step 4 (S4). Coarse Time Interval

In a final phase, the remaining cluster data are corrected for possible rounding errors made by the collector of the data of the activity durations. More precisely, data points with identical values for the test variable RD_i/PD_i are assumed to be mistakenly rounded up or down, as the results of the coarseness of the time scale that is used for reporting the activity durations. For example, when planned values of activity durations are expressed in weeks, it is likely that the real durations are also rounded up to weeks, even if the likelihood that the real duration was an integer number of weeks is relatively low. Therefore, corrections for rounding errors are taken into account when calculating average values of the Blom scores of these so-called tied points. More precisely, these tied points are not merged to a single score value with weight one, but rather to a set of coinciding points to retain their correct composite weight.

In the study of [20], different implementations of S4 have been tested, taking into account rounding error correction with or without including S3 and S4. It has been shown that rounding correction (S4)—although beneficial for calibrating data—is less important for accepting the hypothesis than correcting the data for the Parkinson effect, which is the reason S4 will be taken into account only after S3, as initially proposed in the original calibration procedure.

3.2. Limitations

Although the extended calibration procedure solved the limitations of the original calibration method while retaining its most valuable aspects, the authors still mention some limitations for their extended version, and argue that these limitations cannot be solved by minor adaptations to their procedure only. However, they also have shown that managerial partitioning (S0) adds value to the other four steps, and, hence, it would be wise not to throw away this idea. Therefore, in the current study, we propose a novel methodology that is cast into a more comprehensive and more versatile methodology called the *statistical partitioning heuristic*, which is presented in the next section. The six limitations of the extended calibration method mentioned by [20] in their Section 4.2.4 are summarized along the following lines.

- **Limitation 1.** Only on-time activities can be eliminated in S2. Moreover, if there are no on-time activities in the project, any further analysis is impossible (the proportion x in S3 would per definition also be zero so that no tardy points can be removed either) and no (better) fit can be obtained. Note that early activities are never eliminated from the project in the calibration procedures.
- **Limitation 2.** The p-value is the only measure that is applied to assess the goodness-of-fit, whereas other measures exist that could also be utilized to this end and thus prove useful to calibrate data.
- **Limitation 3.** Partitioning can only be done based on managerial criteria (using the three criteria, i.e., PD, WP and RP) and is thus influenced by human judgement.
- **Limitation 4.** The lognormality hypothesis is not tested for the tardy activities that are removed in S3. This should be done, since these activities do not follow the pure Parkinson distribution like the eliminated on-time activities in S2 do.

- **Limitation 5.** S2 only allows the elimination of *all* on-time activities, whereas removing only a fraction of them could be more optimal (i.e., better fit to the PDLC).
- **Limitation 6.** Although 1000 iterations are performed, the tardy points that are to be removed in S3 are still chosen randomly within every iteration. Deviations in results, however minor, can thus still occur.

In Section 4, the newly developed partitioning heuristic will be discussed, and it will also be shown that the discussed limitations are implicitly taken into account. A summary of the discussed limitations as well as how the new statistical partitioning heuristic has solved them are given in Figure 3.

	Limitation	Solution
1	Only on-time points are eliminated (if no on-time points, no tardy points; never early points)	New selection strategy (on-time, early and tardy points) *New*
2	Goodness of fit restricted to p-value	New stopping strategy (p-value, standard error of regression and adjusted R^2) *New*
3	Partitioning based on managerial criteria (human judgement)	Partitioning based on statistical testing (statistical and human)
4	No hypothesis testing on tardy points	Removed partitions subject to hypothesis testing
5	All on-time points are removed (assumed to be the result of the Parkinson effect)	Only portion of on-time points are removed (some activities assumed to be actually on time)
6	Random selection of portion of tardy points (to be removed)	No random selection of tardy points (always same set of activities)

Figure 3. The limitations of the extended calibration procedure are used to provide solutions in the statistical partitioning heuristic.

4. Partitioning Heuristic

The improved acceptance rate of the extended calibration method as well as its limitations have been the main driving force to develop the new statistical partitioning heuristic. It integrates the hypothesis testing approach of the original calibration method with the human partitioning philosophy of the extended calibration method, and, consequently, follows a similar methodology as both calibration procedures. The main difference is that the statistical partitioning method now partitions the project data not only based on human input, but also using a statistical methodology, and this extension has resulted in a number of significant modifications graphically summarized in Figure 4.

In this section, an overview of the newly developed partitioning method will be given subdivided in three main subsections. Each of these three sections overcome (some of) the limitations that still existed for the extended calibration procedure. We will not run through the solution approach in an explicit stepwise manner as was done in Section 3.1, but rather show where the steps (S0 to S4) have been incorporated and possibly adapted. In Section 4.3.3, the discussed limitations will be addressed chronologically and referred to when and how a particular option of the statistical partitioning heuristic solves them.

4.1. Human Partitioning

The procedure starts with an optional human partitioning step identical to the initialization step S0 of the extended calibration method. Since managerial partitioning has shown to be relevant for the acceptance rate of the extended calibration method, an additional non-human partitioning phase—which is the reason the new procedure is referred to as *statistical* partitioning—will be added to further split the human-based partitions into subpartitions. In the computational experiments of Section 5, results of the statistical partitioning heuristic will be reported with and without the managerial partitioning step S0.

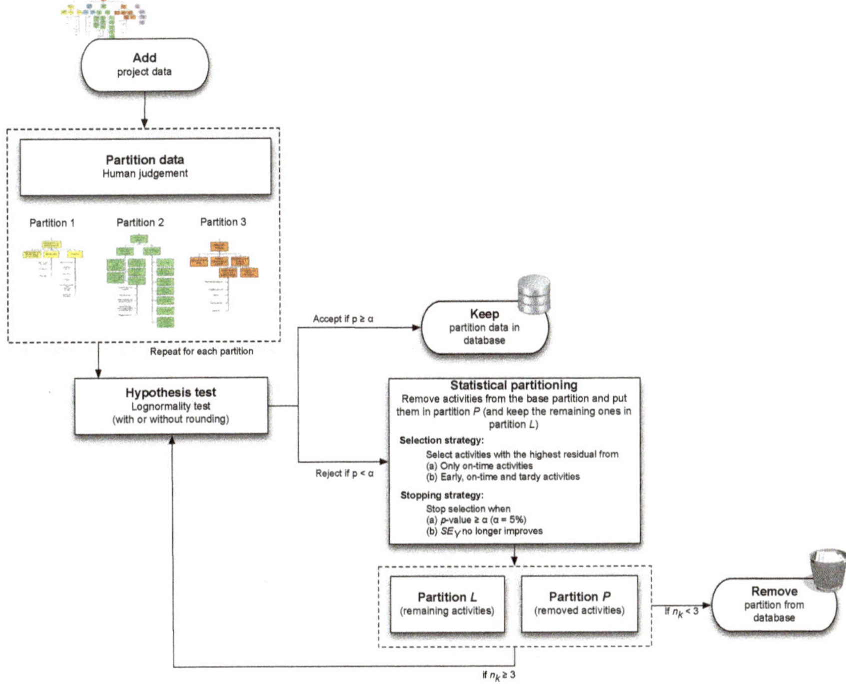

Figure 4. Graphical visualisation of partitioning heuristic.

4.2. Hypothesis Test

The *hypothesis test* (S1) of the statistical partitioning heuristic follows the same methodology as in both calibration procedures, and it can incorporate the data correction for rounding errors or not (S4). The test still assesses whether or not $ln(RD_{ij}/PD_{ij})$ is normally distributed by employing Blom scores and the table of Looney and Gulledge. If the correction for *rounding* errors (S4) is also taken into account, it still corresponds to the averaging of the Blom scores for all clusters of tied points. Therefore, it is not necessary to elaborate on each aspect of the S0 and S4 procedures in detail.

Recall that the hypothesis (S1) was also tested in steps S2 and S3 of the calibration procedures, after the removal of all on-time points and a portion of tardy points to incorporate the effect of Parkinson. As a matter of fact, the major difference between the calibration procedures and the new statistical partitioning method lies exactly in the treatment of the data for the Parkinson's effect (S2 or S3). The (extended) calibration method aims at removing data from the project clusters to be never used again (since it follows the Parkinson effect) and only continues the hypothesis testing on the remaining portion of the data. However, the new statistical partitioning heuristic does not automatically remove data points from the clusters, but, instead, aims at splitting each partition into two separate clusters (subpartitions) and then continues testing the same hypothesis on both partitions. This iterative process of splitting and testing continues until a certain stop criterion is met, and the data of all created subpartitions that pass the test are kept in the database. More precisely, at a certain moment during the search, each subpartition will be either accepted (i.e., the data follow a lognormal distribution) or rejected (i.e., the data do not follow a lognormal distribution or the sample size of the cluster has become too small). As shown in Figure 4, we have set the minimum sample size to 3 since partitions containing too few points may get too easily accepted. The way partitions are split into two subpartitions is defined by two newly developed statistical strategies (selection and stopping), which will be discussed in the next section.

4.3. Statistical Partitioning

In this section, it will be shown how the statistical partitioning heuristic iteratively creates clusters of data with similar characteristics ((sub)partitions) based on statistical testing, similar to the managerial partitioning approach that aims at creating data clusters based on human input. Indeed, the statistical partitioning heuristic iteratively selects data points from a current partition and splits them into two separate clusters, and this process is repeated for each created cluster until a created subpartition can be accepted for lognormality. The specific way how these partitions are split into subpartitions does now no longer require human input but will be done using two new statistical strategies.

The so-called *selection strategy* defines which points of the current partition should be selected for removal when splitting a partition. Each removed point will then be put in a first newly created subpartition, while the remaining non-removed points are put in a second new partition, now with less points than in the original partition. This process of removing data points from the original partition continues until a certain stopping criterion is met as defined by the so-called *stopping strategy*. Once the process stops, the original partition—which we will refer to as the *base partition*—will have been split into two separate subpartitions that will both be subject to the hypothesis test again and—if still not accepted—further partitioning. In the remainder of this manuscript, the term *partition L* will be used to indicate the subpartition with the set of activities that have not been removed from the base partition, while the set of activities that were eliminated from the partition and put in a newly created subpartition is now referred to as *partition P*. It should be noted that the naming of the two partitions P and L found its roots in the testing approach of the previously discussed calibration procedures. Recall that steps S2 and S3 remove all on-time points and a portion of the tardy point from a partition. These removed points are assumed to be a subject of the Parkinson effect (hence, partition P) and are thus removed from the database. The remaining data points in the partition were subject to further testing for the lognormal distribution (hence, partition L) and—if accepted—are kept in the database. A similar logic is followed for the statistical partitioning heuristic, although the treatment of the two partitions P and L now depends on the selection and stopping strategies that will be discussed hereafter.

Both the selection strategy and the stopping strategy can be performed under two different settings (*standard* or *advanced*), which results in $2 \times 2 = 4$ different ways the statistical partitioning heuristic can be performed. Of course, these two strategies cannot work in isolation but will nevertheless be explained separately in Section 4.3.1 and Section 4.3.2. A summary is given in Figure 5.

4.3.1. Selection Strategy

Recall that the partitioning heuristic splits up a partition into two new subpartitions. Partition P contains all the points that are removed from the base partition, while partition L then contains all the non-removed points (but now contains less data points compared to the base partition). The *selection strategy* defines which points will be removed from the base partition and put in partition P, and which points will be kept to create partition L, and can be done in a standard and advanced way.

The *standard selection strategy* does not differ very much from the (extended) calibration method, and defines that only on-time points can be eliminated from the base partition. As a result, partition P with the removed activities will then obviously exhibit a pure Parkinson distribution (since all points are on time), and no further statistical partitioning will be performed for partition P. Partition L can still consist of early, on-time and tardy points, and will be further used by the partitioning heuristic. As shown in Figure 5, no further partitioning will be performed for partition P, and its data are therefore thrown away (cf. STOP in Figure 5), but the specific treatment of partition L (ACCEPT or CONTINUE) depends on the setting of the stopping strategy, which will be discussed in Section 4.3.2.

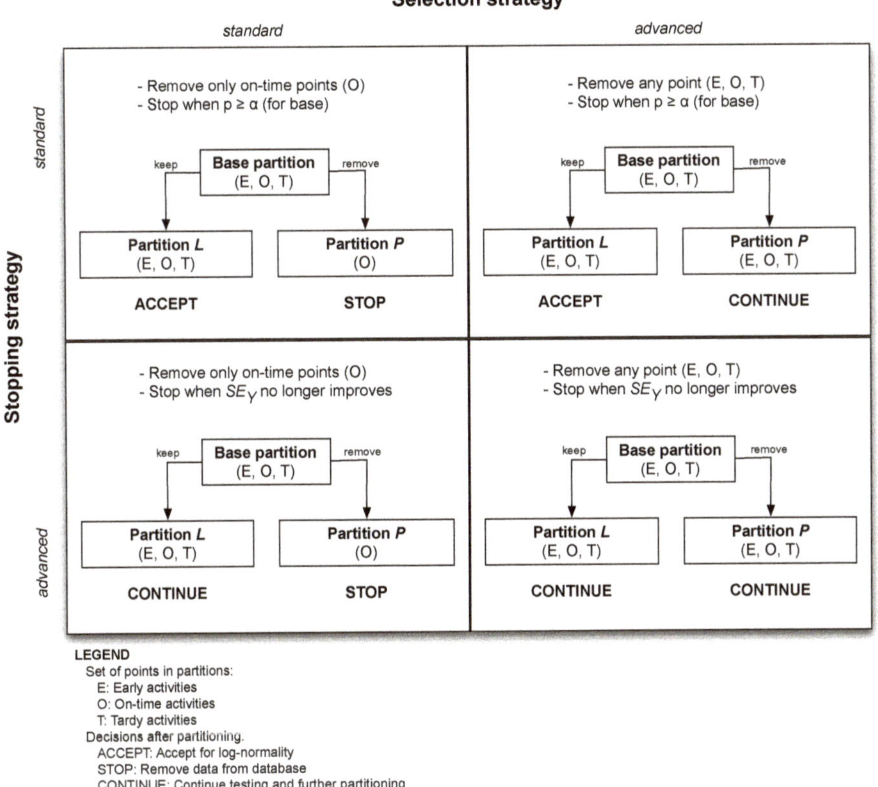

Figure 5. The four settings for the two strategies.

In the *advanced selection strategy*, not only on-time, but rather *all* activities are potential candidates to be selected for removal, and thus both the resulting partitions L and P can now contain early, on-time and tardy points. This approach is called advanced since it is fundamentally different than the approach taken by the calibration procedures (S2 and S3). The most important implication of the advanced setting is that partitions in which not all activities are on time can now be created *automatically*. Indeed, the base partition will be split by eliminating activities from it, put them in partition P and keep the remaining activities in partition L until L attains (optimal) fit (this optimal fit will be defined by the stopping strategy discussed in the next section). The set of removed activities (partition P), however, can now contain both on-time and early/tardy activities (just as partition L) and will thus most likely not exhibit a trivial pure Parkinson distribution (as was the case for the on-time activities of partition P under the standard selection strategy). Therefore, this partition P of removed activities should also undergo a hypothesis test and possibly a partitioning phase, and so should all later partitions that are created as a result of this consecutive application of the partitioning heuristic. In that way, there is an automatic creation of partitions—hence the name statistical *partitioning* heuristic for the method—that should comprise activities that are similar to each other. Unlike the initial managerial partitioning step, no human judgement has interfered with this type of partitioning, which, from now on, we will call it for this reason *statistical* partitioning. Managerial criteria are thus no longer the sole basis for dividing activities into partitions, which addresses limitation 3 in Section 3.2. Nevertheless, managerial partitioning can of course still be performed in combination with the partitioning heuristic, just like for the calibration procedures.

While the set of activities to be removed from the base partition differs between the standard (only on-time points) and advanced (all points) selection strategy, the partitioning heuristic still needs to determine the sequence in which these activities are removed until a stopping criterion is met. Indeed, in contrast to the calibration procedures, the statistical partitioning heuristic needs to select which activity to eliminate in every partitioning step. The term *partitioning step* is used for an iteration of the partitioning heuristic in which one activity is removed. Thus, if there were 10 partitioning steps for a particular project or partition (under certain settings), then 10 activities were eliminated from that project or partition. For this purpose, the procedure calculates the residuals for all activities in the base partition. The residuals e_i are calculated as the deviations between the empirical values $ln(RD_i/PD_i)$ and the linear regression line of those values on the corresponding Blom scores. As a heuristic approach—hence the name statistical partitioning *heuristic*—the activity i with the biggest residual e_i in the base partition is selected for elimination (and put in partition P), since it is expected that this would yield the strongest improvement in the goodness of fit (since the created partitions will be subject to a new hypothesis test again).

4.3.2. Stopping Strategy

The selection strategy defines how the base partition is split into two different partitions by iteratively removing data points (activities) from it to create partitions L and P. Despite the fact that this selection mechanism controls the sequence of points to be removed using the calculation of the residuals, it does not define any stopping criterion during this iterative removal process. To that purpose, the statistical partitioning heuristic also introduces two different versions for the stopping strategy. When the stopping criteria are satisfied, the removal of activities is stopped, and the resulting partitions (L and P) are then the subject to a new partitioning iteration (i.e., they go back to S1 first before they possibly can be split further).

The *standard stopping strategy* employs the p-value to define the stopping criterion. More specifically, the elimination of activities stops when p reaches or exceeds the significance threshold $\alpha = 0.05$ for partition L. Since the p-value is also the condition for accepting the lognormality hypothesis in step S1, this implies that the lognormality test is automatically accepted for this partition L, and all its activities are assumed to follow the lognormal distribution. In this case, no further partitioning is necessary for partition L and all its data points are added to the database (cf. ACCEPT in Figure 5). The data points in partition P are treated differently, and the treatment depends on the option in the selection strategy. Indeed, since the partitioning heuristic is always applied anew to the newly created partitions, every partition P that is created should go back to step S1 and should be tested for lognormality if the advanced selection strategy is chosen. However, under the standard selection strategy, partition P only contains on-time points, and these points will obviously exhibit a pure Parkinson distribution. In this case, no further statistical partitioning will be performed and the data points are removed from the project (cf. STOP in Figure 5).

In the *advanced stopping strategy*, the statistical partitioning is no longer limited to the use of the p-value as the only measure for goodness-of-fit, but the activity removal halts when SE_Y (or R_a^2 as a secondary stopping criterion) does no longer improve. Indeed, it applies the standard error of the regression SE_Y as the main basis for assessing the fit, since SE_Y is the preferred measure for this according to literature. The formula for SE_Y is given below:

$$SE_Y = \sqrt{\frac{\sum_{i=1}^{n} e_i^2}{n-2}}. \tag{1}$$

The denominator is the number of activities in the partition n minus 2 since there are two coefficients that need to be estimated in our case, namely the intercept and the slope of the regression line. SE_Y is also chosen as the primary optimization criterion. By this, we mean that we deem the fit to the PDLC to be improved when the removal of the selected activity has decreased the SE_Y. Obviously, the lower the SE_Y, the better the fit. A perfect fit is obtained when all data points are on the regression line, so then all residuals are per definition zero, which, through Equation (1), implies that SE_Y is also zero in such a case. However, in about 20% of the cases, the partitioning heuristic did not reach the optimal SE_y when *only* that SE_y was considered as optimization criterion; it got stuck in a local optimum. To get out of this local optimum, we added the adjusted R^2 or R_a^2 as a secondary stopping criterion, which—although a very straightforward approach—proved to be a highly effective solution to the problem. Indeed, after adding R_a^2 as a secondary optimization criterion, only 1% of the projects did not attain their optimal SE_y. For completeness, we mention the utilized formula for R_a^2 with respect to the standard coefficient of determination:

$$R_a^2 = 1 - \frac{n-1}{n-2}(1 - R^2). \tag{2}$$

Notice that, unless $R^2 = 1$, R_a^2 is always smaller than R^2. In our context, we need to employ R_a^2 instead of R^2 to allow comparison of regression models with different numbers of observations (activities indeed get removed from the original data set). Just like for the p-value, the higher the R_a^2, the better the fit, with a maximum of 1 to reflect a perfect fit.

As mentioned before, the two settings for the stopping strategy should be used in combination with the two settings for the selection strategy, and it is important to draw the attention to the two fundamental differences with the calibration procedures. First, the treatment of the Parkinson points is fundamentally different. Recall that *all* on-time points are removed in the calibration procedures since they are assumed to be the result of the Parkinson effect. In the standard selection strategy, the procedure also removes on-time points, but it is no longer so that the only possibility is to remove *all* on-time points from the project. The partitioning heuristic allows the elimination of just a fraction of the on-time points in order to get a better fit (defined by the stopping strategy, i.e., p-value or SE_Y). The rationale behind this is that not all on-time points are necessarily the result of the Parkinson effect, as the calibration procedures implicitly assume. Some activities *are* actually on time and should thus effectively be part of partition L. Secondly, not only on-time points are removed, but also early and tardy points are now subject to removal. While the calibration procedures only remove a portion of tardy points to bring the number early, on-time and tardy points back to the original proportions, the statistical partitioning heuristic takes a different approach, and removes both early, and on-time as well as tardy points (under the advanced selection strategy) until the stopping criterion is satisfied. Such an approach creates partitions (L and P) that contain all kinds of activities (early, on-time and tardy) that must be subject to further partitioning, if necessary, and this is fundamentally different than the approach taken by the calibration procedures.

4.3.3. Solutions

In this section, we briefly come back to the discussion of the limitations of the extended calibration procedure of Section 3.2. It will be shown that all the limitations are now solved by using a combination of the two options for the selection and stopping strategies. A summary of these solutions is also given in the right column of Figure 3.

First of all, thanks to the implementation of the selection strategy, three of the six limitations have been solved, as follows:

- **Solution 1.** The calibration procedures only removed on-time (S2) and tardy (S3) activities from the project. This is no longer true in the statistical partitioning heuristic. The advanced selection strategy states that *all* activities are selectable for removal, thus also the early and tardy ones. Early activities could never be eliminated from the project in the calibration procedures.
- **Solution 4.** The calibration procedures never apply the lognormality hypothesis to the removed tardy activities (S3). However, such a test should be performed, since these tardy activities do not follow the pure Parkinson distribution like the eliminated on-time activities in S2 do. Hence, there is no reason why these tardy points should automatically be removed from the database, and, therefore, they are subject to a new hypothesis test in the statistical partitioning heuristic.

- **Solution 6.** Thanks to the use of the e_i criterion, 1000 iterations are no longer necessary (S3). Instead, the statistical partitioning heuristic always selects the exact same set of activities for elimination, since it now relies on the e_i calculations. Since calculations of residuals are invariable, the created partitions would be exactly the same for every simulation run.

Secondly, the stopping strategy has been proposed in the way as described earlier to solve two other limitations:

- **Solution 2.** The p-value is no longer the one and only measure that is applied to assess the goodness-of-fit. Instead, the advanced stopping strategy relies on two other measures—SE_Y and R_a^2—that can also be utilized to assess the goodness-of-fit.
- **Solution 5.** The Parkinson treatment of data points (S2) only allows the elimination of *all* on-time activities, whereas removing only a fraction of them could be more optimal, i.e., leading to a better fit to the PDLC.

Finally, the design of two different options (standard or advanced) for the selection and the stopping strategies is new and solves the last and most important limitation, as follows:

- **Solution 3.** The extended version of the calibration procedure added project data partitioning as a promising feature to accept lognormality, but this new feature could only be performed based on managerial criteria influenced by human judgement. The statistical partitioning heuristic has followed the same logic, but transformed it into a statistical, rather than managerial, partitioning approach. Statistical partitioning is not subject to human (mis-)judgement and not victim to human biases but does not exclude the option of human partitioning as an initialisation step (S0). In the computational experiments of Section 5, it will be shown that human and statistical partitioning lead to a higher acceptance rate of project data.

5. Computational Results

This section shows the results of a set of computational experiments on the same set of projects as used in [20]. All projects are taken from the database of [21] which consisted—at the time of introducing this database—of 51 projects. Additional projects have been added later, and has resulted in a database of 125 projects from companies in Belgium. Twenty-eight projects did not contain authentic time tracking data, and were removed from the analysis (97 left), and 14 projects only contained activities that ended exactly on time (which are assumed to be subject to the Parkinson effect). Hence, 83 remaining projects were used in the extended calibration study and will also be used in the computational experiments of the current paper. The average values for six summary statistics of these 83 projects were published in the extended calibration procedure study and are therefore not repeated here. However, Figure 6 displays a summary of the 83 projects used for the analysis. The top graph shows that more than 70% of the projects come from the construction industry, followed by almost 25% IT projects. The bottom graph displays the real time/cost performance of the projects. The graph shows that the database does not contain projects in the bottom right quadrant (over budget and ahead of schedule), but the three other quadrants contain projects with different degrees of earliness/lateness and budget underruns and overruns.

The results of our computational experiment are divided between three sections. In Section 5.1, all projects are used to test the statistical partitioning heuristic without using managerial partitioning, while Section 5.2 makes use of a subset of these projects, now also adding managerial partitioning to the tests. Finally, Section 5.3 is added with a list of limitations of the statistical partitioning heuristic that can be used as guidelines for future research in this domain.

First of all, it is very important to note that the statistical partitioning heuristic still relies on the p-value to determine whether or not a certain partition follows the PDLC. The reason for this is twofold. First, it allows us to compare the results of the partitioning heuristic to those of the calibration procedures—in which p was the only goodness-of-fit measures that was considered. In addition, second, the only other eligible measure SE_Y does not provide a uniform basis for comparison between projects or partitions, as its numerical value strongly depends on—and can thus vary greatly with—the input values from the data set (i.e., the $ln(RD_i/PD_i)$ values). In other words, no universal fit threshold can be set for SE_Y. This also explains why we will focus more on the p-values than on the SE_Y results in upcoming discussions.

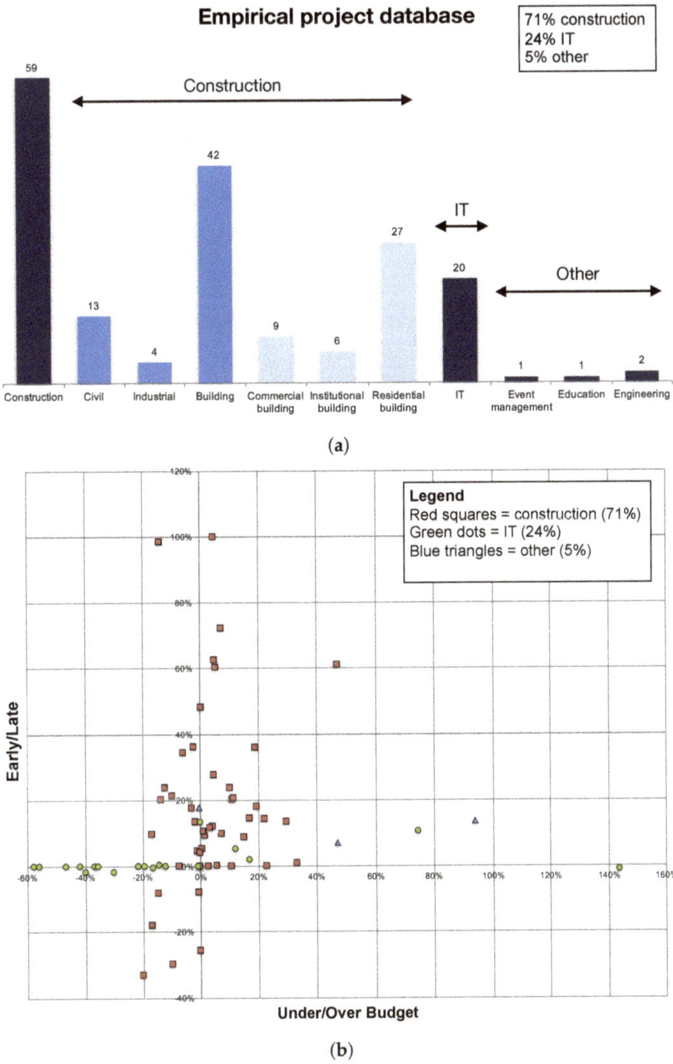

Figure 6. Empirical project database used for the analysis. (**a**) Sector of the 83 projects (mainly construction projects); (**b**) Project time/cost performance.

Secondly, it should also be stressed that SE_Y always remains the main stopping criterion when applying the partitioning heuristic under the advanced stopping strategy. Therefore, we did not include the R_a^2 values in the two tables with computation results, since they are only of secondary importance. Average SE_Y values are mentioned in the tables because of their prime role in the stopping strategy of the statistical procedure.

Finally, we consider eight different settings for the statistical partitioning heuristic, and, since each of them can be performed with or without managerial partitioning, the results had to be divided over two tables. Table 1 shows the outcomes for the application of the statistical partitioning heuristic to our database under the eight different settings without using human partitioning as an initialization step. A second table will show similar results, but now adding a human partitioning step prior to the statistical partitioning steps (Table 2). The eight settings reflect the choices that must be made for hypothesis testing (Section 4.2) and for the selection and stopping strategies of Sections 4.3.1 and 4.3.2. Each choice can be set to either 0 or 1. To represent these different settings in Tables 1 and 2, the code format *rounding–selection–stopping* is introduced as follows:

- The hypothesis test (S1) can be performed with (1) or without (0) rounding (S4), and will further be abbreviated as *rounding* = 0 or 1.
- The selection strategy can be set to standard (0) or advanced (1), and will be abbreviated as *selection* = 0 or 1.
- The stopping strategy can also be set to be standard (0) or advanced (1), abbreviated as mboxemph*stopping* = 0 or 1.

As a result, the eight settings for the parameters (*rounding–selection–stopping*) are then equal to (0-0-0), (1-0-0), (0-0-1), (1-0-1), (0-1-0),(1-1-0), (0-1-1), (1-1-1).

5.1. Without Managerial Partitioning

Table 1 displays the results for the statistical partitioning heuristic without managerial partitioning under the eight different settings. The table is split up in four main rows ((*a*) to (*d*)), and will be explained along the following lines.

Table 1. Results for the partitioning heuristic without managerial partitioning.

| | | Partitioning Setting (Rounding–Selection–Stopping) | | | | | | | |
		(0-0-0)	(1-0-0)	(0-0-1)	(1-0-1)	(0-1-0)	(1-1-0)	(0-1-1)	(1-1-1)
(*a*)	# partitions (total)	83	83	83	83	195	145	249	215
	# partitions (avg/p)	-	-	-	-	2.3	1.7	3.0	2.6
	# partitions (max)	-	-	-	-	5	3	6	5
	1 partition [%]	-	-	-	-	13	36	4	6
	2 partitions [%]	-	-	-	-	51	53	25	42
	3 partitions [%]	-	-	-	-	25	11	47	40
	4 partitions [%]	-	-	-	-	10	0	17	11
	5 partitions [%]	-	-	-	-	1	0	6	1
	6 partitions [%]	-	-	-	-	0	0	1	0
(*b*)	# partitioning steps	2566	2177	2771	2634	1361	365	1705	771
	/project	31	26	33	32	16	4	21	9
(*c*)	% act / partition L	62	73	54	59	-	-	-	-
	% act / partition P	38	27	46	41	-	-	-	-
(*d*)	avg. $\overline{SE}_Y$	0.271	0.229	0.250	0.212	0.257	0.191	0.264	0.139
	avg. p	0.075	0.193	0.280	0.479	0.219	0.362	0.461	0.731
	accepted partitions [%]	61	72	61	72	90	95	86	94

(*a*) **# partitions:** This part displays the number of created partitions (total, average per project and maximum) as well as the percentage of projects with one up to six created partitions. All 83 available projects are considered for every setting of the partitioning heuristic. The total number of activities over these projects amounts to no less then 5068 activities (or an average of 61 activities per project), which can be deemed quite an extensive empirical dataset. Remark that the total number of partitions is equal to the number of considered projects for the settings with *selection* = 0 (shown in the first four (-, 0 ,-) settings). Indeed, when only on-time points can be eliminated, partition P per definition follows a pure Parkinson distribution and should therefore not explicitly be considered. We thus only look at partition L for evaluating the partitioning heuristic with *selection* = 0. When *selection* = 1 (shown in the last four columns), on the other hand, the partitions created by removing any (i.e., not necessarily on-time) activity from the initial project do no longer trivially adhere to the pure Parkinson distribution. Therefore, all created partitions are considered explicitly in these cases. This explains why the number of partitions in Table 1 is bigger than 83 for settings with *selection* = 1.

The row with the average number partitions per project (avg/p) also shows interesting results. In contrast to the situation where *selection* = 0, there can be more (or less) than two partitions when *selection* is set to 1. There is a logical correspondence between the average number of partitions and the average number of partitioning steps per project (part (*b*) of the table). Indeed, the more partitioning steps that are executed, the greater the chance that an extra partition is created. As such, setting (0-1-1), which exhibited the highest number of partitioning steps for *selection* = 1 (1705), also yields the most partitions per project, namely three on average. The minimum is observed for setting (1-1-0) (1.7 partitions per project), which also clearly showed the least partitioning steps (365). Notice that this minimum is less than 2, which means that, under this setting, there are a lot of projects for which the PDLC is accepted (i.e., $p > 0.05$) even without elimination of a single activity, so that all activities fit the proposed distribution as a whole. This is largely due to the beneficial influence of accounting for the rounding effect through the appropriate averaging of Blom scores. When we want to optimize the fit (i.e., further decrease SE_Y), however, activities will need to be eliminated, thus producing at least one extra partition. This explains why, for the setting (1-1-1), there is on average almost one partition more per project than for setting (1-1-0) (2.6 compared to 1.7).

Furthermore, the maximum number of partitions over all projects is also displayed in Table 1, together with the grouping of the projects according to the number of partitions in which they are divided by executing the partitioning heuristic under different settings. A maximum of six partitions—which is in itself still not too much to become inconvenient to work with—only occurs for one project under setting (0-1-1). This is also the only setting for which there are more projects with three partitions than there are with two partitions, the latter clearly being the most common case and in correspondence with the situation where *selection* = 0 (with per definition only one partition L and one partition P).

(*b*) **# partitioning steps:** When further going down the rows in the table, we see that settings with *selection* = 1 require significantly fewer partitioning steps than settings with *selection* = 0. This means that a potential fit can be obtained much faster by allowing all activities (i.e., early, on-time and tardy) to be removed from the base partition, which indicates a first advantage of the partitioning heuristic with respect to the calibration procedures. For setting (1-1-0), for example, an average project only needs four partitioning steps. Obviously, when the advanced stopping strategy is used (*stopping* = 1), the number of necessary partitioning steps increases from 4 to 9. Conversely, accounting for rounding (*rounding* = 1) appears to have a decreasing effect on the required number of partitioning steps, i.e., from 16 to 4 and from 21 to 9, which is assumed to be a positive effect given that a lower number of partitions means bigger clusters of data with similar characteristics.

(*c*) **% activities / partition:** For *selection* = 0, we observe that partition L of an average project comprises between half (54%) and three quarters (73%) of the total activities, depending on the other selected options. This implies that up to about half of the activities (46%) were removed from the base partition and put in partition P (for setting (0-0-1)), which is quite a considerable portion provided that all these eliminated activities had to be on time. This indicates that a great part of the activities of the considered real-life projects were reported as being on time, which supports the existence of the Parkinson effect (and the rounding effect in second instance) and therefore the relevance of the applied methodologies (i.e., the calibration procedures and the partitioning heuristic to validate the PDLC). Note that no values are reported for the settings with *selection* = 1 since, in these cases, even the partition P is subject to further hypothesis testing, possibly resulting in several new partitions.

The division of these partitions into new partitions until the stopping criterion is met is shown by the values for the % activities in each partition under part (*a*) of this table.

(*d*) **Goodness of fit:** More importantly, one can observe that the setting (1-1-1) clearly yields the biggest *p*-value and thus the best fit to the PDLC. This *p*-value is significantly larger than that of the optimum for the extended calibration procedure when no managerial partitioning is executed (0.731 $>>$ 0.385; the latter value is not shown in the table but is the maximum value of the extended calibration procedure found in Table 2 of [20]), and even larger than the overall optimum that occurs when applying initial partitioning according to RP and S4 (0.731 $>$ 0.606; the latter value is the overall maximum *p*-value found in the previously mentioned study). It can thus already be stated that the statistical partitioning heuristic performs better than the extended calibration procedure, also by comparing the percentages of accepted partitions (or projects) without execution of managerial partitioning (maxima: 95% $>$ 81% for the extended calibration procedure). Moreover, accounting for the rounding effect (*rounding* $=$ 1) always appears to be beneficial for the validation chance of the PDLC. Similarly, there is a clear advantage of allowing every activity to be eliminated (*selection* $=$ 1) instead of only the on-time points (*selection* $=$ 0), supported by both *p*-values and accepted partitions' percentages.

We now mention a couple of qualitative reasons why a better performance is observed for *selection* $=$ 1 than for *selection* $=$ 0. First of all, the biggest residual in a certain partitioning step will always be at least as big—and most likely bigger—in the former case than in the latter, since the algorithm can choose from *all* activities when *selection* $=$ 1 and not just from the on-time fraction. Eliminating an activity with a bigger residual means a stronger decrease of SE_Y and thus a faster evolution towards the acceptance of the PDLC. This also explains why *selection* $=$ 1 requires fewer partitioning steps than *selection* $=$ 0, as mentioned earlier.

Secondly, although Table 1 did not yet consider managerial partitioning, there is statistical partitioning when setting *selection* to 1. This means that—in contrast to what is the case for the calibration procedures or when putting *selection* to zero—the early and tardy activities that show very diverse characteristics for their durations can now be assigned to different partitions for which specific distribution profiles can be defined, instead of obstinately trying to fit a single distribution profile to a set of activities that are just too heterogeneous. A good illustrative example is given by the detection of clear outliers in the project data discussed in [20] while validating their extended calibration procedure. These authors propose two straightforward criteria to select outliers, and compare their approach with the approach taken in the empirical validation of the original calibration procedure [19]. In their empirical validation of the original calibration procedure, the authors eliminated 66 activities from the set of projects as clear outliers, but they did not explicitly state how they did this. Using the two proposed criteria to detect outliers for the extended calibration procedure has resulted in the detection of the same 66 activities as being clear outliers, except for one project. This project (ID C2014-03) also had clear outliers when these two new criteria were used, but these outliers were not detected in the first empirical validation study. In the extended calibration study, it was therefore argued that failing to identify and eliminate clear outliers could lead to serious distortions in the results as a motivation for why the two criteria should always be strictly applied. This is, however, is no longer as valid as it was when the statistical partitioning heuristic was used. Using the newly proposed selection and stopping strategies, non-removed outliers would obviously exhibit the biggest residuals and thus automatically be put in a separate partition and could then no longer impede the validation of the PDLC for the other activities (and the resulting partition should be automatically removed from the project database). This also implies that it would no longer be a huge problem to not identify and eliminate the clear outliers beforehand, since the procedure would do this automatically when *selection* $=$ 1. The partitioning heuristic therefore becomes less prone to human error and prevents biased outcomes resulting from such errors, which of course is an advantage of the partitioning heuristic with respect to the calibration procedures and supports the applicability and robustness of the former.

5.2. With Managerial Partitioning

Table 2 presents more similar results than the previous table, but now with the managerial partitioning step as an initialization carried out prior to the statistical partitioning algorithm. The table no longer considers all eight settings for the statistical partitioning heuristic, but fixes the *rounding* value to 1 because this was shown to have a positive effect on both the partitioning efficiency (i.e., fewer partitioning steps) and, foremost, goodness-of-fit (i.e., higher *p*-value). In addition, the *stopping* option is also fixed to 1, since this obviously produces the better *p*-values compared to *stopping* = 0. Moreover, the former setting in fact incorporates the latter, since, up to the point where p becomes greater than 0.05, both approaches run completely parallel. In contrast, the value for the *selection* option is not fixed, since the experiments are set up to assess its influence in combination with managerial partitioning. The settings that are included in Table 2 are thus reduced to (1-0-1) and (1-1-1). Although Table 2 (with managerial partitioning) contains more information than Table 1 (without managerial partitioning), the former will be discussed less extensively than the latter, as many aspects have already been addressed. Rather, we now focus on the most notable results and differences.

Table 2. Results for the partitioning heuristic with managerial partitioning.

| | | Partitioning Setting (Rounding–Selection–Stopping) | | | | | | | |
| | | (1-0-1) | | | | (1-1-1) | | | |
		PD (x4)	PD (x5)	WP	RP	PD (x4)	PD (x5)	WP	RP
(*a*)	# projects	83	83	53	21	83	83	53	21
	avg. # activities	61	61	72	42	61	61	72	42
	tot. # activities	5068	5068	3796	887	5068	5068	3796	887
(*b₁*)	# partitions (human)	232	213	426	65	232	213	426	65
	# partitions (avg/p)	2.8	2.6	8.0	3.1	2.8	2.6	8.0	3.1
	# partitions (max)	4	4	26 *	6	4	4	26 *	6
	1 partition [%]	4	6	36	0	4	6	36	0
	2 partitions [%]	32	40	45	24	32	40	45	24
	3 partitions [%]	45	46	8	52	45	46	8	52
	4 partitions [%]	19	8	7	19	19	8	7	19
	5 partitions [%]	0	0	2	0	0	0	2	0
	6 partitions [%]	0	0	2	5	0	0	2	5
(*b₂*)	# subpartitions (statistical)	-	-	-	-	423	399	631	117
	# subpartitions (avg/p)	-	-	-	-	5.1	4.8	11.9	5.6
	# subpartitions (max)	-	-	-	-	4	4	5	4
	1 subpartition [%]	-	-	-	-	40	37	59	34
	2 subpartitions [%]	-	-	-	-	40	41	35	54
	3 subpartitions [%]	-	-	-	-	18	19	4	11
	4 subpartitions [%]	-	-	-	-	2	3	1	1
	5 subpartitions [%]	-	-	-	-	0	0	1	0
(*c*)	tot. # partitioning steps	2150	2246	835	348	689	751	555	182
	/project	26	27	16	17	8	9	10	9
(*d*)	% act. partition L	79	78	90	77	-	-	-	-
	% act. partition P	21	22	10	23	-	-	-	-
(*f*)	avg. SE_Y	0.161	0.171	0.196	0.101	0.108	0.130	0.146	0.088
	avg. p	0.614	0.589	0.658	0.741	0.774	0.756	0.783	0.811
	accepted (sub)partitions [%]	88	85	92	95	97	94	97	97

* For partitioning criterion WP, a different scale applies for the next six rows: 1 / 2 / 3 / 4 / 5 / 6 partition(s) should be regarded as 1-5/6-10/11-15/16-20/21-25/26-30 partitions, respectively.

(*a*) **# Projects:** A first difference is the number of projects that are considered. This is no longer always 83 because, for some projects, the two of the three criteria for managerial partitioning were not defined by the project manager (i.e., the WPs and/or RPs of the activities were not known, cf. S0 of Section 3.1). The total number of activities that are considered is thus also less than 5068 for WP and RP as partitioning criteria, however, still adequate with a total number of activities of 3796 and 887.

(*b*) **# partitions:** The number of partitions (human) displayed in the table reflects the number of partitions that are created by performing managerial partitioning according to the different criteria. This is the initial partitioning operation (i.e., before executing the actual partitioning heuristic), and obviously yields the same partitions for both *selection* values. On the other hand, subpartitions are created by performing statistical partitioning and are therefore only present when *selection* = 1. In that case, each of the partitions obtained from managerial partitioning is further divided into smaller partitions—therefore called *subpartitions*—using the statistical partitioning heuristic. This means that each project in fact goes through two consecutive partitioning phases when the partitioning heuristic is applied with setting (1-1-1) and including managerial partitioning. The number of subpartitions is obviously larger than the number of partitions, and even reaches 631 over 53 projects for the WP criterion. This comes down to almost 12 subpartitions per project, which might be a bit much to be practical and less relevant since this implies an average of only six activities per subpartition. However, this is not a problem when one of the other managerial criteria is applied, with an average of about five subpartitions per project. The main reason is that project managers apparently define way too much WPs, on average eight per project, with an excessive maximum of 26 WPs for one project. This issue could be resolved by stimulating project managers to limit the number of identified WPs through consideration of higher-level classification criteria.

(*c*) **# partitioning steps:** The number of partitioning steps do not fundamentally differ between the two tables and the table still shows that the setting with *selection* = 1 requires significantly fewer partitioning steps than the setting with *selection* = 0. Furthermore, the introduction of managerial partitioning does not seem to increase the average number of partitioning steps (this remains about 9 (between 8 and 10) for (1-1-1) like in Table 1), which means that the computational effort to partition the data remains just as low.

(*d*) **% activities / partition:** The percentage of activities per partition differs between the two tables. For the setting with *selection* = 0, partition L on average still comprises about 80% (between 77% and 79% as shown in row '% act partition L') of the initial activities, and even 90% for the WP criterion. This is much more than the 59% for (1-0-1) without managerial partitioning from Table 1. Hence, in order to obtain a fit to the PDLC, a far smaller portion of (on-time) activities needs to be removed from the managerial partitions than was the case for the complete project. This indicates that the application of managerial partitioning criteria is indeed relevant and beneficial, and that the definition of them by project managers should thus be stimulated.

(*f*) **Goodness of fit:** The absolute best fit so far in this research is obtained by applying the partitioning heuristic with setting (1-1-1) in combination with managerial partitioning according to the criterion that already proved most profitable in an earlier study, namely RP. The average p-value of 0.811 is significantly higher than the maximum for the extended calibration procedure, which is 0.606 for partitioning step S4 preceded by managerial partitioning according to—also—RP. The percentage of accepted partitions is equal and very high (97%) for both, so we can conclude that the partitioning heuristic outperforms the calibration procedures, regardless even of its qualitative benefits concerning flexibility and robustness. Therefore, we will no longer consider the (extended) calibration procedure in the rest of the discussion.

However, the mentioned p-value of 0.811 is not exceedingly higher than that for partitioning setting (1-1-1) combined with either of the other managerial criteria (p ranging from 0.756 to 0.783) or even without managerial partitioning ($p = 0.731$; see Table 1), and also the partitioning setting (1-0-1) combined with managerial partitioning according to—again—RP comes close with a p of 0.741. The reason for this is that a combination of managerial partitioning and statistical partitioning (which occurs when *selection* = 1) should in fact be seen as a 'double' optimization. Both partitioning approaches already perform very well separately, but combining them takes the distribution fitting another (small) step closer to 'optimal' partitioning. Furthermore, managerial and statistical partitioning do not only perform well on their own; they are mutually also quite comparable. To show this, we need to compare the partitioning heuristic with setting (1-1-1) (so without advanced statistical partitioning) and no managerial partitioning (see Table 1) and that with setting (1-0-1) (so without advanced statistical partitioning) and managerial partitioning according to RP (see Table 2). Remarkably, both exhibit almost identical p-values (0.731 versus 0.741) and accepted partitions percentages (94% versus 95%). This observation is in fact hugely promising, as it indicates that we can just perform the partitioning heuristic with inclusion of the statistical partitioning (i.e., set *selection* to 1) and still obtain very relevant partitions without requiring realistic input for managerial criteria (i.e., WPs or—even better—RPs accurately defined by the project manager). Statistical partitioning is no longer—or at least far less—prone to human judgement and bias than managerial partitioning. In the latter case, project managers indeed need to *accurately* define the WPs or RPs, otherwise the resulting partitions would be faulty and unrealistic anyhow. It might be beneficial to bypass this uncertain human factor, and thus create a more solid and trustworthy methodology for categorizing activities into risk classes and assigning specific distribution profiles to them. The partitioning heuristic developed in this section allows just this. Apart from the discussion of either managerial or statistical partitioning (or both) being preferred, our results clearly show that it is essential to create partitions for a project in order to obtain decent fits of the activity durations to the PDLC.

5.3. Limitations

Notwithstanding the substantial improvements of the statistical partitioning heuristic with respect to the calibration procedures, some extensions to the procedure itself and to the related research could still be made in the future. We now present a few limitations of the current research and propose several potential advances that could be made in these areas.

1. The statistical partitioning heuristic—as the name itself indicates—is still a heuristic and therefore produces good but not (always) optimal results. Indeed, removing the activities with the biggest residuals e_i as long as the SE_Y of the considered base partition (put in partition L) decreases is a plausible and logical approach. However, it is not optimal for multiple reasons. First, it is no certainty that the biggest residual always designates the best activity to eliminate (i.e., which produces the biggest decrease of SE_Y). Second, it is not assessed what the future impact (i.e., over multiple partitioning steps) of this removal would be on the remaining activities in partition L (e.g., maybe it would be more optimal to remove two other high-residual activities instead of that with the biggest residual, but the algorithm does not analyse this). In addition, third, when removing an activity from partition L, it becomes part of another partition (i.e., partition P), but we do not check the influence of this operation on partition P (for all we know, it could deteriorate the SE_Y there). In addition, then there still is the issue of SE_Y being susceptible to lapse into a local optimum, which we now—also heuristically—addressed by considering R_a^2 as a secondary optimization criterion. The ultimate goal would be to develop an algorithm that divides the activities of a project into partitions that all pass the lognormality test (with possible exception of some clear outliers), and moreover, show an average SE_Y over all partitions that is as low as possible (or a p-value that is as high as possible). The advanced algorithm could, for example, contain a fine-tuning stage in which activities can be shifted from one partition to another in order to further improve the overall SE_Y or p. Furthermore, a limit could be set for the minimal allowed partition size, to make the partitions themselves more meaningful and comparison with partitions from similar future projects more workable. We have now set the minimum size of each partition arbitrarily to 3.

2. The employed project database is large for an empirical data set, but still rather limited in comparison to simulation studies using artificial project data. Therefore, the database should ever be further expanded, so that future empirical studies based on it can keep increasing their validity and generalizability.

3. Currently, we only considered the initial partitioning according to one managerial criterion at a time. This could be extended to the application of multiple consecutive criteria. For example, the PD criterion could be performed after the project has already been partitioned according to

RP. In that way, we get even more specific partitions that should exhibit activities that are more strongly related. Furthermore, the extra managerial partitioning could be applied together with or instead of the statistical partitioning (i.e., if *selection* = 1).

4. Furthermore, the managerial partitioning criteria should not stay limited to PD, WP and RP. These are perhaps some of the most obvious and logical criteria, but there can still be others that might show even greater distinctive power for dividing a project into adequate partitions. These extra managerial partitioning criteria could be harvested from more empirical research, for example, into the drivers of project success. If those drivers could be reliably identified for a particular kind of project, they could also provide a good basis for grouping similar activities that thus show similar risks (and should therefore belong to the same partition).

6. Conclusions

Studies have shown that, just like any physical system, projects have entropy that must be managed by spending energy, and this process of energy is called project management. In order to manage the project uncertainty, accurate estimates for activity duration are crucial in order to make informed decisions. This paper presents a new statistical method to better estimate the average and variability of the activity duration distributions in order to help project manager to better manage the project uncertainty (entropy) with the lowest possible effort (energy).

The new statistical calibration method extends two existing calibration methods using an automatic partitioning heuristic. The main objective of such an extension is to improve the ability to define distribution profiles for a project's activity duration that represent as accurately as possible the stochastic nature of the activities. The underlying assumption is that the lognormal distribution is the most appropriate distribution for modelling activity durations, but the parameters for this distribution cannot be easily extracted from empirical data due to hidden earliness and rounded values for the reported activity durations. These procedures were utilized as a starting point for developing a much more extensive calibration procedure, which has programmed in $C++$ and empirically validated on the dataset consisting of more than 5000 activities. These input data come from the real-life project database created by [21] and is freely available at www.or-as.be/research/database.

The previous calibration methods have shown promising results, but also some limitations, and these are also discussed in the current study. First, the original calibration procedure of [19] did not allow the project to be divided into partitions of activities that intrinsically adhere to the same distribution profile. For this reason, [20] have proposed an extended calibration method by introducing the ability of managerial partitioning using human input such as planned duration, the structure of the work breakdown structure or the risk profiles defined for each activity. This extended calibration method proved extremely favourable and confirmed that partitioning is a promising direction for proving the realism of the lognormal distribution for activity duration. Despite this improvement, managerial partitioning is based on criteria defined by the project manager, and, as the project manager is a human being, these criteria are susceptible to bias in human judgement.

To bypass this problem, we developed a completely new approach in the current study which we called the *statistical partitioning heuristic*. It is foremost a *statistical* procedure in contrast to the managerial procedure that requires human input. Moreover, the *partitioning* approach, which was shown to be promising in the extended calibration study, is kept as a *heuristic* tool (i.e., there are other ways of doing the partitioning) in the best possible—but not necessarily optimal—way. Consequently, in statistical partitioning, well-chosen activities that do not fit within a certain partition are eliminated from that partition and assigned to another, which is then also adapted until a fit is reached. The results obtained from this are very good, and almost perfectly match those from performing managerial partitioning in the extended calibration method.

This observation is certainly advantageous, as it suggests that equally adequate partitions can be obtained through the proposed statistical procedure without being susceptible to human bias or, moreover, requiring the definition of managerial criteria. Since project managers are now always able, or willing, to define values for the managerial criteria for all activities, an automatic procedure can replace their cumbersome task. It is therefore advised to perform the statistical partitioning heuristic with the incorporation of advanced selection and stopping strategies for receiving the most appropriate and trustworthy distribution profiles for the activity durations. However, when it is certain that the managerial criteria have been properly defined, managerial partitioning can be executed in combination with (in fact, prior to) the statistical partitioning. Despite the promising results in this study, future research topics can be derived from Section 5.3, since addressing the limitations of the current automatic partitioning heuristic could indeed further advance our research.

Author Contributions: M.V. and J.B. conceived and designed the experiments, performed the experiments and analyzed the data; M.V. wrote the paper.

Funding: This research received no external funding.

Conflicts of Interest: The authors declare no conflict of interest.

References

1. Uyttewaal, E. *Dynamic Scheduling With Microsoft Office Project 2003: The Book by and for Professionals*; International Institute for Learning, Inc.: New York, NY, USA, 2005.

2. Vanhoucke, M. *Project Management with Dynamic Scheduling: Baseline Scheduling, Risk Analysis and Project Control*; Springer: New York, NY, USA, 2012; Volume XVIII.

3. Vanhoucke, M. *Integrated Project Management and Control: First Comes the Theory, Then the Practice*; Management for Professionals; Springer: New York, NY, USA, 2014.

4. Shannon, C.E. A Mathematical Theory of Communication. *Bell Syst. Tech. J.* **1948**, *27*, 379–423, doi:10.1002/j.1538-7305.1948.tb01338.x. [CrossRef]

5. Vanhoucke, M. Using activity sensitivity and network topology information to monitor project time performance. *Omega Int. J. Manag. Sci.* **2010**, *38*, 359–370. [CrossRef]

6. Vanhoucke, M. On the dynamic use of project performance and schedule risk information during project tracking. *Omega Int. J. Manag. Sci.* **2011**, *39*, 416–426. [CrossRef]

7. Martens, A.; Vanhoucke, M. The impact of applying effort to reduce activity variability on the project time and cost performance. *Eur. J. Oper. Res.* **2019**, to appear. [CrossRef]

8. Bushuyev, S.D.; Sochnev, S.V. Entropy measurement as a project control tool. *Int. J. Proj. Manag.* **1999**, *17*, 343–350. [CrossRef]

9. Chenarani, A.; Druzhinin, E. A quantitative measure for evaluating project uncertainty under variation and risk effects. *Eng. Technol. Appl. Sci. Res.* **2017**, *7*, 2083–2088.

10. Tseng, C.C.; Ko, P.W. Measuring schedule uncertainty for a stochastic resource-constrained project using scenario-based approach with utility-entropy decision model. *J. Ind. Prod. Eng.* **2016**, *33*, 558–567. doi:10.1080/21681015.2016.1172522. [CrossRef]

11. Cosgrove, W.J. Entropy as a measure of uncertainty for PERT network completion time distributions and critical path probabilities. *Calif. J. Oper. Manag.* **2010**, *8*, 20–26.

12. Asllani, A.; Ettkin, L. An entropy-based approach for measuring project uncertainty. *Acad. Inf. Manag. Sci. J.* **2007**, *10*, 31–45.

13. Williams, T. A classified bibliography of recent research relating to project risk management. *Eur. J. Oper. Res.* **1995**, *85*, 18–38. [CrossRef]

14. Willems, L.; Vanhoucke, M. Classification of articles and journals on project control and Earned Value Management. *Int. J. Proj. Manag.* **2015**, *33*, 1610–1634. [CrossRef]

15. Trietsch, D.; Mazmanyan, L.; Govergyan, L.; Baker, K.R. Modeling activity times by the Parkinson distribution with a lognormal core: Theory and validation. *Eur. J. Oper. Res.* **2012**, *216*, 386–396. [CrossRef]

16. Malcolm, D.; Roseboom, J.; Clark, C.; Fazar, W. Application of a technique for a research and development program evaluation. *Oper. Res.* **1959**, pp. 646–669. [CrossRef]

17. AbouRizk, S.; Halpin, D.; Wilson, J. Fitting beta distributions based on sample data. *J. Constr. Eng. Manag.* **1994**, *120*, 288–305. [CrossRef]

18. Williams, T. Criticality in stochastic networks. *J. Oper. Res. Soc.* **1992**, *43*, 353–357. [CrossRef]

19. Colin, J.; Vanhoucke, M. Empirical perspective on activity durations for project management simulation studies. *J. Constr. Eng. Manag.* **2015**, *142*, 04015047. [CrossRef]

20. Vanhoucke, M.; Batselier, J. Fitting activity distributions using human partitioning and statistical calibration. *Comput. Ind. Eng.* **2019**, *129*, 126–135. [CrossRef]

21. Batselier, J.; Vanhoucke, M. Construction and evaluation framework for a real-life project database. *Int. J. Proj. Manag.* **2015**, *33*, 697–710. [CrossRef]

22. Blom, G. *Statistical Estimates and Transformed Beta-Variables*; John Wiley & Sons: Hoboken, NJ, USA, 1958.

23. Looney, S.W.; Gulledge, T.R., Jr. Use of the correlation coefficient with normal probability plots. *Am. Stat.* **1985**, *39*, 75–79.

24. Batselier, J.; Vanhoucke, M. Empirical Evaluation of Earned Value Management Forecasting Accuracy for Time and Cost. *J. Constr. Eng. Manag.* **2015**, *141*, 1–13. [CrossRef]

Article

Demand Forecasting Approaches Based on Associated Relationships for Multiple Products

Ming Lei [1,*], Shalang Li [2] and Shasha Yu [1]

[1]   Guanghua School of Management, Peking University, Beijing 100871, China; yuss0505@pku.edu.cn
[2]   Penghua Fund Management Co., Ltd., Shenzhen 518048, China; lishalang_leee@pku.edu.cn
*   Correspondence: leiming@gsm.pku.edu.cn; Tel.: +86-1891-132-5315

Received: 15 August 2019; Accepted: 2 October 2019; Published: 5 October 2019

Abstract: As product variety is an important feature for modern enterprises, multi-product demand forecasting is essential to support order decision-making and inventory management. However, these well-established forecasting approaches for multi-dimensional time series, such as Vector Autoregression (VAR) or dynamic factor model (DFM), all cannot deal very well with time series with high or ultra-high dimensionality, especially when the time series are short. Considering that besides the demand trends in historical data, that of associated products (including highly correlated ones or ones having significantly causality) can also provide rich information for prediction, we propose new forecasting approaches for multiple products in this study. The demand of associated products is treated as predictors to add in AR model to improve its prediction accuracy. If there are many time series associated with the object, we introduce two schemes to simplify variables to avoid over-fitting. Then procurement data from a grid company in China is applied to test forecasting performance of the proposed approaches. The empirical results reveal that compared with four conventional models, namely single exponential smoothing (SES), autoregression (AR), VAR and DFM respectively, the new approaches perform better in terms of forecasting errors and inventory simulation performance. They can provide more effective guidance for actual operational activities.

Keywords: demand forecasting; multiple products; granger causality; correlation; inventory performance

---

1. Introduction

Demand forecasting, a prerequisite for inventory decision-making, plays a vital role in supply chain management. How to improve prediction accuracy has always been the focus of academic circles and enterprises. With the increasingly fierce competition in business, product variety has become an important feature of modern enterprises, which can contribute to meet diverse needs of customers and occupy more market segments [1]. However, many products, where 'many' means hundreds or thousands, bring about a new challenge to demand forecasting. Traditional time series algorithms cannot well adapt to the complex high- or even ultra-high dimensionality, resulting in inferior predictive effectiveness in multi-product scenarios.

It is worth noting that the demand of multiple products is not completely isolated, but rather complex relationships exist between them. According to the relevant literature, there are two common association relationships between different products: correlation and Granger causality. For example, the demand for complementary products is highly correlated, having contemporaneous influence with each other [2]. Materials used in engineering projects have a clear sequence, so Granger causality exists in their demand [3]. Obviously, capturing and making full use of such potential information can be helpful to obtain more accurate prediction results. What's more, when the time series are short, historic trend cannot provide enough information for future demand. Associated relationships can make up for the defects and reduce the bias of prediction. However, as far as we know, there is currently little research taking into account association relationships between products in demand forecasting. In this paper, we incorporate associated relationships among products into the forecasting framework to construct a more accurate prediction approach.

In previous literature, there are mainly three branches of forecasting models for multi-dimensional time series. The first one is a series of statistical methods, represented by the AIRMA model and its extended versions, including VARMA, VARs, BVAR, etc. [4–13]. They treat multi-dimensional time series as an endogenous system. Target variables are regressed by lag items of all series, considering their relations generally. With the development of econometrics, VARs with different settings are widely applied. For example, [8] proposed five types of VAR and utilized industrial production data from OECD countries to test their forecasting effect. The key defect of VAR is that the number of estimated parameters increases exponentially along with the increase in dimensions. For high-dimensional time series, it is easy to cause overfitting, weakening the prediction ability outside the original sample. Some scholars have assumed that estimated parameters obey a specific prior distribution to reduce their number, i.e., BVAR, applied in macro-economic forecasting [14–17], market share forecasting [12] and business forecasting [10]. Some other scholars incorporated some unmodeled predictors from exogenous variables to improve original regression models. For example, [18] integrated intra- and inter-category promotional information to construct multistage LASSO regression to forecast the demand of 3222 products. The results are significantly better than the model only using endogenous variables. Unfortunately, these methods can alleviate but not completely solve the problem of overfitting. Accurate results can be obtained only when the time series is long enough.

The second strand of research concentrates on processing high-dimensional time series through the method of dimension reduction, represented by dynamic factor model (DFM). [19] holds the belief that a small number of latent factors are able to interpret fluctuations of observed macroeconomic indices. As long as these potential factors can be portrayed accurately, the task of forecasting is simplified substantially and precise results are achievable. There are many algorithms for the estimation of dynamic factors, including maximum likelihood [19–22], principle component analysis (PCA) [23–32], and data shrinking methods [33,34]. As for prediction accuracy, [35] collected relevant literature and confirmed that DFM performs better than single time series prediction models through a meta-analysis method. Reference [36] pointed out that a simple AR model may be better than a DFM model when there is a large structural change in the data. Compared with other high-dimensional time series models like shrinkage forecasts, FDM is also superior [37]. What's more, [38] introduced U.S. macroeconomic data to compare two forms of DFM estimation methods [39,40]. The results demonstrated that their forecasting precision is not significantly different. However, DFM has the obstacle to tackle sophisticated high-dimensional time series, due to the existence of some isolated series, and the same is true for ultra-high dimensional time series. More specifically, some unique information may be skipped in the process of selecting a limited number of factors, leading to inefficient estimates. If add more dynamic factors, it will fall into the over parameterized problem again.

As the development of artificial intelligence, various machine learning models have been widely used in the area of forecasting, including neural network [41–43], support vector machine [44–46], nearest neighbor regression [47,48], and so on. They are serious contenders to classical statistical models and form a vital research branch. Different from statistical models, these models construct the dependency between historic data and future values through a black-box and nonlinear process. Reference [49] compared eight types of machine learning models, finding that their rank is unambiguous and does not change much with different features of time series. Reference [50] tested the accuracy of some popular machine learning models. The results demonstrated that their performances were inferior to eight traditional statistical ones. In addition, [50] points out that machine learning models need to become more accurate, reduce their computation load, as well as be less of a black box. Therefore, in this paper, we will continue to optimize the statistics models by associated relationships to get higher accuracy, instead of machine learning models.

By summarizing previous literature related to multi-dimensional time series analysis, we find that these methods all fail to deal with the situation where product dimension is large but time dimension is small. Based on this situation, we innovatively construct an improved forecasting model for the target variable based on its precedent values and endogenous predictors selected from associated relationships. In some scenarios, if there are many time series associated with the object, we adopt two feasible schemes to simplify the variable space. Then we conduct an empirical study by using an actual dataset of a Chinese grid company. The results of forecasting errors and inventory simulation show that new approaches are superior to these conventional time series forecasting models, including SES, AR, VAR and FDM. Generally speaking, the proposed methods have three major advantages. Firstly, the number of estimated parameters is simplified significantly, not depending exponentially on the dimensions. Secondly, each variable has a customized forecasting regression, which can describe isolated time series well. Thirdly, it does not necessary to collect extra data to act as exogenous variables. Therefore, one contribution of our work is that the new approaches innovatively incorporate associated relationships into demand forecasting, getting rid of the transitional dependence on historical data, so it can be applied to forecast short time series with large dimensionality, making up for the void of previous algorithms. In addition, we contribute to solving over-fitting problems, providing a new direction for the subsequent research. Besides the above theoretical implications, the study also has important practical significance. Note that life circles of products especially high-tech products are getting shorter and many new products are born due to the acceleration of technological innovation. Demand forecasting in terms of limited time points is very common and necessary in actual business activities. Therefore, our new approaches have a wide range of application scenarios and can provide more accurate decision-making basis for practitioners.

The remainder of this paper is organized as follows: In Section 2, we give a brief description of two conventional forecasting models for multi-dimensional time series, i.e., VAR and DFM, and then present our new forecasting approaches based on correlation and Granger causality, respectively. Section 3 describes the procurement dataset and analyzes the relationships between the demands for purchased products. An empirical study and its results are discussed in Section 4. Finally, we summarize our conclusions in Section 5.

2. Forecasting Model and Evaluation

In this section, first we review two common models used in multivariable forecasting. Then we detail our new approaches, utilizing correlations and Granger causality among products to improve prediction accuracy respectively. Finally, some indices are introduced to evaluate the forecasting performance.

2.1. VAR and DFM

VAR and DFM are two conventional models used to forecast demand under multi-product scenarios. Both them have specific limitations, struggling with high (or ultra-high) dimensionality and failing to describe evolutions of short time series. Firstly, we introduce the VAR model. Assume that demand of N products at time t is $x_t = (x_{1,t}, x_{2,t}, \ldots, x_{N,t})'$, $t = 1, 2, \ldots, T$, where $x_{j,t}$ represents the demand of jth product at time t. The VAR model is as follows:

$$B(L)x_t = \alpha + \varepsilon_t \tag{1}$$

where $B(L) = I_N - B_1 L - B_2 L^2 - \cdots - B_p L^p$ is a matrix polynomial with p lags in total. B_j is a $N \times N$ parameter matrix of jth lag and L is the lag operator calculated by $L^j x_t = x_{t-j}$. α is a $N \times 1$ constant vector, and ε_t is a $N \times 1$ vector of white noisy process, without contemporaneous correlation. According to (1), it is obvious that there are total $P \times N \times N$ free parameters need to be estimated in VAR model. With the increase of the number of products (i.e., N), the parameters increase quadratically. Therefore, only time series have moderate dimensionality, i.e., the length of data point is long enough relative to the number of products, can VAR obtain efficient estimates.

As for DFM, it extracts some dynamic factors that can explain the most variation of target variables as predictors, turning the curse of dimensionality into a blessing. However, when the number of products is large, there are some isolated products unavoidably. Common factors cannot explain their demand accurately. Keeping the previous assumptions about x_t, the general DFM model is as follows:

$$x_t = \Gamma(L)f_t + \varepsilon_t, \tag{2}$$

$$\Psi(L)f_t = \eta_t, \tag{3}$$

where $f_t = (f_{1,t}, f_{2,t}, \ldots, f_{m,t})'$ is a m-dimensional column vector, representing values for m (m < N) unobserved factors at time t. It can supplant the originally large data. $\Gamma(L) = \Gamma_0 + \Gamma_1 L + \Gamma_2 L^2 + \cdots + \Gamma_p L^p$, $\Psi(L) = I_m + \Psi_1 L + \Psi_2 L^2 + \cdots + \Psi_q L^q$, and the meanings of these parameters are similar to $B(L)$'s in the previous part. ε_t and η_t are residuals, satisfying some idiosyncratic assumptions. Equation (3) aims to get predictive values of dynamic factors, then applied in Equation (2).

We can see that the quality of factors is the key to determine the accuracy of DFM. As mentioned above, there are many methods to extract factors. Among them, PCA is commonly used in forecasting literature [28]. In PCA estimation, assume that $\Gamma(L) = \Gamma_0$, i.e., original time series are only influenced by contemporaneous factors. Because f_t and ε_t are uncorrelated at all lags, we can decompose the covariance matrix of x_t into two parts:

$$\Sigma_{xx} = \Gamma_0 \Sigma_{ff} \Gamma_0' + \Sigma_{\varepsilon\varepsilon} \tag{4}$$

where Σ_{ff} and $\Sigma_{\varepsilon\varepsilon}$ are covariance matrices of f_t and ε_t respectively. Under the assumptions, the eigenvalues of $\Sigma_{\varepsilon\varepsilon}$ is $O(1)$ and $\Gamma_0 \Gamma_0'$ is $O(N)$, the first r eigenvalues of Σ_{xx} are $O(N)$ and the remaining eigenvalues are $O(1)$. Therefore, the first m principal components of x_t can act as dynamic factors. If Γ_0 is known, the estimator of f_t can be calculated by OLS directly, i.e., $\hat{f}_t = (\Gamma_0\Gamma_0')^{-1}\Gamma_0'x_t$. However, Γ_0 is usually unknown for most cases. Similar to regression, the following optimization equation can estimate Γ_0 and f_t:

$$\min_{f_1,f_2,\ldots,f_T,\Gamma_0} \frac{1}{T}\sum_{t=1}^{T} (x_t - \Gamma_0 f_t)'(x_t - \Gamma_0 f_t), \; s.t. \; \Gamma_0\Gamma_0' = I_r. \tag{5}$$

The first order condition for minimizing (5) with respect to f_t shows that $\hat{f}_t = (\hat{\Gamma}_0\hat{\Gamma}_0')^{-1}\hat{\Gamma}_0'x_t$. By substituting this into the objective function, the results demonstrate that $\hat{\Gamma}_0$ equals to the first m eigenvectors of $\hat{\Sigma}_{xx}$, where $\hat{\Sigma}_{xx} = T^{-1}(\sum_{t=1}^{T} x_t x_t')$. More detailed derivation process can refer to [28]. Correspondingly, $\hat{f}_t = \hat{\Gamma}_0'x_t$ is the first m principal components of x_t. It is the final PCA estimator of dynamic factors in DFM. Finally, let $x_{t,forecast} = \hat{\Gamma}_0\hat{f}_{t,forecast}$ to get predictive values of the original time series.

2.2. The Forecasting Approach Based on Correlation

Associated relationships between multiple products can provide rich information for demand forecasting. Mining effective predictors from associated time series, instead of all series, will be helpful to eliminate some irrelevant information and reduce the number of parameters significantly. Based on this believe, we propose new approaches based on two typical association relationships, namely correlation and Granger causality, respectively. It is proved that they have higher accuracy and can work well even if a wide range of products only have limited data points in the time dimension.

We start with the forecasting approach based on correlation between products. If two products are highly correlated, their demand has specific interactions in the contemporaneous period. For example, if the demand for a product increases, its complementary products will also see a rise in demand at the same time, while its substitutes will experience a decline. Therefore, we utilize such hidden information to modify forecasting algorithms and get more accurate results. There are mainly three steps in the forecasting approach based on correlation. Firstly, find a proper variable subset for each product in terms of correlated relationships. To be specific, calculate the correlation coefficients between the target one and all other products. Those highly correlated to the targeted one, i.e., whose correlation coefficient is more than a certain threshold, constitute the proper variable subset. If a product does not have highly correlated ones, its proper variable subset is empty. Secondly, run autoregressive model (AR) for each product to get originally predictive values of its demand. AR only depends on past values of a time series to forecast its future evolution, ignoring useful information hidden in other correlated time series. Therefore, the third step is that reconstruct forecasting model for products whose proper variable subset is not empty. We can add these proper variables into AR to get final results. It is worth noting that in some cases one product may have many highly correlated products, i.e., many proper variables. If the time point is not enough, adding too many proper variables results in the over-fitted problem, similar to VAR. For example, in this paper, the training sample only contains 36 time points in total. In terms of the principle that the number of estimated parameters should be less than 1/10 of that of observations, there are no more than 3 parameters in the forecasting model. Since the autoregressive process of original time series occupies at least two parameters, only one predictor based on correlation can be selected. Therefore, we propose two feasible schemes to control the scale of the forecasting model as follows:

Scheme (I): Only select the product having the highest correlation with the object from the proper variable subset as a predictor added in final model.

Scheme (II): Extract the first principle component of all elements in the proper variable subset as a predictor added in the final model.

More formally, Let $X = [x_1, x_2, \ldots, x_T]$ represent time series of demand for all products during the T periods, and then the correlation matrix ρ_{XX} of X is as follows:

$$\rho_{XX}(i, j) = \frac{cov(X_{i.}, X_{j.})}{\sqrt{var(X_{i.}) \times var(X_{j.})}} \tag{6}$$

where $X_{i.}$ is the ith row of X, i.e., the sophistic demand series of ith product. According to ρ_{XX}, we can pick up products highly correlated to $X_{i.}$, making up for the proper variable subset for ith product. The autoregression $x_{i,t} = \alpha_i + \beta_{i,1}x_{i,t-1} + \beta_{i,2}x_{i,t-2} + \cdots + \beta_{i,p}x_{i,t-p} + \varepsilon_{i,t}$ can get originally predictive demand $\hat{x}_{i,t}$ for ith product in tth period. p is a lag parameter, determined by the Akaike information criterion (AIC). Based on this, $\hat{X}^{[i]}$ is a matrix, containing original prediction values of all proper variables for ith product. Its rows represent time dimension and columns correspond to products, ranked from left to right in terms of their correlation coefficients with $X_{i.}$ in descending order. Assume that $f_i = [f_{i,2}, f_{i,3}, \ldots, f_{i,T}]$ is the effective predictor selected from $\hat{X}^{[i]}$ to improve forecasting. Because it corresponds to prediction values, the first time point is missed. The final model is as follows:

$$x_{i,t} = \alpha_i + \beta_{i,1}x_{i,t-1} + \beta_{i,2}x_{i,t-2} + \cdots + \beta_{i,p}x_{i,t-p} + \beta_{i,p+1}f_{i,t} + \varepsilon_{i,t} \tag{7}$$

Scheme (I) suggests that $f_i = \hat{X}^{[i]}_{.1}$, where $\hat{X}^{[i]}_{.1}$ is the first column of $\hat{X}^{[i]}$, the product having highest correlation with the ith one. According to Scheme (II), f_i is the first principle component of $\hat{X}^{[i]}$. The procedure of calculating principle components is as follows: (i) computing the covariance matrix $\Sigma_{\hat{X}}$ of $\hat{X}^{[i]}$, (ii) determining eigenvalues and eigenvectors $(\lambda_1, e_1), (\lambda_2, e_2), \ldots, (\lambda_n, e_n)$ of $\Sigma_{\hat{X}}$, where $\lambda_1 > \lambda_2 > \cdots > \lambda_n$, (iii) getting the first principle component $f_i = e_1' \hat{X}^{[i]}$.

2.3. The Forecasting Approach Based on Granger Causality

If Granger causality exists between two products, it means that historic observations of one product can explain the future demand of another product (there is a time lag between them). This situation often occurs when the procurement of products has a stable sequence, such as material procurement in engineering projects. The idea of the forecasting approach based on Granger causality is similar to the former one based on correlation, also consisting of three steps. Firstly, find the proper variable subset for every product by doing Granger causal relation test. When the *p*-value of Granger test satisfies a critical condition, the corresponding product can join the proper variable subset of the target one. Secondly, run AR for every product to get its originally predictive demand. Finally, select effective predictors from the proper variable subset to reconstruct the forecasting model, if a product's proper variable subset is not empty. Similarly, there are also two schemes to prevent excessive parameters:

Scheme (I): Only select the product having lowest *p*-value of Granger test with the object from the proper variable subset as a predictor added in final model.

Scheme (II): Extract the first principle component of all elements in the proper variable subset as a predictor added in the final model.

Assume P_{XX}^k is the *p*-value matrix of Granger test for X considering k lags, where k is determined by AIC. The rows of P_{XX}^k describe Granger results while columns are Granger causes. According to P_{XX}^k, we can construct the proper variable subset for the *i*th product, expressed by a matrix $X^{[i,k]}$. The granger cause with the lowest *p*-value of the *i*th product arranges in the first column of $X^{[i,k]}$, and so on. Let $f_i^k = [f_{i,1}^k, f_{i,2}^k, \ldots, f_{i,T}^k]$ represents the effective predictor extracted from $X^{[i,k]}$. According to Scheme (I) and Scheme (II), $f_i^k = X_{1.}^{[i,k]}$ and f_i^k is the first principle component of $X^{[i,k]}$ respectively. The final forecasting model based on Granger causality is as follows:

$$x_{i,t} = \alpha_i + \beta_{i,1} x_{i,t-1} + \beta_{i,2} x_{i,t-2} + \cdots + \beta_{i,p} x_{i,t-p} + \varphi_{i,1} f_{i,t-1}^k + \varepsilon_{i,t}. \tag{8}$$

2.4. The Forecast Accuracy Measures

According to the previous literature, there are two major methods to evaluate the performance for demand forecasting approaches: forecasting errors and inventory performance, from the perspective of forecasting accuracy and actual inventory management, respectively. It is worth noting that the dataset used in this paper has intermittent demand series: the demand of some products is zero in some periods. Therefore, we adopt absolute scaled error (ASE) to measure forecasting errors. It can overcome the drawback of infinities caused by zero division [51]. The formula is as follows:

$$ASE_t = \frac{\left| y_t - \hat{y}_t \right|}{\frac{1}{n-1} \sum_{i=1}^{n-1} \left| y_{i+1} - y_i \right|} \tag{9}$$

Then mean absolute scaled error is $MASE = mean(ASE_t)$. A forecasting approach with lower MASE means that it is more accurate during the whole forecasting period in general. Therefore, we can compare different approaches according to their values of MASE. In addition, we also apply relative error (RE) to measure the accuracy of forecasters, i.e., calculating ratios of their ASE to that of a baseline model. In this paper, we set simple exponential smoothing (SES) model as the benchmark, which can refer to [52]. For multi-period demand forecasting, the overall judgement of RE is usually based on the form of geometric mean instead of arithmetic mean [51,53]. Geometric mean relative absolute scaled error is expressed as $GMRASE = gmean(e_t/e^*_t)$, where e^*_t is errors of the baseline model.

In fact, optimizing forecasting accuracy aims to provide better guidance for order strategy and inventory management, finally reducing inventory costs and improving managerial efficiency. How forecasting results influence inventory performance is also a concern for scholars and enterprises. A lot of studies assess forecasts by means of inventory simulations [54–58]. Therefore, we also introduce inventory performance to evaluate forecasting approaches. The order-up-to-level policy, commonly used in practice, is adopted to control inventory simulation. We set the inventory review period as one month, consistent with the prediction period. The order-up-to level S is $S = \hat{D} + SS$, where $\hat{D}$ is the predictive demand during the lead time (one month), SS is the safety stock related to the desired service level. At the beginning of each period, check the holding stock H. If H is below S, place an order with the ordering quantity H-S. Otherwise, nothing needs to be done. When face out-of-stocks, the demand will be serviced in the next period. To initialize the simulation system, assume that have full stock at the beginning, i.e., H=S. One index of inventory performance is total inventory costs, consisting of two parts: shortage costs and holding costs, i.e., total inventory costs = unit total cost × (mean inventory per month × a + mean stock-out per month × b) [52]. The cost parameters a and b reflect the trade-off between stock-holding and out-of-stock. $b > a$ means that costs of out stocks are more expensive. When $b < a$, by contrast, unit stock-holding costs more dollars. In addition, another index is the inventory ratio. A smaller inventory ratio means higher inventory efficiency. It is calculated by the following expression:

$$\frac{mean\ holding\ stock}{mean\ demand}.$$

3. Data Description

3.1. Data and Pretreatment

In this paper, we obtained a real dataset of material procurement from a large grid company in China over the span from June in 2012 to April in 2016, comprising a total of 47 months. To minimize forecasting errors, we removed trend and seasonal components from the time series, following [38,45].

Because it takes at least three years to estimate seasonal components, we treat the first 36 months as a training set and the remaining 11 months as a test set to evaluate the out-of-sample prediction ability of forecasting approaches. After removing trend and seasonal effect, results of the unit root test suggest that all processed variables are stationary, the subsequent forecasting steps can continue.

Purchased products are mainly infrastructure materials, consisting of cables, transformers, fittings, etc. Note that the demand is intermittent in the dataset. A few products even have zero demand at more than 2/3 of all time points. These products are not suitable to do forecasting and the data have already be cleaned up. Besides, products without procurement in the first 12 months and the last 12 months are also not considered. In total, there are 338 products left. According to product characteristics, they can be aggregated at different levels, forming a hierarchical structure, which is: Family > Category > Subcategory > Product, from the top level to the bottom level. As shown in Table 1, at the most aggregated level, there are two families, namely primary equipment and equipment material respectively, which can be further disaggregated into 15 categories at level 2 and into 59 subcategories at level 3. Besides, it is obvious that the quantity of products and the value of procurement vary significantly within each categories (subcategories).

Table 1. Description of hierarchical structure and procurement scale of products.

Family	Category	Sub-Category	Product	Total Purchase	Average Purchase of Subcategory [1]	Average Purchase of Category [1]
Primary equipment	AC circuit breaker	1	1	2.65	2.65	2.65
	AC transformer	2	18	102.30	51.15	5.68
	AC disconnector	2	2	2.37	1.19	1.19
	Switch cabinet	3	9	71.25	23.75	7.92
	High-voltage fuse	1	1	4.01	4.01	4.01
	Lightning arrestor	1	1	1.70	1.70	1.70
	Load switch	1	3	11.86	11.86	3.95
Equipment material	Tower pole	2	13	112.90	56.45	8.68
	Wire & ground wire	4	40	140.43	35.11	3.51
	Cable	3	49	272.65	90.88	5.56
	Insulator	6	11	11.29	1.88	1.03
	Metal fittings	24	97	16.75	0.70	0.17
	Cable accessory	5	83	11.01	2.20	0.13
	Optical cable accessory	2	5	0.16	0.08	0.03
	Optical cable	2	2	0.87	0.44	0.44

[1] Units of average purchase: Million yuan per month.

3.2. Correlation Analysis

In this sub-section, we compute correlations coefficients between products at different aggregated levels to make clear their dependencies structure, supporting the subsequent forecasting. At the top level, primary equipment and equipment material are highly correlated. The correlation coefficient of the two families is 0.7784.

As for 15 categories, their correlations are shown in Figure 1. A pink dotted line indicates that the correlation value of two linked nodes is in the interval [0.6, 0.7). Similarly, a blue one corresponds to [0.7, 0.8), while a black line means more than 0.8. Seen from Figure 1, there exists high correlations among 8 categories, including AC transformer, insulator, metal fittings, tower pole, AC disconnector, lightning arrester, high-voltage fuse, and wire & ground wire, especially the first four categories. In addition, the correlation network is clustered by a method proposed by [59] and nodes with same colors in Figure 1 represent that they are clustered in a same group. To be specific, AC transformer, AC disconnector, high-voltage fuse and insulator are in G1, while lightning arrester, tower pole, ground wire and metal fittings are in G2, which are mainly consumed in line laying. According to Table 2, a series of centrality indices of metal fittings are almost the biggest in the correlation network, reflecting that its demand has high correlations with demand of all other categories.

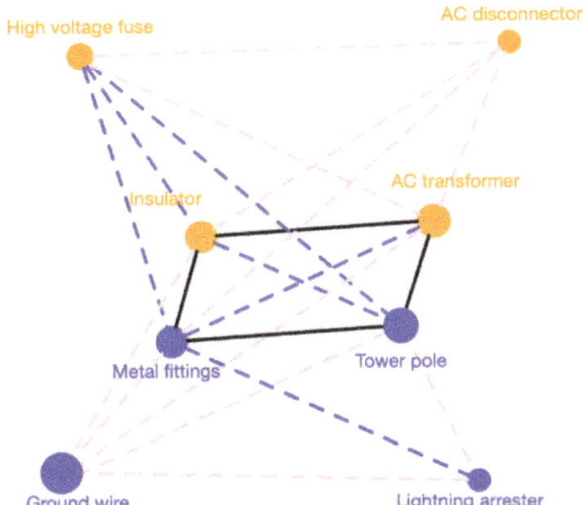

Figure 1. Correlation relationships of categories at level 2.

Table 2. Centrality indices of categories in the correlation network.

Category	Cluster	Degree Centrality	Betweeness Centrality	Eigenvector Centrality	PageRank Centrality	Clustering Coefficient
AC transformer	G1	6	0.833	0.125	1.119	0.800
AC disconnector	G1	4	0.000	0.100	0.787	1.000
High-voltage fuse	G1	5	0.250	0.111	0.952	0.900
Insulator	G1	6	0.833	0.125	1.119	0.800
Metal fittings	G2	7	3.000	0.143	1.299	0.667
Tower pole	G2	6	1.417	0.125	1.299	0.733
Ground wire	G2	5	0.667	0.111	0.964	0.800
lightning arrester	G2	3	0.000	0.091	0.631	1.000

3.3. Granger Causality Analysis

We applied the Granger causal relation test to evaluate the relationship between primary equipment and equipment materials. The result shows that they have no statistically significant causality. In other words, demand of primary equipment does not Granger cause that of equipment materials, with $p = 0.087$. Furthermore, the reverse direction is also not significant, with $p = 0.201$.

In addition, we evaluate the Granger causality relationships among 15 categories at level 2 and visualize the network in Figure 2. Directions of arrows are from Granger causes to Granger results. When the p-value of Granger test locates in $[0.05, 0.01)$ and $[0.01, 0]$, the arrow is drawn by a blue dotted line and a black line respectively. Figure 2 shows that the demands of 14 categories have a significant causal influence on each other, except for cable. It is because cables are widely used in power grid construction, not depending on other products. According to the clustering results of causality network, three groups can be found. G1 contains optical cable accessory, optical cable, load switch and switch cabinet, and procurement of first three categories Granger results in that of switch cabinet. G2 consists of cable accessories, AC transformers, insulator, high-voltage fuse, tower pole and AC disconnector, similar to G1 in correlation analysis, existing complex causality relationships. Overall, high-voltage fuse and tower pole locate at core position, which also have causality relationships with out-group products. Lightning arrester, metal fittings, AC circuit breaker and ground wire are clustered in G3, similar to G2 in correlation analysis. The significant direction of causality from the first three categories to wire & ground wire reflects that a growth of their purchase will increase demand for wire & ground wire later.

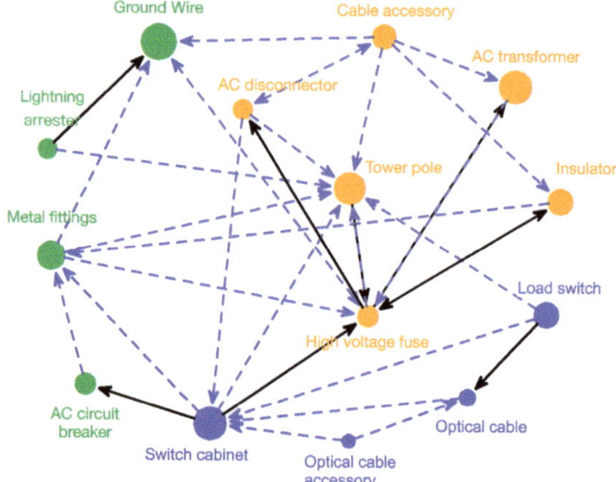

Figure 2. Granger causality relationships of categories at level 2.

To further illustrate Granger causality between categories, we calculate some indices, listed in Table 3. High-voltage fuses and tower poles display the biggest values of in-degree, indicating that their procurement can be greatly explained by lag demand of other categories, while referring to out-degree, high-voltage fuse also ranks first, as well as cable accessory, reflecting that they are strong predictor for follow-up demand of other categories. Besides, switch cabinet, tower pole and high-voltage fuse, with largest centrality indices, are the core nodes in the Granger causality network. what's more, categories in G1 have closer relationship, demonstrated by their higher clustering coefficients.

Table 3. Centrality indices of categories in the causality network.

Category	G	In-Degree	Out-Degree	Betweenes Centrality	Closeness Centrality	Eigenvector Centrality	PageRank Centrality	Clustering Coefficient
Switch cabinet	G1	4	4	52.600	0.056	0.120	1.848	0.179
load switch	G1	0	3	2.667	0.038	0.056	0.768	0.333
Optical cable	G1	2	1	1.000	0.036	0.042	0.808	0.333
Optical cable accessory	G1	0	2	0.000	0.034	0.032	0.575	0.500
AC transfer	G2	2	1	0.400	0.034	0.038	0.543	0.000
High-voltage fuse	G2	6	5	28.967	0.053	0.119	1.578	0.167
Tower pole	G2	7	1	31.067	0.053	0.119	1.583	0.190
Cable accessory	G2	1	5	10.900	0.042	0.074	1.187	0.050
AC disconnector	G2	2	3	3.467	0.045	0.086	0.932	0.417
Insulator	G2	2	2	1.467	0.037	0.059	0.738	0.167
Metal fittings	G3	3	3	19.500	0.050	0.106	1.376	0.300
AC circuit breaker	G3	1	1	0.000	0.037	0.045	0.541	0.500
Lightning arrester	G3	0	2	0.833	0.034	0.037	0.549	0.000
Wire &ground wire	G3	4	1	7.133	0.040	0.067	0.972	0.083

4. Empirical Analysis

4.1. Experimental Setup

In this section, we generate demand forecasting for the above dataset by our new approaches and compare their forecasting performance with four traditional time series models, namely SES, AR, VAR, and DFM, respectively. Except for the enterprise level which only has a single time series, we applied these forecasting approaches on the other four levels of the product hierarchy. Because the number of products varies greatly at the different levels, from several to hundreds, it helps us to investigate whether the new approaches can deal with different data dimensions well.

Nevertheless, DFM is not necessary to do forecasting at the family level. VAR ignores the product level and the subcategory level due to a large number of products at the two levels.

An essential step of our new methods is to set up the criteria for constructing proper variable sets, in other words, to define the critical conditions of high correlation and significant Granger causality. If the critical conditions are too strict, most products may not find proper variables, and then their forecasting demand cannot be corrected by association relationships. Conversely, an excessively loose condition will bring too much disturbing information in proper variable sets, even impairing originally forecasting accuracy. Therefore, a rational critical condition is a key point to obtain satisfying prediction results. Considering that in the forecasting approach based on correlation, originally forecasting values of AR are used as explanatory variables, which may further increase uncertainty, a stricter selection criterion is necessary. We set the correlation coefficient at more than 0.513 as the threshold condition preliminarily. The significance level of the critical value is 0.001 in terms of the size of the training sample. As for the Granger relationship, the standard is that the p-value of Granger test is no less than 0.1. Based on the above settings, we get proper variable sets and then do final demand forecasting. What's more, to further investigate the influence of critical conditions on forecasting performance, we set the critical correlation coefficient as 0.6, 0.7, 0.8, 0.9 as well as the critical Granger significant level as 0.05 and 0.01 separately to repeat prediction process.

Finally, we evaluate the forecasting performance for these models. To begin with we calculate their absolute errors and relative errors in terms of the equations mentioned before. The approach with a smaller average error is considered to be more accurate. Then we introduce t-test to verify whether forecasting errors of our new approaches exist statistically significant differences with the baseline model SES. In addition, demand prediction aims to guide subsequent activities including purchase and inventory management. More accurate forecasting may be helpful to avoid high inventory levels or out of stock, reducing the cost loss of enterprises naturally. Therefore, it is common to assess forecasting approaches by simulating the process of inventory management. we also do inventory simulation for 338 products according to forecasting results of different approaches, to compare their performance from the perspective of inventory management.

4.2. Results and Analysis

4.2.1. Forecasting Accuracy Analysis

Table 4 presents mean values of absolute errors and relative errors for the proposed approaches, as well as four conventional models. CI and CII refer to the approach based on correlations adopting scheme (I) and scheme (II) to control model size respectively. Similarly, GI and GII refer to the approach based on Granger causality and the Roman numerals represent different schemes. The brackets indicate the critical conditions to construct proper variable sets. The approach based on correlations requires the correlation coefficient greater than 0.513. As for the approach based on Granger causality, the p-value of Granger test should be less than 0.1.

Table 4. Forecasting errors of six models at four aggregated levels.

Index	Level	SES [1]	AR	VAR	DFM	CI (0.513)	CII (0.513)	GI (0.1)	GII (0.1)
MASE	Product	0.7837	0.7895	-	0.7712	0.8102	0.5806	0.7001	0.6953
	Subcategory	0.7376	0.7228	-	0.6820	0.7425	0.5593	0.6749	0.6843
	Category	0.7266	0.6692	0.7985	0.7436	0.6813	0.5412	0.6283	0.6448
	Family	0.8107	0.6697	0.6933	-	0.6693	0.6693	0.6749	0.6749
GMRASE	Product	1	1.1006	-	1.1102	1.1068	0.7918	1.0478	1.0004
	Subcategory	1	1.0310	-	1.0031	1.0560	0.7979	1.0012	1.0151
	Category	1	0.9207	1.1981	1.1029	0.9337	0.7479	0.8687	0.8846
	Family	1	0.8332	0.8514	-	0.8289	0.8289	0.8387	0.8387

[1] SES is the baseline model when calculate relative errors.

We can see from Table 4 that for the approaches based on correlations, evaluation results are consistent whether base on MASE or GMRASE. CII performs best among the six models, having a minimum deviation from real demand. Conversely, forecasting accuracy of CI is low, especially at more disaggregated levels. As the dimension of products increase, forecasting errors of CI grows rapidly, even inferior to the original predictive results (AR). This reflects that the forecasting values of the most correlated time series distort the effect of original predictors, leading to lower accuracy, not aligned with our theoretical expectations. However, the first principal component is equivalent to weighted average of all highly correlated time series, which not only contains more effective information but also offsets errors of different correlated series. Therefore, CII can get more accurate results. As for the two types of models based on Granger causality, their accuracy is the same basically at all aggregated levels, superior to that of VAR and DFM. When the product dimension is large, Granger II is more advantageous.

We set different critical values to investigate their influence on accuracy of models. For the approach based on correlation, we set the critical values separately as 0.6, 0.7, 0.8 and 0.9. Table 5 shows the forecasting errors in each situation. With the increase of the critical value, CI becomes more precise while CII is absolutely opposite, witnessing an upward tendency in errors. However, even if the critical value is equal to 0.9, CI performs still worse than AR, let alone CII. For the approach based on Granger causality, forecasting results in terms of different critical conditions are displayed in Table 6. We can see that forecasting errors of GII is always lower than that of GI. When the critical value equal to 0.01, GII has the highest accuracy. In conclusion, scheme (II) can help the approach based on association relationships to get more accurate forecasting results, better than these conventional models. Besides, the critical value should be set in a rationally high level, to ensure that only highly associated time series can be selected to eliminate irrelevant information and the proper variable subset has enough members too to offset errors. In this way, the approaches based on associated relationships can be the most effective.

Table 5. Forecasting errors of the approaches based on correlation under different critical conditions.

Index	Level	CI	CII	CI	CII	CI	CII	CI	CII
		(0.6)		(0.7)		(0.8)		(0.9)	
MASE	Product	0.8095	0.6216	0.7991	0.6926	0.7962	0.7589	0.7933	0.7784
	Subcategory	0.7405	0.5721	0.7400	0.5861	0.7386	0.6112	0.7373	0.6633
	Category	0.6776	0.5346	0.6797	0.5523	0.6651	0.5770	0.6692	0.6692
	Family	0.6693	0.6693	0.6693	0.6693	0.6697	0.6697	0.6697	0.6697
GRMASE	Product	1.1065	0.8516	1.0900	0.9325	1.0990	1.0184	1.1035	1.0760
	Subcategory	1.0528	0.8392	1.0519	0.8403	1.0546	0.8950	1.0499	0.9485
	Category	0.9306	0.7421	0.9353	0.7789	0.9143	0.8094	0.9207	0.9207
	Family	0.8289	0.8289	0.8289	0.8289	0.8332	0.8332	0.8332	0.8332

Table 6. Forecasting errors of the approaches based on Granger causality under different critical conditions.

Index	Level	GI	GII	GI	GII
		(0.05)		(0.01)	
MASE	Product	0.7001	0.6602	0.7007	0.6287
	Subcategory	0.6749	0.6607	0.6841	0.6727
	Category	0.6360	0.6572	0.6336	0.6251
	Family	0.6697	0.6697	0.6697	0.6697
GRMASE	Product	1.0478	0.9601	1.0483	0.9159
	Subcategory	1.0012	0.9851	1.0070	0.9768
	Category	0.8814	0.9059	0.8802	0.8686
	Family	0.8332	0.8332	0.8332	0.8332

To further validate the accuracy of models, we treat the SES model as the benchmark to do *t*-tests for all other models. A negative test statistic means that average ASE of the model is smaller than that of the baseline model, i.e., more accurate than SES model. The smaller the negative t value is, the higher the significance level is. As shown in Table 7, the *t*-test results for absolute errors and relative errors are not consistent. In terms of absolute errors, at the three levels (production, subcategory and category), forecasting errors of GI and GII are all significantly smaller than SES (p-value < 0.01), regardless of the critical values. The models based on correlations outperform SES only when adopt scheme (II) to collect predictors. When it comes to relative errors, CI is still significant superior to the benchmark model only at the category level. GI and GII become insignificant in most cases. In conclusion, significance levels of t-value for relative errors are lower remarkably compared to absolute errors. However, CII still has significantly lower errors, performing best among all models.

Table 7. *T*-test of forecasting errors for different approaches at three aggregated levels.

Model		ASE			RASE		
		Product	Subcategory	Category	Product	Subcategory	Category
AR		0.849	−1.189	−3.301	5.090	0.842	−3.521
VAR		-	-	1.624	-	-	1.780
DFM		−1.392	−2.692	0.398	3.020	0.051	1.154
CI	(0.513)	2.875	0.330	−2.672	6.194	1.564	−2.788
	(0.6)	2.871	0.195	−2.601	6.147	1.471	−2.833
	(0.7)	2.067	0.173	−2.500	5.644	1.458	−2.738
	(0.8)	1.725	0.071	−3.337	5.580	1.404	−3.582
	(0.9)	1.395	−0.021	−3.301	5.317	1.352	−3.521
CII	(0.513)	−14.119	−6.281	−4.799	−11.726	−4.730	−5.256
	(0.6)	−11.002	−5.640	−4.570	−7.693	−3.033	−4.991
	(0.7)	−7.353	−5.008	−3.832	−3.996	−3.271	−4.139
	(0.8)	−2.675	−4.055	−3.221	1.126	−1.915	−3.600
	(0.9)	−0.643	−2.664	−3.301	3.777	−1.016	−3.521
GI	(0.1)	−7.493	−4.178	−3.603	1.405	0.023	−4.114
	(0.05)	−7.493	−4.178	−3.274	1.405	0.023	−3.806
	(0.01)	−7.437	−3.621	−3.450	1.422	0.136	−4.060
GII	(0.1)	−8.197	−3.260	−3.325	0.016	0.202	−3.547
	(0.05)	−10.745	−4.404	−2.282	−1.478	−0.220	−2.564
	(0.01)	−11.906	−3.846	−3.264	−3.566	−0.489	−3.623
Sample size		335	59	15	335	59	15
$p = 0.1$		1.284	1.296	1.341	1.284	1.296	1.341
$p = 0.05$		1.649	1.671	1.753	1.649	1.671	1.753
$p = 0.01$		2.338	2.391	2.602	2.338	2.391	2.602

4.2.2. Inventory Performance Analysis

In our study, we set five desired service levels and three kinds of cost parameters to do inventory simulations for 338 products at the product level. Total inventory costs are shown in Table 8 and Figure 3 illustrates inventory ratios in different scenarios. When satisfying the same service level, CII enjoys the lowest total costs and the inventory ratio and, as expected, CI spends more costs and maintains higher inventory ratios than the original approach (AR) as well as SES and DFM. As for the two approaches based on Granger causality, the results are inconsistent with that of forecasting errors. GI has fewer stock costs and a lower inventory ratio than GII, although GII's forecast accuracy is higher. In addition, compared with AR and DFM, they need higher inventory levels to realize a certain service level.

Table 8. Total inventory costs for different approaches considered satisfying various service levels.

GMRASE	Stock Cost Parameters [1]	SES	AR	DFM	CI (0.513)	CII (0.513)	GI (0.1)	GII (0.1)
0.9	$a = 0.4, b = 0.4$	10.96	10.70	10.33	11.02	8.87	10.14	10.22
	$a = 0.4, b = 0.6$	11.01	10.76	10.38	11.07	8.94	10.19	10.28
	$a = 0.4, b = 0.8$	11.06	10.81	10.43	11.13	9.01	10.25	10.33
0.93	$a = 0.4, b = 0.4$	11.83	11.56	11.21	11.87	9.67	11.01	11.10
	$a = 0.4, b = 0.6$	11.86	11.59	11.24	11.90	9.71	11.04	11.13
	$a = 0.4, b = 0.8$	11.88	11.61	11.26	11.93	9.74	11.07	11.16
0.95	$a = 0.4, b = 0.4$	12.29	12.02	11.69	12.33	10.12	11.48	11.57
	$a = 0.4, b = 0.6$	12.31	12.04	11.70	12.35	10.14	11.50	11.59
	$a = 0.4, b = 0.8$	12.32	12.06	11.71	12.37	10.17	11.52	11.60
0.97	$a = 0.4, b = 0.4$	12.77	12.51	12.18	12.82	10.60	11.97	12.06
	$a = 0.4, b = 0.6$	12.78	12.52	12.19	12.83	10.62	11.98	12.07
	$a = 0.4, b = 0.8$	12.79	12.53	12.20	12.84	10.63	12.00	12.08
0.99	$a = 0.4, b = 0.4$	13.57	13.31	12.98	13.61	11.39	12.76	12.85
	$a = 0.4, b = 0.6$	13.58	13.32	12.98	13.62	11.40	12.77	12.86
	$a = 0.4, b = 0.8$	13.58	13.33	12.99	13.63	11.41	12.78	12.87

[1] Units of total inventory costs: Million.

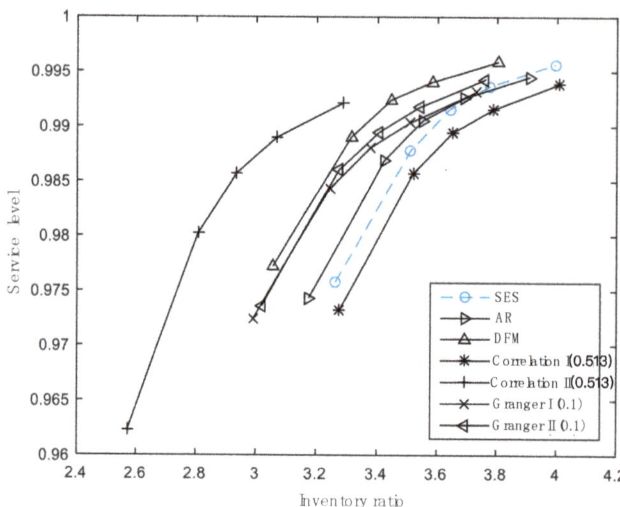

Figure 3. Inventory ratios for different approaches satisfying various service levels.

In addition, we also analyze the inventory performance of the proposed approaches satisfying different threshold values. As shown in Table 9, when the critical value rises, total inventory costs creep up if using CII to forecast demand, and the inventory ratios also reach to the top value (in Figure 4a). CI benefits from high critical values, bringing improvement of inventory performance, but never surpasses that of AR.

Table 9. Total inventory costs for the approach based on correlation with different critical values.

Target Service Level	Stock Cost Parameters [1]	CI	CII	CI	CII	CI	CII	CI	CII
		(0.6)		(0.7)		(0.8)		(0.9)	
0.9	$a = 0.4, b = 0.4$	10.99	9.11	10.88	9.53	10.84	10.57	10.72	10.68
	$a = 0.4, b = 0.6$	11.04	9.18	10.93	9.59	10.90	10.62	10.77	10.74
	$a = 0.4, b = 0.8$	11.10	9.24	10.99	9.65	10.95	10.67	10.82	10.79
0.93	$a = 0.4, b = 0.4$	11.85	9.94	11.73	10.40	11.70	11.43	11.58	11.54
	$a = 0.4, b = 0.6$	11.87	9.97	11.76	10.43	11.72	11.46	11.60	11.57
	$a = 0.4, b = 0.8$	11.90	10.01	11.79	10.45	11.75	11.48	11.63	11.59
0.95	$a = 0.4, b = 0.4$	12.30	10.40	12.19	10.86	12.15	11.89	12.04	12.00
	$a = 0.4, b = 0.6$	12.32	10.42	12.21	10.88	12.17	11.91	12.05	12.02
	$a = 0.4, b = 0.8$	12.34	10.44	12.23	10.90	12.19	11.92	12.07	12.03
0.97	$a = 0.4, b = 0.4$	12.79	10.88	12.68	11.35	12.64	12.38	12.52	12.49
	$a = 0.4, b = 0.6$	12.81	10.90	12.69	11.36	12.66	12.39	12.54	12.50
	$a = 0.4, b = 0.8$	12.82	10.92	12.71	11.38	12.67	12.40	12.55	12.51
0.99	$a = 0.4, b = 0.4$	13.59	11.68	13.47	12.14	13.44	13.17	13.32	13.29
	$a = 0.4, b = 0.6$	13.60	11.69	13.48	12.15	13.45	13.18	13.33	13.30
	$a = 0.4, b = 0.8$	13.61	11.70	13.49	12.16	13.46	13.19	13.34	13.30

[1] Units of total inventory costs: Million.

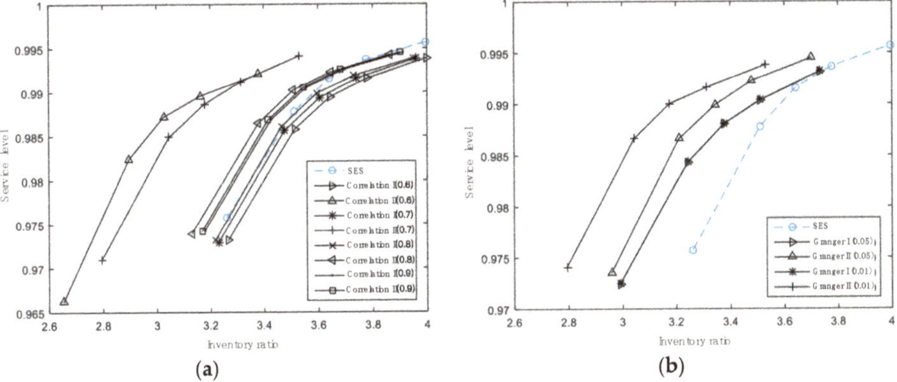

Figure 4. Inventory ratios for approaches with different critical values satisfying various service levels. (a) is for the approach based on correlation and (b) is for the approach based on Granger causality.

For approaches based on Granger causality, their inventory performance is displayed in Table 10 and Figure 4b. We can see that total costs and inventory ratios of GI basically remain the same, regardless of the change of the critical value. GII is more sensitive to the critical value in comparison. A stricter condition to select time series existing Granger causality is helpful to optimize inventory performance for a given service level. As a whole, CII with the critical value equal to 0.513 is the most accurate forecasting model, which can help to reduce inventory costs and improve the service level to the most extent, followed by GII with the critical value equal to 0.01. Therefore, by setting rational thresholds and choosing appropriate methods to mine predictors, the approaches based on associated relationships can achieve higher forecasting accuracy, thereby improving inventory performance. It is helpful to provide effective guidance for actual inventory management activities.

Table 10. Total inventory costs for the approach based on Granger causality with different critical values.

Target Service Level	Stock Cost Parameters [1]	GI	GII	GI	GII
		(0.05)		(0.01)	
0.9	$a = 0.4, b = 0.4$	10.14	10.03	10.14	9.52
	$a = 0.4, b = 0.6$	10.19	10.09	10.20	9.57
	$a = 0.4, b = 0.8$	10.25	10.14	10.25	9.63
0.93	$a = 0.4, b = 0.4$	11.01	10.91	11.02	10.39
	$a = 0.4, b = 0.6$	11.04	10.93	11.05	10.41
	$a = 0.4, b = 0.8$	11.07	10.96	11.07	10.44
0.95	$a = 0.4, b = 0.4$	11.48	11.38	11.48	10.86
	$a = 0.4, b = 0.6$	11.50	11.39	11.50	10.87
	$a = 0.4, b = 0.8$	11.52	11.41	11.52	10.89
0.97	$a = 0.4, b = 0.4$	11.97	11.86	11.97	11.35
	$a = 0.4, b = 0.6$	11.98	11.87	11.98	11.36
	$a = 0.4, b = 0.8$	12.00	11.89	12.00	11.37
0.99	$a = 0.4, b = 0.4$	12.76	12.66	12.76	12.14
	$a = 0.4, b = 0.6$	12.77	12.67	12.77	12.15
	$a = 0.4, b = 0.8$	12.78	12.67	12.78	12.16

[1] Units of total inventory costs: Million.

5. Conclusions

Along with the trend of technological advance, a rapid increase in product diversity and shortening of product life cycles have become important features for enterprises. This means that practitioners need to use limited historical demand to forecast the future, leading to the failure of traditional forecasting approaches. Fortunately, potential relationships between demand time series of multiple products can provide abundant information to improve forecasting accuracy. Therefore, we proposed an improved approach, whose core idea is utilizing the demand of associated products (including highly correlated ones or ones having significantly causality) as predictors to add in AR model to improve its prediction accuracy. Considering that time series may be short, we introduce two feasible schemes to simplify variables to avoid over-fitting. Then monthly procurement data from a Chinese grid company is used to test the forecasting ability of the proposed approaches, as well as four conventional time series approaches SES, AR, VAR, and FDM.

We adopt two types of indicators including forecasting errors and inventory performance to compare these approaches. There are three main findings. Firstly, our new approaches can perform better than those conventional approaches, especially when the product dimensionality is large. Among them, CII with the critical value equal to 0.513 has the best forecasting performance, enjoying the lowest forecasting errors, total inventory costs and inventory ratio, followed by GII with the critical value equal to 0.01. Secondly, Scheme (II) is more effective for the approaches based on associated relationships to achieve higher accuracy, compared with Scheme (I) because it can refine more information and set off the errors of different time series. Finally, it is vital to set a rational threshold condition to select proper variables, which can help to get ideal prediction results. The condition should make sure that most objects have enough proper variables to improve prediction and while the irrelative information is also as less as possible.

However, due to the limitations of the real dataset, we can only pick up one predictor from the associated relationships, so we cannot discuss the influence of the number of selected predictors on the forecasting accuracy. In the future, we can extend our research from the following aspects. Firstly, we can properly test the new approaches through a large wide of diverse datasets, and analyze the most proper number of predictors that should be added in final forecasting models. Secondly, try more methods to dig for predictors from proper variable sets, not limited to PCA.

Then compare their prediction performance to find a better one. In addition, "order-up-to level" is the only inventory strategy used to do inventory simulations in this empirical study. Therefore, we can introduce other widely applied strategies into inventory simulations and explore whether the approaches based on associated relationships are appropriate for different inventory strategies.

Author Contributions: Conceptualization, M.L.; methodology, S.L. and S.Y.; software, S.L.; validation, S.Y.; formal analysis, S.Y. and S.L.; investigation, S.L.; resources, S.L.; data curation, S.L. and S.Y.; writing—original draft preparation, S.Y.; writing—review and editing, S.Y.; visualization, S.L. and S.Y.; supervision, M.L.; project administration, M.L.; funding acquisition, M.L.

Funding: This research received no external funding.

Conflicts of Interest: The authors declare no conflict of interest.

References

1. ElMaraghy, H.; Schuh, G.; ElMaraghy, W.; Piller, F.; Schönsleben, P.; Tseng, M.; Bernard, A. Product variety management. *CIRP Ann.-Manuf. Tech.* **2013**, *62*, 629–652. [CrossRef]
2. Ganesh, M.; Raghunathan, S.; Rajendran, C. The value of information sharing in a multi-product, multi-level supply chain: Impact of product substitution, demand correlation, and partial information sharing. *Decis Support Syst.* **2014**, *58*, 79–94. [CrossRef]
3. Przymus, P.; Hmamouche, Y.; Casali, A.; Lakhal, L. Improving multivariate time series forecasting with random walks with restarts on causality graphs. In Proceedings of the IEEE International Conference on Data Mining Workshops, New Orleans, LA, USA, 18–21 November 2017; pp. 924–931.
4. Riise, T.; Tjozstheim, D. Theory and practice of multivariate ARMA forecasting. *J. Forecast.* **1984**, *3*, 309–317. [CrossRef]
5. Cholette, P.A.; Lamy, R. Mutivariate ARIMA forecasting of irregular time series. *Int. J. Forecast.* **1986**, *2*, 201–216. [CrossRef]
6. Litterman, R.B. Forecasting with Bayesian vector autoregressions—five years of experience. *J. Bus. Econ. Stat.* **1986**, *4*, 25–38.
7. Hafer, R.W.; Sheehan, R.G. The sensitivity of VAR forecasts to alternative lag structures. *Int. J. Forecast.* **1989**, *5*, 399–408. [CrossRef]
8. Funke, M. Assessing the forecasting accuracy of monthly vector autoregressive models: The case of five OECD countries. *Int. J. Forecast.* **1990**, *6*, 363–378. [CrossRef]
9. De Gooijer, J.G.; Klein, A. On the cumulated multi-step-ahead predictions of vector autoregressive moving average processes. *Int. J. Forecast.* **1992**, *7*, 501–513. [CrossRef]
10. Spencer, D.E. Developing a Bayesian vector autoregression forecasting model. *Int. J. Forecast.* **1993**, *9*, 407–421. [CrossRef]
11. Dhrymes, P.J.; Thomakos, D.D. Structural VAR, MARMA and open economy models. *Int. J. Forecast.* **1998**, *14*, 187–198. [CrossRef]
12. Ramos, F.F.R. Forecasts of market shares from VAR and BVAR models: A comparison of their accuracy. *Int. J. Forecast.* **2003**, *19*, 95–110. [CrossRef]
13. Wieringa, J.E.; Horváth, C. Computing level-impulse responses of log-specified VAR systems. *Int. J. Forecast.* **2005**, *21*, 279–289. [CrossRef]
14. Ashley, R. On the relative worth of recent macroeconomic forecasts. *Int. J. Forecast.* **1988**, *4*, 363–376. [CrossRef]
15. Artis, M.J.; Zhang, W. BVAR forecasts for the G-7. *Int. J. Forecast.* **1990**, *6*, 349–362. [CrossRef]
16. Holden, K.; Broomhead, A. An examination of vector autoregressive forecasts for the UK economy. *Int. J. Forecast.* **1990**, *6*, 11–23. [CrossRef]
17. Brave, S.A.; Butters, R.A.; Justiniano, A. Forecasting economic activity with mixed frequency BVARs. *Int. J. Forecast.* **2019**, *35*, 1692–1707. [CrossRef]
18. Ma, S.; Fildes, R.; Huang, T. Demand forecasting with high dimensional data: The case of SKU retail sales forecasting with intra- and inter-category promotional information. *Eur. J. Oper. Res.* **2016**, *249*, 245–257. [CrossRef]

19. Sargent, T.J.; Sims, C.A. Business cycle modeling without pretending to have too much a priori economic theory. *New Meth. J. Bus. Cycle Res.* **1977**, *1*, 145–168.

20. Geweke, J. The dynamic factor analysis of economic time series. In *Latent Variables in Socio-Economic Models*; North-Holland Publishing Company: Amsterdam, The Netherlands, 1977; pp. 365–383.

21. Engle, R.; Watson, M. A one-factor multivariate time series model of metropolitan wage rates. *J. Am. Stat. Assoc.* **1981**, *76*, 774–781. [CrossRef]

22. Stock, J.H.; Watson, M.W. New indexes of coincident and leading economic indicators. *NBER Macroecon. Annu.* **1989**, *4*, 351–394. [CrossRef]

23. Brillinger, D.R. *Time Series: Data Analysis and Theory*; Holden-Day: San Francisco, CA, USA, 1981.

24. Connor, G.; Korajczyk, R.A. Performance measurement with the arbitrage pricing theory: A new framework for analysis. *J. Financ. Econ.* **1986**, *15*, 373–394. [CrossRef]

25. Ding, A.A.; Hwang, J.T.G. Prediction intervals, factor analysis models, and high-dimensional empirical linear prediction. *J. Am. Stat. Assoc.* **1999**, *94*, 446–455. [CrossRef]

26. Forni, M.; Hallin, M.; Lippi, M.; Reichlin, L. The generalized dynamic factor model consistency and rates. *J. Econom.* **2004**, *119*, 231–255. [CrossRef]

27. Bai, J. Inferential theory for factor models of large dimensions. *Econometrica* **2003**, *71*, 135–171. [CrossRef]

28. Stock, J.H.; Watson, M.W. Forecasting with many predictors. In *Handbook of Economic Forecasting*; Elsevier: Amsterdam, The Netherlands, 2006; Chapter 10; pp. 515–554.

29. Boivin, J.; Ng, S. Are more data always better for factor analysis? *J. Econom.* **2006**, *132*, 169–194. [CrossRef]

30. Otrok, C.; Whiteman, C.H. Bayesian leading indicators: Measuring and predicting economic conditions in Iowa. *Int. Econ. Rev.* **1998**, *39*, 997–1014. [CrossRef]

31. Kim, C.J.; Nelson, C.R. Business cycle turning points, a new coincident index, and tests of duration dependence based on a dynamic factor model with regime switching. *Rev. Econ. Stat.* **1998**, *80*, 188–201. [CrossRef]

32. Kose, M.A.; Otrok, C.; Whiteman, C.H. International business cycles: World, region, and country-specific factors. *Am. Econ. Rev.* **2003**, *93*, 1216–1239. [CrossRef]

33. Cepni, O.; Guney, I.E.; Swanson, N.R. Forecasting and nowcasting emerging market GDP growth rates: The role of latent global economic policy uncertainty and macroeconomic data surprise factors. *J. Forecast.* **2018**, *1*, 1–19. [CrossRef]

34. Kim, H.H.; Swanson, N.R. Mining big data using parsimonious factor, machine learning, variable selection and shrinkage methods. *Int. J. Forecast.* **2018**, *34*, 339–354. [CrossRef]

35. Eickmeier, S.; Ziegler, C. How successful are dynamic factor models at forecasting output and inflation? A meta-analytic approach. *J. Forecast.* **2008**, *27*, 237–265. [CrossRef]

36. Stock, J.H.; Watson, M.W. Generalized shrinkage methods for forecasting using many predictors. *J. Bus. Econ. Stat.* **2012**, *30*, 481–493. [CrossRef]

37. D'Agostino, A.; Giannone, D. Comparing alternative predictors based on large-panel factor models. *Oxf. Bullet. Econom. Stat.* **2012**, *74*, 306–326. [CrossRef]

38. Stock, J.H.; Watson, M.W. Forecasting using principal components from a large number of predictors. *J. Am. Stat. Assoc.* **2002**, *97*, 1167–1179. [CrossRef]

39. Forni, M.; Hallin, M.; Reichlin, L.L. The Generalized Dynamic Factor Model: One-Sided Estimation and Forecasting. *J. Am. Stat. Assoc.* **2005**, *100*, 830–840. [CrossRef]

40. Khandelwal, I.; Adhikari, R.; Verma, G. Time series forecasting using hybrid ARIMA and ANN models based on DWT decomposition. *Procedia Comput. Sci.* **2015**, *48*, 173–179. [CrossRef]

41. Kandananond, K. Forecasting electricity demand in Thailand with an artificial neural network approach. *Energies* **2011**, *4*, 1246–1257. [CrossRef]

42. Singh, P. Rainfall and financial forecasting using fuzzy time series and neural networks based model. *Int. J. Mach. Learn. Cyber.* **2018**, *9*, 491–506. [CrossRef]

43. Wang, J.; Wang, J. Forecasting stochastic neural network based on financial empirical mode decomposition. *Neural Netw.* **2017**, *90*, 8–20. [CrossRef]

44. Fu, S.; Li, Y.; Sun, S.; Li, H. Evolutionary support vector machine for RMB exchange rate forecasting. *Physica A* **2019**, *521*, 692–704. [CrossRef]

45. Villegas, M.A.; Pedregal, D.J.; Trapero, J.R. A support vector machine for model selection in demand forecasting applications. *Comput. Ind. Eng.* **2018**, *121*, 1–7. [CrossRef]

46. Martínez, F.; Frías, M.P.; Pérez, M.D.; Rivera, A.J. A methodology for applying k-nearest neighbor to time series forecasting. *Artif. Intell. Rev.* **2019**, *52*, 2019–2037. [CrossRef]

47. Dell'Acqua, P.; Bellotti, F.; Berta, R.; De Gloria, A. Time-aware multivariate nearest neighbor regression methods for traffic flow prediction. *IEEE Tran. Intell. Transport. Syst.* **2015**, *16*, 3393–3402. [CrossRef]

48. Ahmed, N.K.; Atiya, A.F.; Gayar, N.E.; EI-Shishiny, H. An Empirical Comparison of Machine Learning Models for Time Series Forecasting. *Econom. Rev.* **2010**, *29*, 594–621. [CrossRef]

49. Makridakis, S.; Spiliotis, E.; Assimakopoulos, V. Statistical and Machine Learning forecasting methods: Concerns and ways forward. *PLoS ONE* **2018**, *13*, e0194889. [CrossRef]

50. Prestwich, S.; Rossi, R.; Tarim, S.A.; Hnich, B. Mean-based error measures for intermittent demand forecasting. *Int. J. Prod. Res.* **2014**, *52*, 6782–6791. [CrossRef]

51. Moon, S.; Hicks, C.; Simpson, A. The development of a hierarchical forecasting method for predicting spare parts demand in the South Korean Navy—A case study. *Int. J. Prod. Econom.* **2012**, *140*, 794–802. [CrossRef]

52. Syntetos, A.A.; Boylan, J.E.; Croston, J.D. On the categorization of demand patterns. *J. Oper. Res. Soc.* **2005**, *56*, 95–503. [CrossRef]

53. Babai, M.Z.; Ali, M.M.; Nikolopoulos, K. Impact of temporal aggregation on stock control performance of intermittent demand estimators: Empirical analysis. *Omega* **2012**, *40*, 713–721. [CrossRef]

54. Kourentzes, N. Intermittent demand forecasts with neural networks. *Int. J. Prod. Econom.* **2013**, *143*, 198–206. [CrossRef]

55. Kourentzes, N.; Petropoulos, F.; Trapero, J.R. Improving forecasting by estimating time series structural components across multiple frequencies. *Int. J. Forecast.* **2014**, *30*, 291–302. [CrossRef]

56. Petropoulos, F.; Kourentzes, N.; Nikolopoulos, K. Another look at estimators for intermittent demand. *Int. J. Prod. Econom.* **2016**, *181*, 154–161. [CrossRef]

57. Petropoulos, F.; Wang, X.; Disney, S.M. The inventory performance of forecasting methods: Evidence from the M3 competition data. *Int. J. Forecast.* **2018**, *35*, 251–265. [CrossRef]

58. Chatfield, D.C.; Kim, J.G.; Harrison, T.P.; Hayya, J.C. The bullwhip effect—impact of stochastic lead time, information quality, and information sharing: A simulation study. *Prod. Oper. Manag.* **2004**, *13*, 340–353. [CrossRef]

59. Wakita, K.; Tsurumi, T. Finding community structure in Mega-scale social networks [extended abstract]. In Proceedings of the International Conference on World Wide Web, Banff, AB, Canada, 8–12 May 2007.

Article

An Entropy-Based Machine Learning Algorithm for Combining Macroeconomic Forecasts

Carles Bretó [1], Priscila Espinosa [2], Penélope Hernández [3] and Jose M. Pavía [4,*]

[1] Department of Economic Analysis, Universitat de València, Avda. Tarongers s/n, 46022 Valencia, Spain; carles.breto@uv.es

[2] Department of Applied Economics, University of Valencia, Avda. Tarongers s/n, 46022 Valencia, Spain; priscila.espinosa@uv.es

[3] ERI-CES, UMICCS, Department of Economic Analysis, University of Valencia, Calle Serpis 29, 46022 Valencia, Spain; penelope.hernandez@uv.es

[4] UMICCS, Department of Applied Economics, Universitat de València, Avda. Tarongers s/n, 46022 Valencia, Spain

* Correspondence: pavia@uv.es; Tel.: +34-963828404

Received: 30 August 2019; Accepted: 18 October 2019; Published: 19 October 2019

Abstract: This paper applies a Machine Learning approach with the aim of providing a single aggregated prediction from a set of individual predictions. Departing from the well-known maximum-entropy inference methodology, a new factor capturing the distance between the true and the estimated aggregated predictions presents a new problem. Algorithms such as ridge, lasso or elastic net help in finding a new methodology to tackle this issue. We carry out a simulation study to evaluate the performance of such a procedure and apply it in order to forecast and measure predictive ability using a dataset of predictions on Spanish gross domestic product.

Keywords: maximum-entropy inference; Kullback–Leibler; combining predictions; GDP; averaging

---

1. Introduction

This paper applies a Machine Learning approach with the aim of providing a single aggregated prediction from a set of individual predictions. Departing from the well-known maximum-entropy inference methodology, a new factor capturing the distance between the true and the estimated aggregated predictions presents a new problem. To tackle the issues posed by this additional factor, one can look at machine learning (ML) algorithms like ridge regression, lasso or elastic nets. By doing so, the main contribution of this paper is a novel algorithm that combines classic maximum-entropy inference with machine learning and regularization principles by applying a penalty when the aggregated forecast fails to match the forecast target. Via a simulation exercise, we assess the performance of the algorithm and compare it against the naive approach in which aggregated predictions are built as averages of individuals predictions. We also apply this algorithm to a dataset of predictions on Spanish gross domestic product to produce optimal weights that are then used to produce predictions, the predictive ability of which is also evaluated.

Nowadays, there is an increasing number of prospective sources and methods stating a wide variety of forecasts for a given economic variable. The traditional methods for combining forecasts are based on the relative past performance of the forecasters to be combined. However, the number of forecasters has increased considerably over recent years, with the new ones not having had enough time to sufficiently demonstrate their predictive ability, an issue relevant in Economics.

The convenience of combining individual results to obtain a single aggregated prediction is not only problematic in Economics. In Physical Theory, understood as Statistical Mechanics, the seminal works of Jaynes ([1,2]) provide the connection with Information Theory that suggests a constructive method for setting up probability distribution with partial knowledge. Another reason why an Information Theory approach could be a more appropriate way of tackling the problem of the prediction aggregation is an informational matter. Rational expectation says that experts should converge eventually to the true prediction. After a long but successful learning process, experts should make similar predictions. Therefore, a uniform distribution over the set of predictions should be the ultimate combination of predictions. Such a distribution maximizes its entropy.

The machine learning literature on combining forecasts is vast and includes among others the approaches of bagging [3], boosting [4,5] or neural network blending [6]. In the field of economics, combining forecasts has a long tradition and is still an active area (see, e.g., Refs. [7–10]). Prediction combination in order to forecast gross production represents also an active subfield of research (e.g., Refs. [11–16]). The ASA/NBER business outlook surveys started producing composite economic forecasts on 1968 shortly after Ref. [17] commented on the advantages of averaging several forecasts of gross production (as pointed out in Ref. [18]).

From the classic theory, the combination of individual results to obtain a single aggregated prediction consists on a vector of weights that calibrates different degrees of expert ability. Several alternatives can be considered for the combination of forecasts involving different degrees of sophistication. For instance, Ref. [19] considers a minimization of variance-covariance; Ref. [20] offer a method to compute the weights in order to minimize the error variance of the combination. Another method called the regression methods by Ref. [21] interprets the coefficient vector of a linear projection as the corresponding weights. This line of research takes into account the same optimization problem by changing the restriction conditions. We present the benchmark model for the optimization problem of the aggregation of prediction under the perspective of Information Theory. This model activates the criterium of Kullback–Leibler distance to determine the weights of the aggregation of prediction. The nature and objectives of the above problem consists of combining the predictions trying to keep constant (uniform) the knowledge provided by each of them and verifying the true prediction.

Under this perspective, a second approach, the Machine Learning technique, presents a second optimization problem. We draw inspiration from some machine learning algorithms to suggest a specification that combines both objectives: the relative distance expression and the constraints part related to the true prediction. We propose a new specification that also introduces temporal parameters related to an arbitrary temporal structure. Parameters that weight each of the divergences between the aggregation of the predictions and the true predictions. The resulting optimization problem resembles that of regression with regularization [22] and we propose solving it using nested cross-validation [23].

Empirical features of the proposed algorithm are illustrated using a dataset of predictions on Spanish gross domestic product (GDP). The dataset used in this application comes from Fundación de las Cajas de Ahorro, FUNCAS. This is a rich dataset with a sufficient number of institutions making predictions to allow the use of the proposed algorithm. Using this dataset, the proposed algorithm produces optimal weights which are then used to produce both predictions and the predictive ability. Although the dataset does not allow us to disentangle clear differences between the proposed algorithm and a naive forecast, the algorithm is robust in the sense that selecting predictions made in either July or December leads to similar results and interpretations.

The differences between the proposed algorithm and the naive forecast are further explored in a simulation study. Such a study reveals that the proposed algorithm becomes more suitable than the simpler, naive overall average as the length of the target time series increases, as the number of forecasting institutions decreases and as the institutions with predictions sharper than the rest become fewer in number and depart more from the rest.

The paper is organized as follows. In Section 2 we present the model. In Section 3 we introduce the Machine Learning algorithm applied to the maximum-entropy inference problem. In Section 4 the above algorithm is applied to a dataset of predictions on Spanish gross domestic product and in Section 5 assessed via a simulation exercise. Section 6 presents the concluding remarks.

2. Model

This section presents first the benchmark model for the optimization problem of the aggregation of predictions under the perspective of Information Theory. This model activates the criterium of Kullback–Leibler distance to determine the optimal weights of the aggregation of predictions. A second approach, the machine learning technique, provides the second model. Finally, the relationship between both approaches is described.

2.1. Benchmark Maximum-Entropy-Inference (MEI)

Given a set of agents I, let $\{y_{i,t}\}_{i\in I, t\geq 0}$ be forecasts for an economic variable at time t made at a prior time. We consider the combination of the individual results or weighted by a vector of parameters for each possible forecast denoted by ω_i. The weights ω_i are interpreted as the degrees of expertise for every agent $i \in I$. By assuming a non-degenerate distribution of weights, the true prediction at time t is denoted a_t, which verifies $\sum_{i\in I} \omega_i y_{i,t} = a_t$. The first problem we tackle is to find out the weights ω_i such that the true prediction fits the aggregation of predictions.

A parallel problem that we consider is the entropy maximization of the distribution of $\{\omega_i\}_{i\in I}$ subject to the true value coinciding with the aggregation of predictions for all possible temporal horizon t. This optimization problem is expressed as follows:

$$\max_{\omega_i} \sum_{i\in I} \omega_i \log \omega_i^{-1}$$

$$\text{subject to} \qquad \sum_{i\in I} \omega_i = 1 \qquad \omega_i \geq 0,$$

$$\sum_{i\in I} \omega_i y_{i,t} = a_t \qquad \text{for } t \geq 0$$

This methodology known as maximum-entropy inference is equivalent to the problem of finding out a non-negative distribution of weights $\{\omega_i\}_{i\in I}$ that minimizes the Kullback–Leibler-distance between such a distribution and the uniform distribution over the set of agents, that is, $\frac{1}{|I|}$. The Kullback–Leibler distance between two distributions p and q is defined as $K(p,q) = \sum_x p(x) \log \frac{p(x)}{q(x)}$. Notice that the Kullback–Leibler distance is always non-negative but it is not a proper distance since it neither verifies the symmetric nor the triangular properties. This approach based on the Kullback–Leibler-distance assures a non biased outcome over the set of agents.

Formally,

$$\min_{\omega_i} \sum_{i\in I} \frac{1}{|I|} \log(\omega_i |I|)^{-1} \tag{1}$$

$$\text{subject to} \qquad \sum_{i\in I} \omega_i = 1 \qquad \omega_i \geq 0,$$

$$\sum_{i\in I} \omega_i y_{i,t} = a_t \qquad \text{for } t \geq 0$$

To solve this, we start with the Lagrangian of the above problem. It should be noted that the cardinality of variables compared with the set of restrictions may be not enough to guarantee a unique solution, even the existence of a solution. Despite being able to characterize a set of possible weights that minimize the relative distance, it may not fit the true prediction condition.

This issue is well recognized in the literature, since the complexity of finding a proper solution increases with the cardinality of the parameters and conditions; hence, it is necessary to use numerical algorithms to find out (if it exists) the set of candidates of solution.

In order to reduce the complexity of the problem, we can consider a new parametrization of $\{\omega_i\}_{i\in I}$. In particular, one can reparameterize $\omega_i = \frac{e^{x_i}}{\sum e^{x_i}}$ for $x_i \in \mathbf{R}$. This guarantees that $\sum_i \omega_i = 1$ with $\omega_i > 0$ while simplifying the optimization problem by reducing the number of constraints from three to one.

Another way to approach the original problem is to allow a balance between the two restrictions written in the Lagrangian. On the one hand, the distribution of weights should minimize the relative entropy, on the other hand, this system should generate an aggregation of predictions as close as possible to the true prediction.

2.2. Machine-Learning-Inference (MLI)

The nature and objectives of problem (1) consists of (i) combining the predictions of a set of institutions, (ii) trying to keep constant (uniform) the knowledge (the information) provided by each of them and (iii) verifying (approaching as much as possible) a set of restrictions. Under this perspective, we draw inspiration from some machine learning algorithms, such as ridge, lasso or elastic net (in ridge, lasso or elastic net the goal is to minimize a distance, keeping under control the number of parameters of the model to avoid overfitting and all this is controlled by a parameter that allow to rescale or determine the relative importance of each source error function) to propose a specification that combines both objectives: the relative distance expression and the constraints part related to the true predictions. We propose a new specification, which also introduces the parameters δ_t related to an arbitrary temporal structure (for example, each of those parameters may depend on the time distance of the restriction to the forecasted period or on the certainty available about the corresponding constraint value), with a parameter λ that weights the restrictions imposed by the distance between the aggregation of the predictions and the true predictions. It is possible to consider a family of norms since the problem is in $\mathbf{R}^n$ and we are under a normed vector space.

Putting together both expressions and being mindful that the parameters $\{\omega_i\}$ are parametrized, we formalize the minimization problem under a machine learning perspective as:

$$\min_{x_i} \sum_{i \in I} \frac{1}{|I|} \log(\omega_i|I|)^{-1} + \lambda \sum_t \delta_t \left\| \sum_{i \in I} \omega_i y_{i,t} - a_t \right\| \tag{2}$$

The connection of the proposed specification to the machine learning literature stems from the form of the objective function (Equation (2)) and its two summands. The first one refers to the divergence of Kullback–Leibler: $\sum_{i \in I} \frac{1}{|I|} \log(\omega_i|I|)^{-1}$. The second one corresponds (resembles) to a flexible regularization term: $\lambda \sum_t \delta_t \left\| \sum_{i \in I} \omega_i y_{i,t} - a_t \right\|$.

Lambda (λ, hereafter) is a penalty parameter to choose weights that minimize the divergence of Kullback–Leibler to a uniform distribution and penalize the magnitude of the deviation of the weighted prediction from the observed value. On the one hand, when λ is equal to 0 there is no past prediction penalty and the result is equivalent to the classic model without temporal restrictions. On the other hand, when λ grows the breach of the temporal restrictions is gaining weight and dominates Equation (2). In this latter case, the problem may be thought of as a weighted regression problem but with the coefficients restricted to being positive and to adding up to one and without showing the drawbacks of traditional procedures when the number of forecasters is larger than the number of temporal restrictions.

The delta parameters (δ, hereafter) are an improvement measure for the magnitude of the importance that λ gives to the breach of the restrictions. In other words, δ weights the relative importance to the restriction from one year to another.

2.3. From Maximum-Entropy Inference to Machine Learning Inference

Problem (1) indeed shares the same essence as the minimization of problem (2). The first problem is a constrained optimization problem and the second one incorporates this restriction to the objective function. The methodology of solving problem (1) is by the method of the Lagrange multipliers. Specifically, the constrained problem is converted into a structural form with both the objective and the constrained conditions together multiplied by parameters depending on the set of restrictions. Solving the first order conditions of the Lagrangian function, the optimum is derived. The Lagrangian for (1) is written as:

$$\mathcal{L} = \sum_{i \in I} \frac{1}{|I|} \log(\omega_i|I|)^{-1} + \sum_t \lambda_t \left(\sum_{i \in I} \omega_i y_{i,t} - a_t \right)$$

It should be noted that the solution $\{(\omega_i^*, \lambda_t^*)\}_{(i \in I, t)}$, if it exists, pushes down to 0 the second part of the Lagrangian since the restrictions must hold and moreover minimize the relative distance.

Let us now assume a family of problems denoted by $\mathcal{P}(\lambda)$. Fixing λ we have the following minimization problem:

$$\min_{x_i} \sum_{i \in I} \frac{1}{|I|} \log(\omega_i|I|)^{-1} + \lambda \sum_t \delta_t \left\| \sum_{i \in I} \omega_i y_{i,t} - a_t \right\| \tag{3}$$

When the norm is the absolute distance and $\lambda_t = \lambda \delta_t$, both problems, (3) and (1) coincide. If a solution in the former problem (3) exists, then such a solution is a candidate for the later problem (1) for the specific λ_t. Only the restrictions may not be satisfied in problem (3) if this distortion allows the reduction of (if it is possible) the relative entropy with the uniform distribution. Therefore, under the assumption of existence of solution, both problems will offer the same class of solution.

The consideration in the optimum allows us to consider addressing problem (3) from another perspective when in fact problem (1) has no solution or it is too complex to find. The algorithms and structural forms borrowed by machine learning could be a way to approach the solution from a machine learning framework.

3. Algorithm

This section proposes an algorithm to deal with problem (3). The proposed algorithm finds a solution to problem (3) analogously as they do well-understood regularization, machine-learning algorithms. The main steps of the algorithm are splitting the data into training, validation and test sets and choosing the penalty coefficient, λ, via cross-validation on validation sets. The parameters δ_t are exogenous. Following this, the algorithm prediction error can be computed on the test sets and the x_i values estimated using the whole set after addressing the λ and δ parameters, bearing in mind that the parameters $\{\omega_i\}$ are parametrized. The estimated values are finally used as weights to combine the individual predictions.

For cross-validation, we follow the time-series machine-learning literature and propose the use of rolling-origin evaluation [24], also known as rolling-origin-recalibration evaluation [25]. These are forms of nested cross-validation, which should give an almost unbiased estimate of error [23]. Once the number of institutions (forecasters) that we could be used to properly define the training, validation and test sets are selected, we can start to solve the optimization problem. As we will have already noticed, the institutions must be the same in the training, testing and validation sets. If this condition is not fulfilled, the problem will not be well defined. To solve this issue, in our application (see Section 4), the dimensionality of the initial data bank was reduced from 21 to around 10 forecasters satisfying the condition of existence of data for the three phases. This gives us three sets of data sampling with around 10 institutions for each phase.

As a possible specification we select one of the possible options, we consider the quadratic norm, a ridge regression, in the objective function and add a parameter λ and the δ's that characterize the slackness of the process. For simplicity we use $\delta_t = 1$, where we give equal importance to all restrictions. Different values of λ, from a grid of values, are tested to find the optimum that minimizes the divergence and penalizes the combinatorial prediction with respect to the observed value.

The steps of the proposed algorithm are described in detail in Algorithm 1. The output of the algorithm is a prediction for period $T+1$ denoted by $\hat{a}_{T+1}$. The requirements to apply this algorithm are: (i) the three dataset splits mentioned above (training, validation and test), (ii) a set of discrete values of λ between 0 and infinity, (iii) a set of discount values δ emphasizing the λ parameter, and (iv) a prediction error function. The algorithm solves the optimization problem on the training subset for each of the different values of λ and δ. Once the optimization problem is solved, we get a set of prediction errors on the validation set, as many as values for λ. Subsequently, through cross-validation, we make the selection of the λ that minimizes this prediction error. Thanks to this selection, it is possible to obtain the best penalty in terms of prediction error. Once the best λ is obtained, we apply the algorithm on the test set and evaluate its performance. We get the ω_i that minimize the objective function and a measure of its prediction error.

Algorithm 1: Machine learning based entropy

---

1 **input:**

2 Forecast data made by institution i for year n, $\{y_{i,n};\ i \text{ in } 1:I,\ n \text{ in } 1:N+1\}$ $(N \geq 2)$

3 Realized values, $a_{1:N}$

4 Set of penalty coefficients, $\{\lambda_j,\ j \text{ in } 1:J\}$

5 Set of discount coefficients $\{\delta_{t,T},\ t \text{ in } 1:(N-1),\ T \text{ in } 2:N\}$

6 Forecast error function f

7 **output:**

8 Prediction $\hat{a}_{N+1}$

9 **Pseudocode:**

10 For n in $2:N$

11   For j in $1:J$

12   Solve for weights using the training subset $y_1,\ldots,y_{n-1}$:

13   Set $\omega_{i,n,j} = \text{argmin}_{\{\omega_i\}} \sum_{i \in I} \frac{1}{|I|} \log(\omega_i|I|)^{-1} + \lambda_j \sum_{t=1}^{n-1} \delta_{t,n} || \sum_{i \in I} \omega_i y_{i,t} - a_t ||$

14   Determine the forecast error using the validation set y_n:

15   Set $e_{n,j} = f(a_n, \sum_{i \in I} \omega_{i,n,j} y_{i,n})$

16   End For

17 End For

18   Set $j^* = \text{argmin}_j (N-1)^{-1} \sum_{t=2}^{N} e_{t,j}$

19   Set $\lambda^* = \lambda_{j^*}$

20 Solve for weights using λ^* and the full data set:

21   Set $\omega_i^* = \text{argmin}_{\{\omega_i\}} \sum_{i \in I} \frac{1}{|I|} \log(\omega_i|I|)^{-1} + \lambda^* \sum_{t=1}^{N} \delta_{t,N} || \sum_{i \in I} \omega_i y_{i,t} - a_t ||$

22 Set $\hat{a}_{N+1} = \sum_{i \in I} \omega_i^* y_{i,N+1}$

---

4. Data Analysis

A dataset of predictions on Spanish gross domestic product is used to illustrate empirical features of the proposed algorithm. The proposed algorithm produces optimal weights ω_i^* (Table 1) that are used to produce predictions $\hat{a}_{T+1}$ (Table 2), the predictive ability of which can be assessed. The predictive ability of the proposed algorithm for this dataset is similar to that of alternative naive forecast algorithms, in agreement with the simulation exercise of Table 3.

The dataset used in this application comes from the Fundación de las Cajas de Ahorro, FUNCAS. The sample covers the economic predictions of different institutions from 2000 to 2018. The selected sample contains a total of 21 institutions: Analistas financieros, Asesor, Bankia, BBVA, Caixabank, Cámara de Comercio de España, CatalunyaCaixa, CEEM-URJC, Cemex, CEOE, CEPREDE-UAM, ESADE, Funcas, ICAE-UCM, IEE, Instituto de Macroeconomía y Finanzas (Universidad CJC), Instituto Flores de Lemus, Intermoney, Repsol, Santander, Solchaga Recio & asociados). Each agency makes two predictions a year, in July and December for both the current and the following year. Therefore, each year is predicted by each agency up to 4 times. FUNCAS prediction panels are very well known with a prominent experience in economic research and for their thorough work in collecting forecasts at the regional and national levels. In addition, FUNCAS provides such information for free (see www.funcas.es).

For this data analysis, a quadratic forecast error function $f(x,y) = (x - y)^2$ and the following algorithmic parameter values have been used: $\lambda \in \{1 \times 10^{-4}, 2 \times 10^{-4}, \ldots, 8 \times 10^{15}, 9 \times 10^{15}\}$ and $\delta_{t,T} = 1$ for all t, T. The optimization problems have been solved using the free software R version-3.6.1 [26] and the optimization algorithms available in the nloptr library, which serves as an interface for the NLOPT library [27]. NLOPT algorithms can be global or local and based on derivatives or gradient free and include, for example, the augmented Lagrangian algorithm, which uses subsidiary local optimization algorithms. All optimizations have been initialized with a uniform starting point.

To help illustrate the application of the algorithm, Tables 1 and 2 focus on the subset of the full dataset that only includes forecasts for each given year made in July of that same year. Alternative restrictions of the full dataset are possible, for example, forecasts for each year made in December of that same year or forecasts for each year made in July of the previous year. Such alternative restrictions lead to similar key features regarding predictive ability and optimal weights as are described below.

Key features of the optimal weights ω_i^* output by the proposed algorithm included in Table 1 are weight variation across years and across institutions, variations that can be substantial but also reveal some consistencies. The years for which Table 1 reports optimal weights are 2002 through 2018. For the first two prediction years, all weights are negligible except for one, with that single key institution representing about 10% of the number of institutions. For the remaining fifteen years, weights spread out producing 20% to 60% of key institutions. Institutions range from those receiving large optimal weights (e.g., CatalunyaCaixa with 100% on 2002–2003 or IEE and ICO with about 75% on 2004 and 2006 respectively) to those receiving negligible weights. Some institutions are not considered in some years. Of the initial 21 institutions in the full dataset, only 13 produced forecasts from 2000, of which only 9 were still producing forecasts by the end of the sample. Considering years and institutions jointly gives two institution groups: institutions with strikes of substantial weights (e.g., using 25% as threshold: CatalunyaCaixa, IEE and ICO) and the rest of institutions.

Some key factors to assess the predictive ability of predictions $\hat{a}_{T+1}$ made using the proposed algorithm included in Table 2 have been varied as parameters in the simulation study. The simulation study considers multiple combinations of different parameters (Table 3). A combination of parameters that resembles the features in the data could be: (i) 40% of key agents, given that about half of the estimated optimal weights in Table 1 are non-negligible , i.e., $\geq 4\%$ (note however that the fraction of non-negligible weights grows substantially over time in the data while it remains constant in the simulation study); (ii) 10 forecasting agents, given that the number of agents decreases from 13 on 2000 to 9 on 2018; and (iii) a sample size of $T = 20$ years, with the data covering nineteen years (2000–2018). According to the simulation study, such combination of parameters seems to have potential for favoring either the naive or the proposed algorithm depending on the degree of variability between predictions. A variability of $SD = 0.2$ might be reasonable for the data, since predictions for a year are made in July of that same year. This amount of variability produced an average increase of 3.21% in the root mean square prediction error relative to the naive algorithm in the simulation study. This is consistent with the differences reported in Table 3 for the data.

Table 1. Optimal weights ω_i^* output by the proposed algorithm.

Institucion	2002	2003	2004	2005	2006	2007	2008	2009	2010	2011	2012	2013	2014	2015	2016	2017	2018
Analistas Financieros	0.00	0.00	0.01	0.01	0.01	0.04	0.04	0.01	0.01	0.05	0.08	0.09	0.09	0.10	0.10	0.16	0.16
Bankia	0.00	0.00	0.01	0.01	0.01	0.04	0.04	0.01	0.01	0.05	0.08	0.08	0.08	0.07	0.08	0.11	0.11
BBVA	0.00	0.00	0.01	0.01	0.01	0.04	0.04	0.01	0.01	0.04	0.08	0.07	0.07	0.04	0.04	0.06	0.06
Caixabank	0.00	0.00	0.01	0.01	0.01	0.05	0.05	0.01	0.01	0.05	0.08	0.09	0.09	0.11	0.09	0.13	0.14
CatalunyaCaixa	1.00	1.00	0.03	0.04	0.06	0.17	0.17	0.36	0.37	0.33	0.08	0.10	0.10				
CEPREDE-UAM	0.00	0.00	0.01	0.01	0.01	0.05	0.05	0.01	0.01	0.04	0.08	0.07	0.07	0.04	0.04	0.05	0.05
Funcas	0.00	0.00	0.01	0.01	0.01	0.04	0.04	0.01	0.01	0.04	0.08	0.08	0.08	0.06	0.07	0.10	0.10
ICAE-UCM	0.00	0.00	0.01	0.01	0.01	0.03	0.03	0.01	0.01	0.04	0.08	0.08	0.08	0.06	0.05	0.07	0.07
ICO	0.00	0.00	0.13	0.64	0.74	0.26	0.26										
IEE	0.00	0.00	0.75	0.22	0.11	0.13	0.13	0.52	0.51	0.13	0.08	0.10	0.10	0.25	0.27		
Instituto Flores de Lemus	0.00	0.00	0.01	0.01	0.01	0.04	0.04	0.01	0.01	0.04	0.08	0.08	0.08	0.06	0.06		
Intermoney	0.00	0.00	0.01	0.01	0.01	0.04	0.04	0.01	0.01	0.08	0.08	0.08	0.08	0.08	0.09	0.12	0.12
Santander	0.00	0.00	0.02	0.02	0.02	0.07	0.07	0.03	0.03	0.12	0.08	0.09	0.09	0.12	0.12	0.19	0.19

Table 2. Gross Domestic Product (GDP), forecasts $\hat{a}_{T+1}$ and corresponding sample forecast root mean square errors (RMSE) for the time period 2000–2018 using different methods: the arithmetic average of predictions made by all institutions (naive); the proposed algorithm (machine); and the arithmetic average of the subset of predictions used to make the predictions with the proposed algorithm (naive2).

Year	GDP	Naive	Naive2	Machine
2000	5.20	4.01		
2001	4.00	3.00		
2002	2.90	2.09	2.01	2.30
2003	3.20	2.28	2.22	2.10
2004	3.20	2.81	2.77	2.74
2005	3.70	3.28	3.29	3.28
2006	4.20	3.37	3.39	3.30
2007	3.80	3.85	3.88	3.84
2008	1.10	1.74	1.69	1.76
2009	−3.60	−3.64	−3.57	−3.65
2010	0.00	−0.59	−0.52	−0.72
2011	−1.00	0.79	0.80	0.86
2012	−2.90	−1.69	−1.56	−1.60
2013	−1.70	−1.49	−1.48	−1.50
2014	1.40	1.19	1.20	1.18
2015	3.60	3.05	3.03	3.09
2016	3.20	2.85	2.83	2.88
2017	3.00	3.15	3.16	3.16
2018	2.60	2.79	2.82	2.84
Sample RMSE		0.76	0.73	0.74

5. Simulation Study

5.1. Simulation Set-Up

The simulation study covers a wide range of scenarios, each evaluated using 30 replicates. Each replicate is constructed using the following process. Initially, the Spanish gross domestic product actual data to be predicted in our data analysis is used to obtain parameter estimates ($\hat{\mu}$, $\hat{\phi}$ and $\hat{\sigma}$), for a standard autoregressive process of order one, $y_t = \mu + \phi y_{t-1} + \sigma_\epsilon \epsilon_t$. These estimates are used in each replicate to generate a preliminary simulated target time series, $\{\tilde{y}_t^*\}$. This preliminary target is then used to generate simulated predictions for each institution, $\{\tilde{y}_{i,t}\}$, by adding noise (all noises considered, i.e., ϵ_t, η_t and ε_t), are independent standard Gaussian noises) with different intensities parameterized by its standard deviation, that is, $\tilde{y}_{i,t} = \tilde{y}_t^* + \sigma_\eta \eta_t$. These simulated predictions are then aggregated using simulated weights, $\{\tilde{\omega}_i\}$. Simulated weights depend on the number of key agents (institutions) considered. For a 100% of key agents, simulated weights are set to equal weights. For 40% and 10% of key agents, that percentage of the total of institutions is randomly selected and randomly assigned uniform weights between 0.5 and 1. The other institutions are assigned a negligible weight and all weights are rescaled to add up to one. These simulated weights are used to produce the final simulated target time series $\tilde{y}_t = \sum_i \tilde{\omega}_i \tilde{y}_{i,t} + \sigma_\varepsilon \varepsilon_t$ (with σ_ε fixed at 0.1 to introduce some but not much deviation from the direct aggregate). Algorithm performance for different such simulated target time series is analyzed by varying the following parameters: the number of institutions, the sample size, the percentage of key agents and the noise standard deviation.

The number of institutions or agents takes values 10, 20 and 40. The first two values are slightly under and slightly over the number of institutions in our data analysis (Section 4, Table 1). The third value corresponds to an ideal, large number of institutions. Sample size (T) takes values 6, 10 and 20. The first value matches the observations available in our data analysis and the other values consider reasonable and desirable horizons respectively. The percentages of key agents considered are 10%, 40% and 100%, with the latter corresponding to all institutions weighting equally in the generating the target time series. The noise standard deviation (SD), σ_η, takes values 0.1, 0.2 and 0.3. While the first two values are appropriate for near-future forecasts (e.g., forecasts for a given year made in December of that same year), the last value corresponds to forecasts further into the future (e.g., for a given year made in July of the preceding year).

5.2. Simulation Results

The results from the simulation study are as expected (Tables 3 and 4). The proposed algorithm becomes preferable to the simpler, naive overall average as the length of the target time series increases and as the number of both institutions and key institutions decreases. The simulation study reveals that the root average square error can more than double when using the naive algorithm instead of the proposed one. Also, while the results show a good number of improvements of relative error over 20%, negative results seem to stop at around 12%.

The results in terms of weight recovery are shown in Table 4. We assess weight recovery via the Kullback–Leibler divergence between true and recovered weights. A small Kullback–Leibler divergence between these weights is linked to the improvements identified by the simulation study in forecast error resulting from applying the proposed algorithm. The results from Table 4 are in agreement with those from Table 3.

The so-called *forecast combination puzzle* consists in the realization that simple combinations of point forecasts have been found to outperform elaborated weighted combinations in repeated empirical applications [28]. Smith and Wallis [28] pointed out at finite-sample errors in weight estimation as a likely culprit. More recently, Genre et al. [13] establish that "we would not conclude that there exists a strong case for considering combinations other than equal weighting as a means of better summarizing the information collected as part of the regular quarterly rounds" of the Survey of Professional Forecasters. Our findings are in agreement with this literature. The agreement is both from the empirical perspective and from that of the simulation study. This agreement complements the main contribution of this paper in connecting the information theory literature with the machine learning literature in the context of forecast combination. The success of equal weighting for forecast combination can also be linked to the fact that forecasting institutions tend to form a well-informed consensus, which benefits simultaneously from a herd effect [29] and a wisdom-of-the-crowds effect [30].

Table 3. Relative changes (in %) of root average square error (averaging over years) of the arithmetic average of simulated institution predictions ("naive2" in Table 2) with respect to the the proposed algorithm. The parameters are the number of institutions or agents, sample size T (inner subtable dimensions), key agents and noise standard deviation (outer dimensions).

Key Agents		Noise $SD = 0.1$ Sample Size (T)			Noise $SD = 0.2$ Sample Size (T)			Noise $SD = 0.3$ Sample Size (T)		
	Agents	$T = 6$	$T = 10$	$T = 20$	$T = 6$	$T = 10$	$T = 20$	$T = 6$	$T = 10$	$T = 20$
10%	10	9.054	4.412	18.990	30.949	38.425	54.184	47.125	104.228	103.081
	20	−10.597	−4.592	−1.951	0.189	6.892	9.293	11.791	21.052	34.796
	40	−2.586	−5.273	−3.923	−2.806	−1.578	−0.214	−1.774	0.725	2.796
	Agents	$T = 6$	$T = 10$	$T = 20$	$T = 6$	$T = 10$	$T = 20$	$T = 6$	$T = 10$	$T = 20$
40%	10	−7.664	−7.808	−3.638	−1.210	−3.582	−3.214	2.117	0.619	4.131
	20	−9.622	−0.997	−4.204	−10.501	−7.674	−6.759	−8.960	−7.862	−4.530
	40	−6.705	−3.98	−6.479	3.712	−1.271	−6.835	−4.716	−5.989	−3.499
	Agents	$T = 6$	$T = 10$	$T = 20$	$T = 6$	$T = 10$	$T = 20$	$T = 6$	$T = 10$	$T = 20$
100%	10	−5.369	−5.737	−6.963	1.667	−10.27	−6.739	−9.857	−9.534	−8.856
	20	−11.972	−8.850	−3.873	−9.304	−12.018	−7.438	−11.344	−8.470	−6.108
	40	−10.179	−9.636	−5.620	−11.407	−8.597	−3.475	−9.157	−7.703	−7.479

Table 4. Kullback–Leibler divergence between true and recovered weights.

Key Agents		Noise $SD = 0.1$ Sample Size (T)			Noise $SD = 0.2$ Sample Size (T)			Noise $SD = 0.3$ Sample Size (T)		
	Agents	$T = 6$	$T = 10$	$T = 20$	$T = 6$	$T = 10$	$T = 20$	$T = 6$	$T = 10$	$T = 20$
10%	10	2.653	1.197	0.362	0.815	0.376	0.154	0.686	0.166	0.116
	20	2.964	2.378	1.438	1.958	1.316	0.574	1.663	0.930	0.286
	40	3.118	2.623	2.253	2.483	2.314	1.832	2.122	1.960	1.295
	Agents	$T = 6$	$T = 10$	$T = 20$	$T = 6$	$T = 10$	$T = 20$	$T = 6$	$T = 10$	$T = 20$
40%	10	2.495	1.613	1.229	1.429	1.015	0.702	1.229	0.734	0.438
	20	2.638	2.232	1.143	1.497	1.083	0.960	1.208	0.986	0.871
	40	2.053	1.462	1.178	1.281	1.220	1.022	1.039	1.163	0.960
	Agents	$T = 6$	$T = 10$	$T = 20$	$T = 6$	$T = 10$	$T = 20$	$T = 6$	$T = 10$	$T = 20$
100%	10	1.647	1.346	0.357	0.639	0.356	0.068	0.371	0.253	0.035
	20	1.173	0.860	0.613	0.637	0.259	0.098	0.222	0.135	0.092
	40	0.856	0.731	0.392	0.423	0.304	0.325	0.294	0.294	0.155

6. Concluding Remarks

According to prediction and sampling theories, forecasting errors and variances of single forecasts can be reduced by combining individual predictions. The traditional methods for combining forecasts are based on assessing the relative past performance of the forecasters to be combined. The problem, however, becomes indeterminate as soon as the number of forecasters is larger than the number of past results. To overcome this issue, an alternative is to assume some set of a priori weights and to apply the principle of maximum entropy to obtain a set of a posteriori weights, subject to the constraint that the combined predictions equal the realized values. Unfortunately, this is a complex problem that grows with the cardinality of the variables and the possibility of finding a solution is not guarantee.

In order to reach a solution within the information theory framework we propose a fresh approach to the problem and, inspired in the machine learning literature, we suggest a new specification based on regularization regression and an algorithm to solve it. The new approach always produces a solution, being moreover quite flexible. It permits the use of different norms to measure the discrepancies among the combined predictions and the realized values and to weight the relative importance of the discrepancies. Our regularization approach also has the advantage of producing, as a by-product, the weights assigned to the different forecasters. These weights could be understood as a measure of the forecasters' ability and be used as a tool to decide the methodologies deserving more credit.

Further flexibility could be introduced in our model. For instance, by substituting in Equation (2) the single prediction values by prediction functions (for example, regression equations). In this case, the parameters of such prediction functions would be estimated simultaneously, during the cross-validation step. This will enable us to apply our proposal in one step when, for instance, we try to obtain, from a set of national forecasts, a prediction for a regional economy where single forecasts are not available. We could substitute the (unavailable) single regional forecasts for a parametrized function (e.g., a dynamic regression equation) of the national values.

In our algorithm, we have considered a quadratic norm (a ridge penalty) and a rolling-origin evaluation as cross-validation strategy. Obviously, other penalties (e.g., lasso or elastic net) are also possible and, likewise, there is also room for implementing other methods of cross-validation. For instance, we can explicitly omit the temporal order of the data in the training sets and carry out leave-one-out cross-validation. At the end, the relative importance of the most recent predictions can be implicitly included in our specification through the δ's coefficients.

Regarding our application, as it is a common practice we have used the last reliable GDP available figures (all the countries elaborate several vintages of GDP. National accounts are regularly revised as statistical information is enlarged. For instance, in the case of Spain, the estimates from each year undergo three revisions until they are considered definite [31]) as realized values, a_t. In our opinion, this is not however the best strategy to be followed for a "combiner" of macroeconomic forecasts. Instead, flash estimates should be used. Flash estimates (the most provisional and least reliable figures, though) are the most appealing, getting a strong attention (on the one hand, they occupy the front pages of the media and are the ones more analysed, debated and commented on. Revised and definitive data, published three to four years later, attract little public opinion interest. On the other hand, and more importantly, the flash estimates serve as a framework for decision-making by economic stakeholders. Decisions which may give rise to rights and obligations: budgetary stability commitments in the EU, ceilings on general government expenditure, size of deficit or government debt allowed). This may entail marked consequences on the weights each forecaster receives.

The key contribution of this paper is to link the maximum-entropy inference methodology from the information theory literature with regularization from the machine learning literature with the ultimate goal of combining forecasts. Although one might envisage linking forecast combination algorithms other than regularization (e.g., boosting or bagging) with the information theory literature, it does not seem immediatly clear how this could be done. Such immediacy seems to be one of the advantages of regularization over alternative algorithms when it comes to connecting the machine learning and information theory literature.

Author Contributions: Conceptualization, C.B., P.E., P.H. and J.M.P.; Methodology, C.B., P.E., P.H. and J.M.P.; Software, C.B., P.E., P.H. and J.M.P.; Formal analysis, C.B., P.E., P.H. and J.M.P.; Investigation, C.B., P.E., P.H. and J.M.P.; Resources, C.B., P.E., P.H. and J.M.P.; Data curation, P.E. and J.M.P.; Writing–original draft preparation, C.B., P.E., P.H. and J.M.P.; Writing–review and editing, C.B., P.E., P.H. and J.M.P.; Supervision, C.B., P.H. and J.M.P.; Project administration, J.M.P.; Funding acquisition, P.H. and J.M.P.

Funding: The authors acknowledge the support of Generalitat Valenciana through the aggreement "Desarrollo y mantenimiento de las previsiones macroeconómicas de la Comunitat Valenciana" (Consellería de Economía Sostenible, Sectores Productivos, Comercio y Trabajo) and the project AICO/2019/053 (Consellería d'Innovació, Universitats, Ciència i Societat Digital). The authors also thank the support of the Spanish Ministry of Science, Innovation and Universities and the Spanish Agency of Research, co-funded with FEDER funds, project ECO2017-87245-R.

Acknowledgments: The authors wish to thank two anonymous reviewers for their valuable comments and suggestions and the Guest Editors and Journal Editors for their help and kindness. They also like to thank Marie Hodkinson for revising the English of the paper.

Conflicts of Interest: The authors declare no conflict of interest.

References

1. Jaynes, E.T. Information Theory and Statistical Mechanics I. *Phys. Rev.* **1957**, *106*, 620–630. [CrossRef]
2. Jaynes, E.T. Information Theory and Statistical Mechanics II. *Phys. Rev.* **1957**, *108*, 171–190. [CrossRef]
3. Breiman, L. Bagging Predictors. *Mach. Learn.* **1996**, *24*, 123–140. [CrossRef]
4. Freund, Y. Boosting a Weak Learning Algorithm by Majority. *Inf. Comput.* **1995**, *121*, 256–285. [CrossRef]
5. Schapire, R.E. The Strength of Weak Learnability. *Mach. Learn.* **1990**, *5*, 197–227. [CrossRef]
6. Shnarch, E.; Alzate, C.; Dankin, L.; Gleize, M.; Hou, Y.; Choshen, L.; Aharonov, R.; Slonim, N. Will it Blend? Blending Weak and Strong Labeled Data in a Neural Network for Argumentation Mining. In Proceedings of the 56th Annual Meeting of the Association for Computational Linguistics (ACL 2018), Melbourne, Australia, 15–20 July 2018; Volume 2: Short Papers, pp. 599–605.
7. Fernández-Vázquez, E.; Moreno, B.; Hewings, G.J. A Data-Weighted Prior Estimator for Forecast Combination. *Entropy* **2019**, *21*, 429. [CrossRef]
8. Chan, F.; Pauwels, L.L. Some Theoretical Results on Forecast Combinations. *Int. J. Forecast.* **2018**, *34*, 64–74. [CrossRef]
9. Timmermann, A. Forecast combinations. *Handb. Econ. Forecast.* **2006**, *1*, 135–196.
10. Armstrong, J.S. Combining Forecasts. In *Principles of Forecasting: A Handbook for Researchers and Practitioners*; Springer: Boston, MA, USA, 2001; pp. 417–439.

11. Fernández-Vázquez, E.; Moreno, B. Entropy Econometrics for Combining Regional Economic Forecasts: A Data-Weighted Prior Estimator. *J. Georgr Syst.* **2017**, *19*, 349–370. [CrossRef]

12. Hsiao, C.; Wan, S.K. Is There an Optimal Forecast Combination? *J. Econ.* **2014**, *178*, 294–309. [CrossRef]

13. Genre, V.; Kenny, G.; Meyler, A.; Timmermann, A. Combining Expert Forecasts: Can Anything Beat the Simple Average? *Int. J. Forecast.* **2013**, *29*, 108–121. [CrossRef]

14. Moreno, B.; Lopez, A.J. Combining Economic Forecasts Through Information Measures. *Appl. Econ. Lett.* **2007**, *14*, 899–903. [CrossRef]

15. Moreno, B.; López, A.J. Combining Economic Forecasts by Using a Maximum Entropy Econometric Approach. *J. Forecast.* **2013**, *32*, 124–136. [CrossRef]

16. Capistrán, C.; Timmermann, A. Forecast Combination with Entry and Exit of Experts. *J. Bus. Ecol. Stat.* **2009**, *27*, 428–440. [CrossRef]

17. Zarnowitz, V. An Appraisal of Short-Term Economic Forecasts. In *NBER Books*; number zarn67-1; National Bureau of Economic Research, Inc.: Cambridge, MA, USA, September 1967.

18. Clemen, R.T. Combining forecasts: A Review and Annotated Bibliography. *Int. J. Forecast.* **1989**, *5*, 559–583. [CrossRef]

19. Bates, J.M.; Granger, C. The Combination of Forecasts. *Oper. Res. Q.* **1969**, *20*, 451–468. [CrossRef]

20. Newbold, P.; Granger, C.W. Experience with Forecasting Univariate Time Series and the Combination of Forecasts. *J. R. Stat. Soc. Ser. A Gener.* **1974**, *137*, 131–146. [CrossRef]

21. Granger, C.W.; Ramanathan, R. Improved Methods of Combining Forecasts. *J. Forecast.* **1984**, *3*, 197–204. [CrossRef]

22. Hastie, T.; Tibshirani, R.; Friedman, J. *The Elements of Statistical Learning: Data Mining, Inference and Prediction*, 2nd ed.; Springer: New York, NY, USA, 2009.

23. Varma, S.; Simon, R. Bias in Error Estimation when Using Cross-Validation for Model Selection. *BMC Bioinform.* **2006**, *7*, 91. [CrossRef]

24. Tashman, L.J. Out-of-Sample Tests of Forecasting Accuracy: An Analysis and Review. *Int. J. Forecast.* **2000**, *16*, 437–450. [CrossRef]

25. Bergmeir, C.; Benítez, J.M. On the Use of Cross-Validation for Time Series Predictor Evaluation. *Inf. Sci.* **2012**, *191*, 192–213. [CrossRef]

26. R Core Team. *R: A Language and Environment for Statistical Computing*; R Foundation for Statistical Computing: Vienna, Austria, 2019. Available online: https://www.r-project.org/ (accessed on 5 July 2019).

27. Steven, G.J. The NLopt Nonlinear-Optimization Package. 2019. Available online: http://github.com/stevengj/nlopt (accessed on 1 June 2019).

28. Smith, J.;Wallis, K.F. A Simple Explanation of the Forecast Combination Puzzle. *Oxf. Bull. Econ. Stat.* **2009**, *71*, 331–355. [CrossRef]

29. Pons-Novell, J. Strategic Bias, Herding Behavior and Economic Forecasts. *J. Forecast.* **2003**, *22*, 67–77. [CrossRef]

30. Surowiecki, J. *The Wisdom of Crowds: Why the Many Are Smarter Than the Few and How Collective Wisdom Shapes Business, Economies, Societies and Nations*; Doubleday: New York, NY, USA, 2004.

31. Cabrer, B.; Serrano, G.; Pavía, J.M. Evaluación del Sesgo en las Estimaciones de Contabilidad Nacional Trimestral: Estudio de las Añadas en España. *Estudios de Economía Aplicada* **2017**, *35*, 271–298. (In Spanish)

Article

Time Series Complexities and Their Relationship to Forecasting Performance

Mirna Ponce-Flores [1,*,†], **Juan Frausto-Solís** [1,*,†], **Guillermo Santamaría-Bonfil** [2,*,†], **Joaquín Pérez-Ortega** [3] and **Juan J. González-Barbosa** [1]

[1] Graduate Program Division, Tecnológico Nacional de México/Instituto Tecnológico de Ciudad Madero, Cd. Madero 89440, Mexico; jjgonzalezbarbosa@gmail.com

[2] Information Technologies Department, Consejo Nacional de Ciencia y Tecnología—Instituto Nacional de Electricidad y Energías Limpias, Cuernavaca 62490, Mexico

[3] Computing Department, Tecnológico Nacional de México/Centro Nacional de Investigación y Desarrollo Tecnológico, Cuernavaca 62490, Mexico; jpo.cenidet@gmail.com

* Correspondence: mirna_poncef@hotmail.com (M.P.-F.); juan.frausto@gmail.com (J.F.-S.); guillermo.santamaria@ineel.mx (G.S.-B.)

† These authors contributed equally to this work.

Received: 17 December 2019; Accepted: 7 January 2020; Published: 10 January 2020

Abstract: Entropy is a key concept in the characterization of uncertainty for any given signal, and its extensions such as Spectral Entropy and Permutation Entropy. They have been used to measure the complexity of time series. However, these measures are subject to the discretization employed to study the states of the system, and identifying the relationship between complexity measures and the expected performance of the four selected forecasting methods that participate in the M4 Competition. This relationship allows the decision, in advance, of which algorithm is adequate. Therefore, in this paper, we found the relationships between entropy-based complexity framework and the forecasting error of four selected methods (Smyl, Theta, ARIMA, and ETS). Moreover, we present a framework extension based on the Emergence, Self-Organization, and Complexity paradigm. The experimentation with both synthetic and M4 Competition time series show that the feature space induced by complexities, visually constrains the forecasting method performance to specific regions; where the logarithm of its metric error is poorer, the Complexity based on the emergence and self-organization is maximal.

Keywords: classical forecasting methods; complexity; entropy; error measures; symbolic analysis; M4 competition

---

1. Introduction

Presently, time series forecasting is applied to many areas such as weather, finance, ecology, health, electrochemical reactions, computer networks, and so on [1]. Among the most popular and effective methods stand the classical time series models such as the Simple Exponential Smoothing (SES) and the Autoregressive Integrated Moving Average (ARIMA). Also, forecasting methods of machine learning such as Neural Networks have gained popularity after the results of the Smyl winning method of the M4 Competition, and the benchmark forecasting methods of Theta, ARIMA, and ETS [2]. In the forecasting area, researchers agree that it is too difficult to identify a suitable forecasting method for a particular time series beforehand, even knowing its specific statistical characteristics [3]. For instance, time series (TS) complexity [3] is a widely debated measure, which it is supposed to quantify the *intricacy* of the time series, allowing choice of the forecasting methods to be applied. Shannon's entropy has been used to measure the complexity of discrete systems [4]. Although the entropy formula was conceived in the thermodynamic area, the entropy concept has spread to different disciplines adapting its meaning in regard to the applied area and making tools for many applications [5–7]. For example, in [8] a package with functions to measure emergence, self-organization, and complexity applied to discrete and continuous data is presented as a framework; the present study is based on them. However, to the best of the authors' knowledge, these formulae have not been applied to assess the *forecastability* of time series. Furthermore, this framework is extended with other measures. We present four complexity measures based on entropy and a methodology for determining the relationships between these complexity measures and the forecasting error of the Smyl [9], Theta [10], ARIMA [11,12], and ETS [13] methods; all of them were participants of the M4 Competition [14]. This study was made for a dataset with some synthetic time series [15] and more than 20,000 time series taken from M4 Competition [16], which is a reference point used by many researchers. We obtain the prediction error with the forecasting values of each one of the four selected methods, and we determine four complexity measures based on the relationship between Entropy and Mean Absolute Scaled Error (MASE) error [17], but for functionality we use the logarithm values of MASE error $log(MASE)$. We present a complexity $log(MASE)$ analysis, and we apply a visualization method [18] for the time series of the dataset. Finally, the experimentation shows that the permutation and 2-regimen complexities are the measures that identify patterns of the distribution of TS on the two-dimensional space; also we found a relationship between the permutation complexity and the $log(MASE)$ values and finally we make a comparison between the four forecasting methods reinforcing the known *No-free lunch* theorem.

This paper is organized as follows. Section 2 presents the materials used in this research; Section 3 describes the methods, parameters settings, methodology and the dataset used in the experimentation. In Section 4, we provide the results of the experimentation. Finally, Section 5 presents the conclusion for this work.

2. Materials

The complete dataset of the time series used in this paper is divided into two subsets: Synthetic and M4 Competition. Each of these is described in the following subsections.

2.1. Synthetic Time Series

Three generating mechanisms were considered for the construction the of synthetic TS: (a) sine waves; (b) logistic map; and (c) a time series tool, namely *GRATIS* [15]. It is worth noting that in order to generate time series belonging to the same mechanism type, either the parameters of the generating function are modified or noise is introduced at a certain specific Signal-to-Noise Ratio (SNR). The synthetic TS considered are *Sine Wave corrupted by uniform noise, Sine Wave corrupted by Gaussian noise, 1-D logistic Map*, and the *GRATIS tool*.

2.1.1. Sine Waves TS

A stationary and seasonal TS is generated using a sine wave of the form:

$$x_t = \alpha \cdot sin(\omega t), \tag{1}$$

where x_t is the observation at time t, α corresponds to the wave amplitude parameter, and ω to the wave frequency. A family of time series is spawned from Equation (1) by corrupting the wave at specific SNRs. In the case of the latter, the contaminated series is defined as

$$X = f(x) + k \cdot \epsilon,$$

where $f(x)$ is the sine wave, k is an increasing constant, and $\epsilon \in P(X)$ is a shock which follows a uniform or Gaussian distribution. In these cases, the SNR is determined by

$$SNR = \frac{Var(f(x))}{Var(\epsilon)},$$

where larger values of SNR imply that it is easier to detect the signal, and smaller values otherwise.

2.1.2. Logistic Map TS

The logistic Map is a 1-dimensional chaotic dynamic system which is commonly employed as benchmark to study tools and methods used to characterize chaotic dynamics [19–21]. The logistic map function is defined as

$$x_{t+1} = r \cdot x_t(1 - x_t), \tag{2}$$

where x_0 is a random number within $0 < x_t < 1$, and r is a constant. In fact, this last parameter is the one that defines the behavior of Equation (2). More precisely, when $r < 1$ the system always collapses to zero, for $1 \leq r \leq 3$ the system tends to a single value, for $3 < r < 3.6$ the system is fixed to period-doubling points, and from $r \sim 3.6$ the system exhibits a chaotic behavior.

2.1.3. GRATIS TS

The last subset of time series was generated using the GRATIS tool [15] that is based on Gaussian Mixture Autoregressive (MAR) models. These models contain multiple stationary or non-stationary autoregressive components, non-linearity, non-Gaussianity, and heteroscedasticity. To tune the parameters for MAR models, the GRATIS' authors use a Genetic Algorithm when the distance between the target feature vector and the feature vector is close to zero. This tool generates time series with diverse parameters such as length, frequency, and behavior features.

2.2. M4 Competition TS

The complete set is composed of 100,000 real-life series divided into sets named "periods" (Yearly, Quarterly, Monthly, Weekly, Daily, and Hourly) and subdivided into subsets or types (Demographics, Finance, Industry, Macro, Micro, and Other). Our criterion for selecting time series was: (1) To choose time series with a minimum of two hundred and 50 observations; (2) The frequency group should have more than one set type. Consequently, TS from the Hourly group were not selected since it only contains time series from the type *Other*. The complete dataset is shown in Table 1, which has two final columns named size and percentage (%); the former refers to the number of time series selected from each frequency group, and the latter is the correspondent percentage of selected time series concerning that group. According to the last criterion, in our dataset, we consider only the subsets Yearly, Quarterly, Monthly, Weekly, and Daily; the total number of the TS in our dataset is 22,610.

Table 1. M4 Competition time series.

Frequency	Demographic	Finance	Industry	Macro	Micro	Other	Total	Selected Series Size	%
Yearly	1088	6519	3716	3903	6538	1236	23,000	56	0.24%
Quarterly	1858	5305	4637	5315	6020	865	24,000	256	1.07%
Monthly	5728	10,987	10,017	10,016	10,975	277	48,000	18,406	38.35%
Weekly	24	164	6	41	112	12	359	293	81.62%
Daily	10	1559	422	127	1476	633	4227	3599	85.14%
Hourly	0	0	0	0	0	414	414	0	0.00%
Total	8708	24,534	18,798	19,402	25,121	3437	100,000	22,610	22.61%

3. Methods

To analyze the relationship between the prediction errors of classical forecasting methods such as ARIMA, we build a feature space [18] based on Shannon entropy (H) features that presumably can be used to identify those TS instances where ARIMA forecasting errors are expected to be higher or lower, accordingly. These features are based on four entropy-based complexity measures, namely the frequentist binning approach (H_{dist}) [22]; 2-Regimes (H_{2reg}) [19] and Permutation (H_{perm}) entropy [23].

These three built upon the notions of symbolic dynamics, and the Spectral entropy (H_{spct}) [24] based on the analysis of the spectrum of a time series. The main difference between these measures is the *discretization* or *symbolization* followed to describe the states of dynamical systems, which has a deep impact on the quantification of entropy [25]. For instance, consider a TS with hundreds of points coming from a sine wave; with the frequentist binning approach, we will be studying a system whose probability distribution follows an arc-sine distribution, whereas if we represent it by symbols that correspond to 1-period waves, we will be studying a system which follows a Dirac delta probability distribution.

On the other hand, H_{dist} has been used to study a dynamical system in terms of the rate of Emergence (E) of new states or information, the rate of Self-organization (S) displayed as discernible patterns, and the interplay between these two called Complexity (C), hereafter *ESC* for short [8]. In particular, systems with higher C concentrates its dynamics into a few highly probable states with many less frequent states [8]. In this work, the *ESC* framework is extended to study the interplay between E and S for H_{spct}, H_{2reg}, and H_{perm}.

Therefore, first, the Shannon-based complexity measures and TS symbolization for each is presented; second, the *ESC* framework is introduced along with the Complexity Feature Space; third, the forecasting methods are briefly defined; finally, the proposed analysis methodology is detailed.

3.1. A Background on Entropies

Entropy is a term with many meanings, but in the information theory area it usually refers to the average ratio of uncertainty a process produces, which is measured by the well-known discrete Shannon Entropy equation [4,26].

$$H = \sum_{i=1}^{n} p_i log_a p_i, \tag{3}$$

where H stands for Shannon Entropy, n is the total number of TS observations, a is the logarithm base, and p_i is the probability for each symbol of the TS alphabet. It is worth noting that *information* may refer to the capacity of a channel for transmitting messages, the consequence of a message, the semantic meaning conveyed by it, and so on, all regardless of its specific meaning [27]. Entropy-based measures are the first option when the task at hand is the quantification of the *complexity* of a time series [28]. However, what does *complexity* stand for?

Complexity science is a multidisciplinary field in charge of studying dynamical systems composed of several parts, whose behavior is nonlinear, and that cannot be studied neither by the laws of linear thermodynamics nor by modelling parts in isolation [29,30]. A key aspect of these systems is that individual parts' interactions will heavily determine the future states of the overall system, and shall induce spatial, functional, or temporal structures all alone (i.e., self-organizing) [27]. Similarly, these systems are considered *open* since they exchange matter, energy, and information with their environments [27,30]. These are observed in a multitude of disciplines such as biology, ecology, economy, linguistics, and so on; it is common to study their dynamical behavior through the observation of one or more of its variables in the form of TS [31].

There are several measures of *complexity*, but at its core remains the notion of information altogether to some form of Shannon entropy formulation [8]. These two notions spawn a myriad of complexity measures; among them stand out H_{perm}, the Kolmogorov–Sinai (KS) complexity, H_{spct}, H_{2reg}, Transfer entropy, LMC complexity, ϵ-complexity, *ESC*, and so on [8,27,31–33]. The diversity of such measurements is given by the inexorable subjectivity of what shall be considered *complex*, which is translated into a specific *quantization* of a TS regarding an observer point of view [25,27].

In this work, *quantization* stands for the procedure to estimate the discrete probability distribution from a TS; in other words, how we describe the states of the system. In the classical H, continuous measurements are typically transformed into discrete states by binning measurements into non-overlapping ranges. To emphasize this form of entropy estimation *per se*, it will be referred as H_{dist} and will reserve H for the concept. However, there are other ways in which we can discretize a time series into a probability distribution, which needs to be hand in hand with the properties of the time series that are analyzed. In Figure 1, a cartoon of the four symbolizations used in this work is shown.

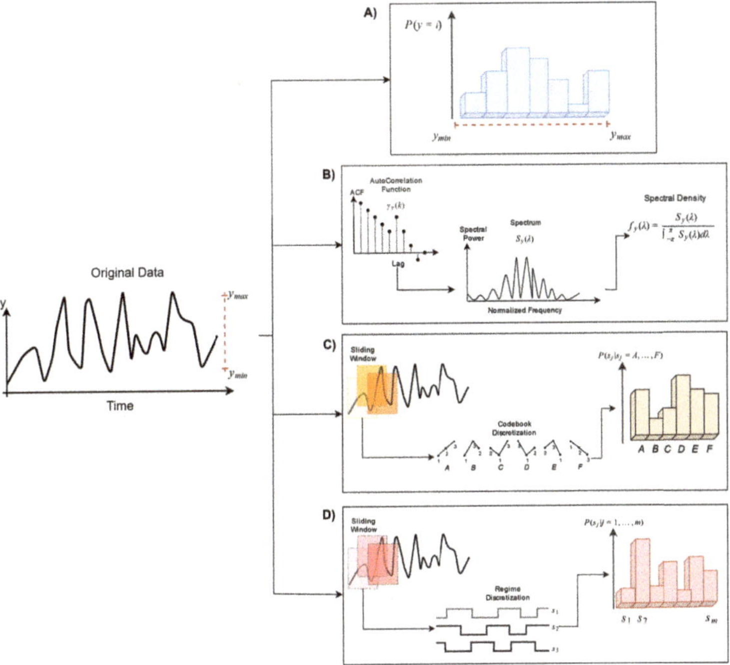

Figure 1. Four possible characterizations of the states of a dynamical system. On (**A**) the frequentist binning approach; on (**B**) the spectral probability density of the TS is estimated by the classical Fourier transform of the Auto Correlation Function (ACF); on (**C**,**D**) symbolic transformations define the alphabet by ordinal rank patterns and sequences of the first derivative sign, respectively.

3.1.1. Spectral Entropy

Power Spectral Density (PSD) estimation is commonly used in signal-processing literature. By transforming a given time series x_t from the time domain to the frequency domain using the discrete Fourier transform, the latter provides information about the power of each frequency component. These frequencies describe a spectral probability distribution which can be used to assess the uncertainty about a future prediction, namely spectral entropy H_{spct} (a cartoon of this process is shown in Figure 1B). To calculate this from a TS (assumed to be stationary), we first require its Autocovariance Function (ACVF). This is defined as

$$\gamma_x(k) = \mathbb{E}[(x_t - \mu_x)(x_{t-k} - \mu_x)], \qquad k \in \mathbb{Z}, \tag{4}$$

where μ_x is the TS mean value and k corresponds to the lag. With the ACVF, the spectrum of the TS is obtained through the Fourier transform such as

$$S_x(\lambda) = \frac{1}{2\pi} \sum_{j=-\infty}^{\infty} \gamma_x(j)e^{ij\lambda}, \qquad \lambda \in [-\pi, \pi], \tag{5}$$

where $i = \sqrt{-1}$ and $S_x : [-\pi, \pi] \to \mathbb{R}^+$. To understand the implications of Equation (5) consider a white noise TS ω. In such case, $\gamma_\omega(k) = 0$ for $k \neq 0$, thus, the spectrum is constant for all $\lambda \in [-\pi, \pi]$ [24]. Then, if we define $\sigma_x^2 = \int_{-\pi}^{\pi} S_x(\lambda)$ for $k = 0$, the spectral density of x_t is

$$f_x(\lambda) = \frac{S_x(\lambda)}{\sigma_x^2} = \frac{1}{2\pi} \sum_{j=-\infty}^{\infty} \rho_x(j)e^{ij\lambda}, \tag{6}$$

where $\rho(k) = \frac{\gamma(k)}{\gamma(0)}$ corresponds to the Autocorrelation Function (ACF). The $f_x(\lambda)$ can be used as a probability density function of a random variable such that it is ascribed to the unit circle. For instance, in the case of ω, $f_x(\omega) = \frac{1}{2\pi}$, which corresponds to the spectral density of the uniform distribution [24].

Using Equation (6), the spectral entropy H_{spct} is defined as

$$H_{spct} = \int_{-\pi}^{\pi} f_x(\lambda)log_a f_x(\lambda)d\lambda. \tag{7}$$

If the value obtained by Equation (7) is relatively small, x_t is more *forecastable* since it contains more signal than noise, whereas a larger value stands otherwise [18]. The previous analysis can be simplified by normalizing H_{spct} within $0 \leq H_{spct} \leq 1$ by dividing it by the uniform distribution entropy i.e., $log_a(2\pi)$ (which has the maximum entropy for a finite support). In this sense, the uncertainty about a required prediction x_{t+h} is given by the characteristics of the process itself [24].

3.1.2. Permutation Entropy

Permutation Entropy (H_{perm}) was conceived by Bandt and Pompe as an entropy-based measure for measuring the *complexity* of a TS [23]. H_{perm} is based on the concepts of H and Symbolic Dynamics (SD). In contrast to H_{dist} and H_{spct}, H_{perm} does not ignore temporal information.

The SD carried to obtain H_{perm} consists of transforming TS data into a sequence of discrete symbols, i.e., length-L Ordinal Patterns (OP). These are produced by encoding consecutive observations contained in a sliding window of size L, $L \geq 2$, into *permutations* determined by observations rank order in each window [20,34]. However, to determine the L-window, a Phase Space Reconstruction (PSR) needs to be carried out [28,35]. Such reconstruction employs two parameters—the embedding dimension d_e and the time delay τ. Formally, given a TS of the form $X = x_1, x_2, \ldots, x_t | x_i \in \mathbb{R}$, a point mapped to the reconstructed d_e-dimensional space is of the form $\vec{x}_j = \{x_t, x_{t-\tau}, \ldots, x_{(d_e-1)\tau}\} | \vec{x}_j \in X_R$, thus, $L = (d_e - 1)\tau$.

Once TS is mapped into this space, portrait permutations are obtained. A permutation $\pi_j \in \Pi$ is given by the permutation of indices (from 0 to $d_e - 1$), which puts the d_e values of a given $\vec{x}_j$ into ascending sorted order. Notice that there are $d_e!$ different permutations. Afterwards, the Permutation Distribution (PD), also known as *codebook*, is calculated by counting the relative frequency of each symbol. Analyzing the behavior of a TS by its PD has several advantages: it is invariant to monotonic transformations of the underlying TS, requires low computational effort, is robust to noise, and does not require knowledge of the data range beforehand [20,35].

Once the PD is estimated, H_{perm} is obtained such that

$$H_{perm} = - \sum_{\pi_j \in \Pi} \pi_j log_a \pi_j, \tag{8}$$

where Π is the set of all different d_e permutations. A cartoon of this process is shown in Figure 1C.

For convenience, H_{perm} can be normalized by dividing it by $log_a(d_e!)$ to constrain it to $0 \leq H_{perm} \leq 1$. In this sense, a lower value of the normalized H_{perm} corresponds to more regular and deterministic process, whereas a value closer to 1 is observed in more random and noisier TS. Notice that H_{perm} is closely related to the Kolmogorov–Sinani (KS) entropy and equal to it when the TS is stationary. However, in contrast to KS entropy, H_{perm} it is computationally feasible to calculate H_{perm} for long L-windows [21,28,34].

3.1.3. 2-Regimes Entropy

Under SD, a *regime* stands for a qualitative behavior defined as a growth model or dynamical rule with its own state space that allows the existence of multiple attractors in equilibrium or not, at the same time [19]. This *symbolization* (transforming TS values into symbols) allows study, for instance, of structural changes in a TS such as sudden changes in trend, such as changes in the governing rules of a dynamical system (e.g., switching between trends). An adequate symbolization allows the highlighting of temporal patterns, to improve the signal-to-noise ratio, improve computation efficacy and efficiency, to mention a few [36]. To understand this, consider a time series of the form $X = x_1, x_2, \ldots, x_t | x_i \in \mathbb{R}$. Typically, the *symbolization* of a TS is carried out by dividing $\mathbb{R}$ into $q \geq 1 \mid q \in \{1, 2, \ldots\}$ non-overlapping bins. Such bins represent the states of the system; hence, each x_i is mapped to its corresponding partition, mapping from a sequence of points X into a sequence of symbols $Z = z_1, z_2, \ldots, z_t$. In the simplest case when $q = 2$, the original TS is mapped into a sequence of two symbols (i.e., 0 or 1) using a threshold that partitions $\mathbb{R}$ into two intervals, namely *2-regimes* symbolization. This representation is useful to study trends of growth (expansion) or fall (contraction) in a TS, for instance the *bear* and *bull* regimes in an economic market. In this work, 2-regime symbolization is carried out by employing the sign of the first difference such that

$$z_t = \begin{cases} 1 & 1 \cdot sgn(x_t - x_{t-1}) > 0, \\ 0 & \text{otherwise,} \end{cases} \tag{9}$$

where *sgn* stands for the sign function. An example of this codification is presented in Table 2, where the first row displays the original observations, and the second the corresponding 2-regimes codification. Notice that due to Equation (9), the length of the codified TS is n-1.

Table 2. Example of a converted time series for codifying TS values.

49	52	53	61	71	67	72	52	48	...	54
–	1	1	1	1	0	1	0	0	...	1

Once the TS is symbolized into Z, Equation (3) can be employed to calculate a two-regime entropy (H_{2reg}). It is worth mentioning that H_{2reg} can be considered to be a special case of H_{perm}, i.e., during the symbolization step of a TS, a permutation with $d_e! = 2$ will produce equivalent symbols to those obtained by Equation (9). However, by considering sequences of contiguous symbolized observations $z_t, z_{t+1}, \ldots, z_{t+d}$, an alphabet of size 2^d can be explored, allowing study of richer 2-regimes alphabets. In this work, the alphabet size is set to $2^d = 256$, $d = 8$. Finally, notice that H_{2reg} can be normalized by $log_a(2^d)$ to constrain it within $0 \leq H_{2reg} \leq 1$. In this sense, a lower value is obtained when there is a predominant regime (e.g., trend or drift); a value closer to one stands for a more random and noisier TS.

3.2. ESC and the Complexity Feature Space

Regarding the *forecastability* (i.e., determining a system future states) of a TS $\Omega(x_t)$, some complexity measures such as H_{spct}, defines it as the complement of the average uncertainty of the process (given by its spectral density) such that $\Omega(x_t) = 1 - H*$, where $H*$ corresponds to the normalized version of H_{spct} [24]. On the other hand, others indicate that the *forecastability* shall be in terms of existing and new patterns. Thus, complexity may be defined as the relationship between stability and instability [30], Information and Disequilibrium [32], redundancy and new information [21], or Emergence and Self-organization [29]. In particular, it has been established that among the basic properties of complex systems stand the emergence, self-organization, and complexity [27]. Therefore, here we decided to extend the *ESC* paradigm using different entropy functions, namely H_{spct}, H_{perm}, and H_{2reg}. In this sense, it is possible to measure (1) the average uncertainty given by a probability distribution considering multiple quantizations, (2) estimate their compliments associated with the *forecastability*, and (3) analyze the interplay between these two.

Formally, the Emergence (E), Self-organization (S), and Complexity (C) for a TS, irrespectively of the entropy of choice, is given by the following

$$E = -K \cdot H^p(x_t) \tag{10}$$

$$S = 1 - E, \tag{11}$$

$$C = 4 \cdot E \cdot S, \tag{12}$$

where $H^p(x_t)$ is the normalized version of H_{dist}, H_{spct}, H_{perm}, or H_{2reg}, such that $0 \leq E, S, C \leq 1$. This normalization is carried by the constant $K = \frac{1}{log_a(U_b)}$, which corresponds to the entropy of uniform distribution with an alphabet of size b. It is worth mentioning that when required, E, S, and C for a particular entropy mentioned above will be referred to with the entropy ID underscored. For instance, if we refer to (E, S, C) tuple for H_{spct}, these may be referred as $(E_{spct}, S_{spct}, C_{spct})$, respectively.

The feature space conformed by these 12 measures is called the Complexity Feature Space (CFS). Hence, any TS is now mapped to a 12-D space and given the aforementioned definitions of complexity, and is expected that in the CFS it will be grouped into a specific region in accordance with its *forecastability*. Notice that such a region will depend, in part, of the model used to forecast [21] as well as the forecasting horizon. However, to obtain any information from the CFS regarding the relationship between forecastability and complexity, it is a necessary tool for its analysis. For that matter, visual tools based on a dimensional reduction technique such as Principal Components Analysis (PCA) can be employed [18]. In this sense, any TS from the CFS are now displayed as 2-D points whose dimensions correspond to two principal component axes. Although PCA leads to a loss of information due to its linear nature, the topological distribution of points is mostly preserved [18]. Finally, this feature space can be improved by considering other entropy-based complexity measures such as Transfer Entropy [37] or Tsallis Entropy [38] or different characteristics such as the trend, frequency, or seasonality [2,18,39,40].

3.3. Forecasting Methods: Smyl, Theta, ARIMA and ETS

On M4 Competition, 61 forecasting methods participated, the sharing dataset contains in addition the forecast values for the better 25 methods. We select four of them, considering the Smyl winning method and three classical benchmark methods; each of them is described in the next paragraph, and ordered according to the final position in the competition.

- Smyl: This is a hybrid method that combines exponential smoothing (ES) with recurrent neural network (RNN); this method is called ES-RNN [9] and is the winning method for M4 Competition.
- Theta: was one of winning methods on M3, the previous competition, and in the past was indicated to be a variant of the classical exponential smoothing method [10].
- ARIMA (Autoregressive Integrated Moving Average): It is one of the most widely used by the Box & Jenkins methodology [41], mainly applied for nonlinear patterns in TS.
- ETS (exponential smoothing state space [13]): This method is especially used in forecasting for TS that presents trends and seasonality.

The ARIMA method is used to forecast all complete datasets, including synthetic and M4 Competition TS, and the other three methods are used only for M4 Competition TS.

3.4. Analyzing the Forecasting Performance in the CFS

A global view of the executed steps to build the CFS for analyzing the relationship between TS forecastability and complexities is presented in Figure 2.

The first step consists of gathering TS. In our case, we tested the CFS using two types of data sets: synthetic, and M4 Competition TS. Afterward, parameters of TS complexity measures, such the alphabet size, is determined. The third step consists of the calculation of the *ESC* for every type of

Entropy function. Recall that these produce a total of 12 measures per TS. The latter is repeated for each TS that belongs to the set (either Synthetic or selected M4 Competition TS). The fourth step is to make the forecast for each TS in accordance with its corresponding forecasting horizon, and measure its error using a performance measure. In the fifth step the *ESC* measures of the dataset along with PCA are used to build the CFS to visually display TS in 2-D. Finally, in the last step the performance metric is displayed in the 2-D CFS to assess its relationship with the complexity measures. To enhance this step, the relationship between forecastability and complexity is assessed by plotting quartiles of the performance metric.

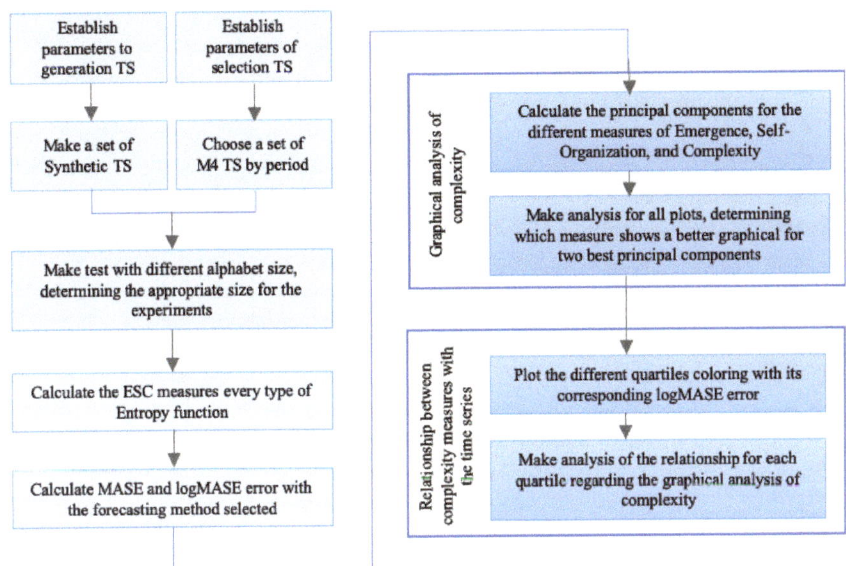

Figure 2. Proposed method.

3.4.1. Parameters Settings

So far, we have neglected some details regarding the parameters to build the CFS to analyze forecasting performance. First, we detail the parameters used to generate synthetic TS data. Then, entropy-based parameters and the performance metric are presented.

All synthetic TS have 10^4 observations, all sine waves have an amplitude of $\alpha = 1$, and frequency of $\omega = 2$. SNRs for both TS corrupted by uniform and Gaussian noise are $SNR = 10^{-3}, 10^{-2}, 10^{-1}, 10^{-0.9}, 10^{-0.5}, 10^{-0.1}, 1, 10$. In particular, for the Gaussian noise, we used a standard deviation of $\sigma^2 = 1$. Thus, for each sine wave, we generated 8 TS, giving a total of 16 sine waves. Regarding the logistic map, we employed an $r \in [3.1, 4]$ such that $\Delta r = 0.005$ with a starting point of $x_0 = 0.1$ which produces 181 TS. The last subset of synthetic TS is 16 TS generated with the tool GRATIS described in a previous section. We choose parameters like length equal to 600, spectral entropy between 0.25 and 1.00 rank, and Monthly frequency. In total, the Synthetic dataset is composed of 215 TS. On the other hand, we selected a subset of the M4 Competition data composed of 22,610 TS. Regarding the forecasting horizon, for the M4 Competition TS, the dataset contains for each TS a training and test observation part, and a defined horizon as well, thus, the Yearly period has an horizon of 6, Quarterly 8, Monthly 18, Weekly 13, and Daily 14, considering that the synthetic TS generated corresponds to Monthly period, following the same scheme of M4 Competition, the horizon selected was of 18.

Regarding the entropy-based complexities, there are some parameters to be established beforehand. In the case of H_{spct} we employed the implementation of the package *forecast* [42], which is an already normalized version of entropy i.e., $0 \leq H_{spct} \leq 1$. In contrast, H_{dist}, H_{2reg}, and H_{perm} require selection of an alphabet size. It is worth mentioning that the selection of the alphabet length was rather arbitrary, and perhaps it is a parameter for tuning or taking advantage. In the case of H_{dist} and H_{2reg} we used an alphabet size of $d = 8$; thus, the alphabet size was $2^d = 256$. The case of H_{perm} is special, since it will depend on the time series and the reconstructed phase space. For such purpose, the Mutual Information method and the Approximate Nearest Neighbor method are employed to estimate τ and d_e, respectively, in accordance with Cao's practical method [43]. In particular, for the logistic map $\tau = 2, d_e = 8$ in accordance to [28], in this case the alphabet size is $8! = 40,320$ permutations (however only those $P(\pi_j) > 0$ are considered). To estimate the PD required by H_{perm} the package *pdc* is employed [35]. The H_{2reg} and ESC measures from the entropy-based complexities were calculated using a self-implementation in R based on [8]. To estimate the error of forecasting methods, we use forecast values of 4 methods provided in *M4comp2018* package. The MASE is calculated using the *forecast* package. Furthermore, during the experimentation we noticed a logarithm relationship between the MASE and some ESC values, thus, MASE values are scaled log_{10} to highlight this relationship.

4. Results

To understand the relationship between the CFS and prediction performance for a given model, experiments were carried out using two different datasets: synthetic and M4 TS. For the first case, the purpose is to understand the relationship between ESC complexities against data whose underlying mechanisms can be controlled. In the second case, the value of the CFS in a real-world setting is explored to obtain a better idea of its potential in identifying regions (perhaps groups) of forecastability.

4.1. Complexities and Forecastability of the Synthetic TS

This section is divided into two subsections that are described below. The forecastability is analyzed only with the ARIMA forecasting method, which was executed from the *forecast* package. However, its parameters *p, r, and q* are tuned by following the procedure in [44]. This method is executed with *ARIMA* function with different combinations of $p \in [0,10]$, $d \in [1,3]$, $q \in [0,10]$, and selecting those that obtained the smallest Akaike Information Criterion (AIC) value, all of these trying to obtain the better forecast for each TS that belongs to the subset of synthetic TS.

4.1.1. The Logistic Map

To start the discussion of synthetic TS results, the logistic map is analyzed. This is a common benchmark used for the elucidation of the relationship between entropy-based complexities and forecastability [20,21,28]. Hence, the well-known Feigenbaum diagram along with its corresponding ESC measurements (from top to bottom) obtained for the logistic map are shown in Figure 3. Recall that the Feigenbaum diagram is a visual summary of the values (x_t) visited by a system as a function of a bifurcation parameter. Thus, in this case, as the parameter r grows the logistic map transitions from permanent oscillations between fixed-point pairs to the chaotic regime. Colors in the Feigenbaum diagram correspond to the $log(MASE)$ obtained by an ARIMA model: lower errors are shown in dark blue whereas those with higher values are displayed in bright yellow. Hence, as the logistic map dynamics becomes more chaotic, the TS become less forecastable by the ARIMA model.

For the ESC plots, colors correspond to different entropy-based complexities: red for H_{dist}, green for H_{spct}, blue for H_{perm}, and purple for H_{2reg}. Observe that all entropy-based complexities are constant when oscillating between two values ($r \leq 3.44$), except for H_{perm}. In this case, ESC values using a binary alphabet shall be ($E = 1, S = 0, C = 0$) for H_{dist}, H_{2reg}, and H_{perm}, however, by forcing an arbitrary large alphabet size, the self-organization is revealed. In fact, for H_{perm} and $r \leq 3.44$, ($E = 0, S = 1, C = 0$) for most cases consequence of a Dirac delta PD, with the exception of some spikes in which new ordinal patterns emerge. Observe that several of these spikes have worse $log(MASE)$ than those obtained by contiguous r values. H_{dist} grows immediately after $r = 3.44$ due to doubling of the limit cycle, but remains steady until $r = 3.54$, this contrasts to H_{2reg} which does not grow, and H_{spct} and H_{perm} which increases slower. In fact, H_{dist} seems to be a more sensitive measure to the alphabet size and not necessarily TS intricacy, since its E becomes very high between $[3.54, 3.63]$ in comparison to the rest of the complexities. Eventually, H_{spct} and H_{dist} concur in that the emergence of new states ($E \sim 1$) (or complementary, the reduction in self-organization $S \sim 0$) is similar to a random process. Conversely, H_{2reg} and H_{perm} increase slower as r grows, although the former does not change until $r \sim 3.68$ indicating that regimes displayed by the logistic map for $r \leq 3.68$ are constant.

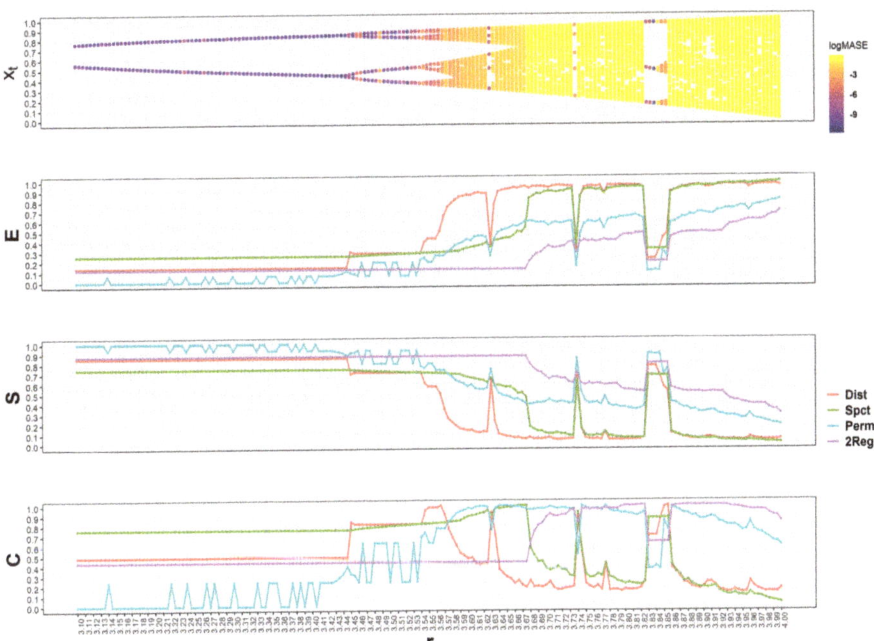

Figure 3. The Logistic Map and its ESC (Emergence, Self-Organization, and Complexity). The top plot shows the bifurcation diagram, whereas below the corresponding ESC for different entropy measures is showed.

The interplay between new states and the self-organization of the system, displayed by C is very interesting. For H_{dist}, when the logistic map has 2 fixed points ($r \leq 3.44$) a $C = 0.5$ is obtained, when the period doubles it increases to $C \sim 0.8$, and it shows maximal complexity ($C \sim 1$) at points between double-periods and chaotic regimes (*at the edge of chaos*). Hence, for obtaining a lower $log(MASE)$ it is necessary that $0.5 \leq C_{dist} \leq 1$, $S \geq 0.5$, and $E \leq 0.5$. For H_{spct} a similar relationship is observed in the sense that high C values are associated with lower $log(MASE)$ due to $E \leq 0.5$ and $S \geq 0.5$ proportions. Notice that this C separates ARIMA performance into two performance regions, with the worst $log(MASE)$ corresponding to complexities below 0.6, dropping even to $C \sim 0$ as the logistic map becomes more chaotic. In contrast, C for H_{perm} and H_{2reg} have larger complexity values for worse forecasting performance; C_{2reg} separates ARIMA performance into two performance regions similar to C_{spct}.

4.1.2. The CFS of All Synthetic Data

All the synthetic data were mapped as 2D point in the CFS which is displayed in Figure 4.

In Figure 4A *ESC* variables are projected into the CFS plane to display its loadings. Notice that the first two Principal Components (PCs) explain a large amount of the variance in the data ($PC1 \sim 73\%$, $PC2 \sim 10.6\%$), due to most of the series in the data set belonging to the logistic map. C_{perm}, E_{dist}, E_{perm}, and E_{2reg} have positive loadings on the PC1, whereas S_{dist}, S_{perm}, and S_{2reg} have negative loadings. E_{spct} and S_{spct} are parallel to its H_{dist} counterpart, but with lower loadings on the PC1. The rest of the complexities have lower loadings in these two PCs. Convex hulls are used to denote each TS source; however, note that these are constrained to specific regions in the CFS. Hulls corresponding to the corrupted sine waves mostly overlap each other, and share a large portion with GRATIS data.

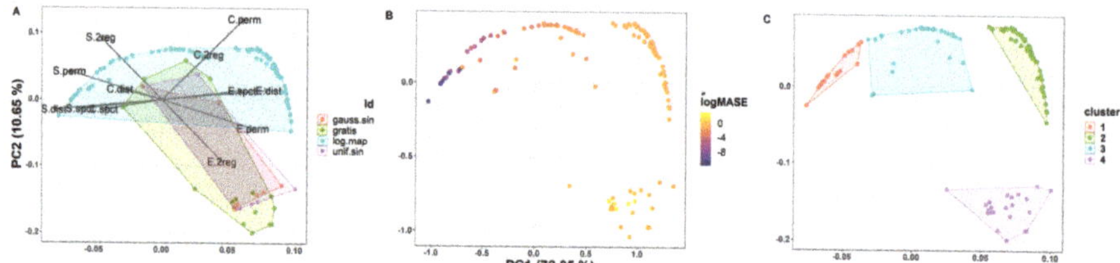

Figure 4. The Logistic Map and its ESC. The top plot shows the bifurcation diagram, whereas below the corresponding ESC for different entropy measures its showed. (**A**) *ESC* variables are projected into the Complexity Feature Space (CFS) plane to display its loadings; (**B**) Two dimension Time Series are colored in accordance to its $log(MASE)$; (**C**) K-means clustering algorithm results using four centroids.

In Figure 4B 2D TS are colored in accordance to its $log(MASE)$. Observe that a clockwise relationship between forecasting performance is displayed: ARIMA best performance lies in the upper left quadrant and its worst results on the lower right. It is interesting that the worst $log(MASE)$ correspond to noisy time series, instead of the chaotic source, and that they are *conveniently* confined to specific regions in the CFS. By *convenient* we meant that a clustering algorithm may be used to cluster TS characterized by the *ESC* variables to obtain performance clusters, employed to determine if a forecasting method is useful or not for a given TS. Encouraged by the latter, the results, obtained by the popular K-means clustering algorithm using four centroids, are shown in Figure 4C. Notice that the resulting clusters correspond to the performance regions mentioned before.

4.2. Complexities and Forecastability of the M4 Competition TS

Before we delved into the analysis of M4 Competition results, we display the relationships of different *ESC* measures of the M4 set in the CFS. In Figure 5 all TS (Yearly, Quarterly, Monthly, Weekly and Daily) are displayed as 2-D points; we focus on the C measure for each entropy measure (Figure 5a 2-regimen, Figure 5b distribution, Figure 5c permutation, and Figure 5d spectral); colors range from brighter (corresponding to higher values $C \sim 1$) to darker (corresponding to lower values $C \sim 0$). Observe that both C_{2reg} and C_{perm} achieves the line gradient behavior with high complexity values as the PC1 becomes more negative, and lower complexity values as it becomes more positive. Interestingly, regarding PC2, they are on opposite sides. On the other side, C_{dist} visual gradient is perceived more on the y-axis (lower values are positive and higher values are negative), in contrast to the PC1 where no clear relationship between high and low C values is observed. Similarly, C_{spct} shows high values over most of the two PCs plane. However, for both C_{dist} and C_{spct} this behavior can be product of the reduction of dimensionality by the linear method.

These intuitions are corroborated by the loadings of these variables on the four most important PCs, which are presented in Table 3. Notice that the PC1 and PC2 are mainly represented by Permutation and 2-regimen complexities. On the other hand, for PC3 the most significant variable is C_{dist} which has a negative loading, whereas for PC4, the most significant variable is C_{spct}. In particular, the C part of the *ESC* measures will be used for the analysis in Section 4.3. Table 4 shows results for the explained variance proportion corresponding to each principal component. Observe that the first two PCs account for most of the variance ($\approx$77%) in data, and with only 4 PCs we account for the 100% of the variance.

In Figure 6a selected M4 TS are shown in the CFS color-coded by the period that corresponds to its frequency. Observe that Daily and Monthly TS are readily identifiable in the 2d projection, while the former is restrained to a specific region of the CFS, and the latter is spread across the CFS. Weekly TS are constrained to the middle section of the CFS, while Yearly and Quarterly TS are barely noticeable. On the other hand, in Figure 6b M4 TS are shown colored in accordance to the winning method, where 6838 points correspond to ARIMA, 6384 correspond to the SMYL, 5064 to the ETS, and 4324 to the Theta. It is worth mentioning that even when ARIMA wins in more TS than the Smyl algorithm, error magnitudes of the former are larger in comparison to the latter. Moreover, there are no specific regions in which any of the tested methods obtain better performance than the rest, which is consistent with the *No-Free Lunch* theorem.

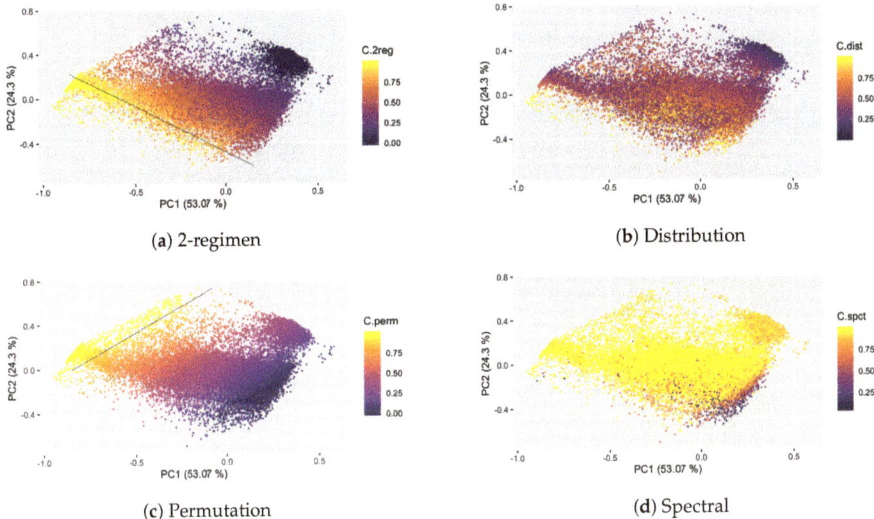

(**a**) 2-regimen   (**b**) Distribution

(**c**) Permutation   (**d**) Spectral

Figure 5. Four complexity measures and the Principal Components Analysis (PCA) of 12 features (ESC).

Table 3. Principal Components Analysis (PCA) results.

	PC1	PC2	PC3	PC4
C.2reg	−0.6768	−0.5947	0.4336	0.0142
C.dist	−0.2003	−0.4150	−0.8776	−0.1323
C.perm	−0.7057	0.6777	−0.1757	0.1086
C.spct	−0.0607	0.1219	0.1047	−0.9851

Table 4. Proportion of variance for the principal components.

	PC1	PC2	PC3	PC4
Standard deviation	0.2923	0.1978	0.1592	0.1052
Proportion of Variance	0.5308	0.2431	0.1574	0.0687
Cumulative Proportion	0.5308	0.7739	0.9313	1.0000

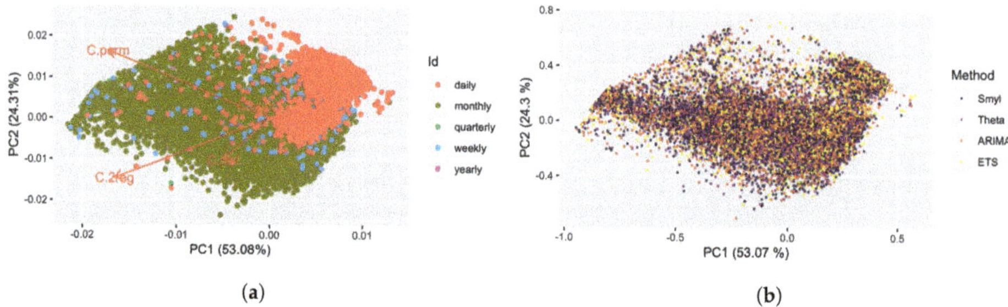

(a)   (b)

Figure 6. Analysis of TS regarding its Period frame and Winning method by TS. (**a**) Selected M4 Time Series are shown in the Complexity Feature Space (CFS), and each one is colored according to the period of its frequency; (**b**) M4 Time Series are colored according to the winning method.

Continuing with the experiments on M4 dataset, one of our main interests is to determine the forecastability of the M4 Competition through the complexity measures of TS. Therefore, we consider four methods of M4-Competitions in order to establish whether there exists or not a relationship between the MASE error ($log(MASE)$ to effects of functionality) by forecasting method (Smyl, Theta, ARIMA, and ETS). The first activity was to divide the complete dataset into four quartiles, in Figure 7, with each gray point representing one TS that belongs to the complete dataset of M4 Competition, and the dark green point representing the TS whose ($log(MASE)$ value is found of the first quartile; specifically, in Figure 7a, the ($log(MASE)$ values corresponding to the Smyl forecasting method, the Figure 7b, the ($log(MASE)$ values corresponding to the Theta forecasting method, and so on, this figure shows that the TS with low $log(MASE)$ value are concentrated in the negatives values of the second principal component and has a high value for the first principal component, according to Figure 6a; this kind belongs mainly to the Monthly period.

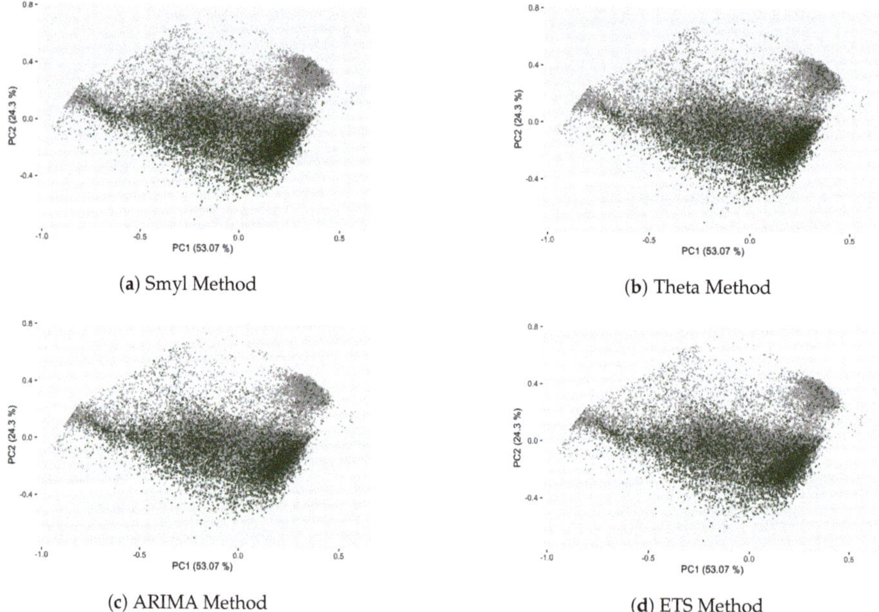

(a) Smyl Method   (b) Theta Method

(c) ARIMA Method   (d) ETS Method

Figure 7. Relationship between $log(MASE)$ and Complexity measures of the first quartile.

In the same way, Figure 8 represents the TS that integrates the four quartiles according to the $log(MASE)$ values, where the purple point corresponds to the TS that belongs to this quartile. Making a comparison between Figure 8a–d it is noted that the $log(MASE)$ values for each one of forecast methods is closer between them, and in terms of distribution area for these TS, we determine that when the complexity measures are higher, the $log(MASE)$ value is higher too; moreover, compared to the distribution of TS by periods (Yearly, Quarterly, Monthly, Weekly and Daily), the major part of Daily TS belongs to this quartile.

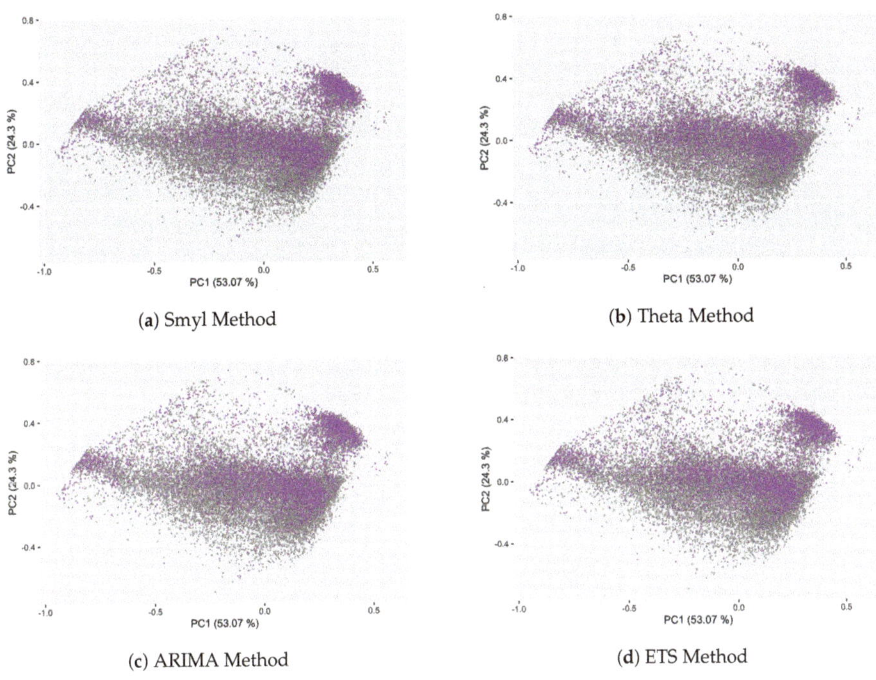

(a) Smyl Method   (b) Theta Method

(c) ARIMA Method   (d) ETS Method

Figure 8. Relationship between $log(MASE)$ and Complexity measures of the fourth quartile.

4.3. Regression Results

After analyzing the TS applying Principal Component Analysis, we used the Principal Components Regression method (PCR) to adjust a model of linear regression by least squares using the four components generated on a previous subsection. For this test, we select the TS by period and divide them in a subset of training and another subset of test with 80% and 20%, respectively. The estimated error of prediction was calculated with the Mean Square Error (MSE), and the results are presented in Table 5, the best prediction values were obtained in the Quarterly and Weekly periods, and the high prediction error was obtained in the Yearly period; it is important to remember that the subset of Yearly period is composed only of 56 TS.

Table 5. Mean Square Error of prediction with a linear regression model.

	Yearly	Quarterly	Monthly	Weekly	Daily
MSE	115.0187	6.8431	21.1561	4.3047	56.2699

5. Conclusions

In this work, we proposed four possible characterizations of the state of a dynamic system based on Shannon entropy: a frequentist binning approach (distribution), the spectral probability density of the TS (spectral), and symbolic transformations (permutation and 2-regimes) defining the alphabet by ordinal rank patterns, and sequences of the first derivative sign. These characterizations are the measures of complexity, and they are bounded between zero (i.e., minimal Entropy/Complexity) and one (i.e., maximal Entropy/Complexity). One important feature for these measures is that Entropy is maximal when TS states are equiprobable. In contrast, Complexity is maximal when the system tends to high Self-Organization or high Emergence (i.e., discernible patterns with some noise or high noise with some discernible patterns). From those measures, we determined the principal components, and through its loadings, we found that C_{perm} and C_{2reg} are those measures that represent patterns that identify TS groups with similar features. Also, by plotting the TS by its $log(MASE)$ in different quartiles, we observed that the TS with low $log(MASE)$ are concentrating along with the first principal component. Moreover, comparing the four forecasting methods, the behavior is very similar between them; it is important to emphasize that for every TS, the $log(MASE)$ values displayed in this space are very close among each other. Thus, these plots only corroborate the supposition that the winning method is the best for the quantity of TS where the winning is individually. Another important result is that we found that from the four forecasting methods identified as the winner of each TS dispersed over the complete TS, we see that the two principal components are consistent with the *No-free lunch* theorem. Finally, we determine that the TS with complexity measures closer to zero correspond to a low $log(MASE)$ error, whereas when complexities measures are high, the $log(MASE)$ tends to be high.

Author Contributions: In this paper, M.P.-F., J.F.-S., and G.S.-B. made the conceptualization for obtaining Entropy, Complexity, and prediction error measure; M.P.-F., J.F.-S., and G.S.-B. made the formal analysis; G.S.-B. conceptualized the ESC entropy extensions and the CFS; M.P.-F. and G.S.-B. performed the experiments; M.P.-F., J.F.-S., G.S.-B., J.P.-O. and J.J.G.-B. validated the experiments and wrote the paper. All authors have read and agreed to the published version of the manuscript.

Funding: This research received no external funding.

Acknowledgments: The authors acknowledge CONACYT, Cátedras CONACYT program, and TecNM/ITCM for the use of its installations.

Conflicts of Interest: The authors declare no conflict of interest. The dataset collection, analyses, and results interpretation were completely made by the authors, as the writing of the manuscript as well. The decision to publish the results prepared for this paper was completely taken by the authors. The are no funders that had a role in the design of the study, in the collection, analyses, or interpretation of data; in the writing of the manuscript, or in the decision to publish the results.

References

1. Montgomery, D.C.; Jennings, C.L.; Kulahci, M. *Introduction to Time Series Analysis and Forecasting*; John Wiley & Sons: Chichester, UK, 2008; p. 469.
2. Makridakis, S.; Spiliotis, E.; Assimakopoulos, V. Predicting/hypothesizing the findings of the M4 Competition. *Int. J. Forecast.* **2019**, *36*, 29–36. [CrossRef]
3. Wang, X.; Smith, K.; Hyndman, R. Characteristic-based clustering for time series data. *Data Min. Knowl. Discov.* **2006**, *13*, 335–364. [CrossRef]
4. Shannon, C.E.; Weaver, W.; Blahut, R.E. The mathematical theory of communication. *Urbana Univ. Ill. Press* **1949**, *117*, 379–423. [CrossRef]
5. Ribeiro, H.V.; Jauregui, M.; Zunino, L.; Lenzi, E.K. Characterizing time series via complexity-entropy curves. *Phys. Rev. E* **2017**, *95*. [CrossRef] [PubMed]
6. Mortoza, L.P.; Piqueira, J.R. Measuring complexity in Brazilian economic crises. *PLoS ONE* **2017**, *12*, e0173280. [CrossRef] [PubMed]
7. Mikhailovsky, G.E.; Levich, A.P. Entropy, information and complexity or which aims the arrow of time? *Entropy* **2015**, *17*, 4863–4890. [CrossRef]

8. Santamaría-bonfil, G. A Package for Measuring emergence, Self-organization, and Complexity Based on Shannon entropy. *Front. Robot. AI* **2017**, *4*, 10. [CrossRef]

9. Smyl, S. A hybrid method of exponential smoothing and recurrent neural networks for time series forecasting. *Int. J. Forecast.* **2020**, *36*, 75–85. [CrossRef]

10. Assimakopoulos, V.; Nikolopoulos, K. The theta model: A decomposition approach to forecasting. *Int. J. Forecast.* **2000**, *16*, 521–530. [CrossRef]

11. Brockwell, P.; Davis, R. *Introduction to Time Series and Forecasting*; Springer: Berlin/Heidelberger, Germany, 2002; p. 437.

12. De Gooijer, J.G.; Hyndman, R.J.; Gooijer, J.G.D.; Hyndman, R.J. 25 Years of Time Series Forecasting. *Int. J. Forecast.* **2006**, *22*, 443–473. [CrossRef]

13. Hyndman, R.J.; Koehler, A.B.; Snyder, R.D.; Grose, S. A state space framework for automatic forecasting using exponential smoothing methods. *Int. J. Forecast.* **2002**, *18*, 439–454. [CrossRef]

14. Makridakis, S.; Spiliotis, E.; Assimakopoulos, V. The M4 Competition: 100,000 time series and 61 forecasting methods. *Int. J. Forecast.* **2020**, *36*, 54–74. [CrossRef]

15. Kang, Y.; Hyndman, R.J.; Li, F. GRATIS: GeneRAting TIme Series with diverse and controllable characteristics. *arXiv* **2019**, arXiv:stat.ML/1903.02787.

16. Makridakis, S.; Spiliotis, E.; Assimakopoulos, V. The M4 Competition: Results , findings , conclusion and way forward The M4 Competition: Results, findings, conclusion and way forward. *Int. J. Forecast.* **2018**, *34*. [CrossRef]

17. Hyndman, R.J.; Koehler, A.B. Another look at measures of forecast accuracy. *Int. J. Forecast.* **2006**, *22*, 679–688. [CrossRef]

18. Kang, Y.; Hyndman, R.J.; Smith-Miles, K. Visualising forecasting algorithm performance using time series instance spaces. *Int. J. Forecast.* **2017**, *33*, 345–358. [CrossRef]

19. Brida, J.G.; Punzo, L.F. Symbolic time series analysis and dynamic regimes. *Struct. Chang. Econ. Dyn.* **2003**, *14*, 159–183. [CrossRef]

20. Amigó, J.M.; Keller, K.; Unakafova, V.A. Ordinal symbolic analysis and its application to biomedical recordings. *Philos. Trans. R. Soc. A Math. Phys. Eng. Sci.* **2015**, *373*. [CrossRef]

21. Pennekamp, F.; Iles, A.C.; Garland, J.; Brennan, G.; Brose, U.; Gaedke, U.; Jacob, U.; Kratina, P.; Matthews, B.; Munch, S.; et al. The intrinsic predictability of ecological time series and its potential to guide forecasting. *Ecol. Monogr.* **2019**, *89*. [CrossRef]

22. Verdú, S. Empirical Estimation of Information Measures: A Literature Guide. *Entropy* **2019**, *21*, 720. [CrossRef]

23. Bandt, C.; Pompe, B. Permutation entropy: A natural complexity measure for time series. *Phys. Rev. Lett.* **2002**, *88*, 174102. [CrossRef] [PubMed]

24. Goerg, G. Forecastable Component Analysis. In Proceedings of the 30th International Conference on Machine Learning (ICML-13), Atlanta, GA, USA, 16–21 June 2013; Volume 28, pp. 64–72.

25. Zenil, H.; Kiani, N.A.; Tegnér, J. Low-algorithmic-complexity entropy-deceiving graphs. *Phys. Rev. E* **2017**, *96*, 012308. [CrossRef] [PubMed]

26. Balzter, H.; Tate, N.J.; Kaduk, J.; Harper, D.; Page, S.; Morrison, R.; Muskulus, M.; Jones, P. Multi-scale entropy analysis as a method for time-series analysis of climate data. *Climate* **2015**, *3*, 227–240. [CrossRef]

27. Haken, H.; Portugali, J. Information and Self-Organization. *Entropy* **2017**, *19*, 18. [CrossRef]

28. Riedl, M.; Müller, A.; Wessel, N. Practical considerations of permutation entropy: A tutorial review. *Eur. Phys. J. Spec. Top.* **2013**, *222*, 249–262. [CrossRef]

29. Gershenson, C.; Fernández, N. Complexity and information: Measuring emergence, self-organization, and homeostasis at multiple scales. *Complexity* **2012**, *18*, 29–44. [CrossRef]

30. Atmanspacher, H. On macrostates in complex multi-scale systems. *Entropy* **2016**, *18*, 426. [CrossRef]

31. Zunino, L.; Olivares, F.; Bariviera, A.F.; Rosso, O.A. A simple and fast representation space for classifying complex time series. *Phys. Lett. Sect. A Gen. At. Solid State Phys.* **2017**, *381*, 1021–1028. [CrossRef]

32. López-Ruiz, R.; Mancini, H.L.; Calbet, X. A statistical measure of complexity. *Phys. Lett. A* **1995**, *209*, 321–326. [CrossRef]

33. Piryatinska, A.; Darkhovsky, B.; Kaplan, A. Binary classification of multichannel-EEG records based on the ϵ-complexity of continuous vector functions. *Comput. Methods Programs Biomed.* **2017**, *152*, 131–139. [CrossRef]

34. Amigó, J.M. The equality of Kolmogorov-Sinai entropy and metric permutation entropy generalized. *Phys. D Nonlinear Phenom.* **2012**, *241*, 789–793. [CrossRef]
35. Brandmaier, A.M. pdc: An R Package for Complexity-Based Clustering of Time Series. *J. Stat. Softw.* **2015**, *67*, 1–23. [CrossRef]
36. Alcaraz, R. Symbolic entropy analysis and its applications. *Entropy* **2018**, *20*, 568. [CrossRef]
37. Lizier, J.T. JIDT: An Information-Theoretic Toolkit for Studying the Dynamics of Complex Systems. *Front. Robot. AI* **2014**. [CrossRef]
38. Fernández, N.; Aguilar, J.; Piña-García, C.; Gershenson, C. Complexity of lakes in a latitudinal gradient. *Ecol. Complex.* **2017**, *31*, 1–20. [CrossRef]
39. Box, G.E.; Jenkins, G.M.; Reinsel, G.C.; Ljung, G.M. *Time Series Analysis Forecasting and Control*; John Wiley & Sons: Chichester, UK, 2015.
40. Contreras-Reyes, J.E.; Canales, T.M.; Rojas, P.M. Influence of climate variability on anchovy reproductive timing off northern Chile. *J. Mar. Syst.* **2016**, *164*, 67–75. [CrossRef]
41. Box, G.E.P.; Cox, D.R. An Analysis of Transformations. *J. R. Stat. Soc. Ser. B (Methodol.)* **1964**. [CrossRef]
42. Hyndman, R.J.; Khandakar, Y. Automatic time series forecasting: The forecast package for R. *J. Stat. Softw.* **2008**. [CrossRef]
43. Cao, L. Practical method for determining the minimum embedding dimension of a scalar time series. *Phys. D Nonlinear Phenom.* **1997**, *110*, 43–50. [CrossRef]
44. Hyndman, R.J.; Athanasopoulos, G. *Forecasting: Principles and Practice*; OTexts: Melbourne, Australia, 2018.

MDPI
St. Alban-Anlage 66
4052 Basel
Switzerland
Tel. +41 61 683 77 34
Fax +41 61 302 89 18
www.mdpi.com

Entropy Editorial Office
E-mail: entropy@mdpi.com
www.mdpi.com/journal/entropy